Brain Functional Connectivity: Prediction, Dynamics, and Modeling

Guest Editor
Alexander N. Pisarchik
Center for Biomedical
Technology
Universidad Politécnica
de Madrid
Madrid
Spain

Editorial Office
MDPI AG
Grosspeteranlage 5
4052 Basel, Switzerland

This is a reprint of the Special Issue, published open access by the journal *Applied Sciences* (ISSN 2076-3417), freely accessible at: https://www.mdpi.com/journal/applsci/special_issues/V7P02B9C03.

For citation purposes, cite each article independently as indicated on the article page online and as indicated below:

Lastname, A.A.; Lastname, B.B. Article Title. *Journal Name* **Year**, *Volume Number*, Page Range.

ISBN 978-3-7258-6612-0 (Hbk)
ISBN 978-3-7258-6613-7 (PDF)
https://doi.org/10.3390/books978-3-7258-6613-7

Contents

About the Editor

Alexander N. Pisarchik

Alexander N. Pisarchik is a Distinguished Researcher and Isaac-Peral Professor at the Center for Biomedical Technology of Universidad Politécnica de Madrid. His research interests center on nonlinear dynamics, complexity, and synchronization in biological systems, with a particular focus on brain networks, neuroengineering, and computational neuroscience. He leads international projects aimed at developing novel methods for diagnostic and therapeutic tools through the analysis of neurophysiological signals and the modeling of brain dynamics. With a prolific career spanning over five decades, his significant achievements include over 300 peer-reviewed publications, several monographs, and patents. He has been honored with the CONACYT Chair of Excellence in Mexico, is Level III in the National System of Researchers, and serves as an Evaluator of European projects. His editorial leadership stems from his deep commitment to advancing the interdisciplinary frontier where theoretical physics meets applied biomedical research.

Preface

This reprint presents a curated selection of cutting-edge research dedicated to mapping and understanding the functional architecture of the brain. Its scope encompasses theoretical, experimental, and analytical advances in brain connectivity, with a unifying aim: to translate the dynamic interplay of neural networks into deeper mechanistic insights and tangible applications. The motivation for this collection stems from a critical juncture in neuroscience. While the brain's complexity remains profound, new methodologies are empowering us to move beyond mere observation toward predictive and explanatory models. By integrating tools from signal processing, electrophysiology, and artificial intelligence, we are now poised to decode the neural signatures of cognition, behavior, and their disruption in disease. This reprint is addressed to a broad audience of neuroscientists, clinicians, and engineers. For researchers, it offers a snapshot of methodological innovation and conceptual shifts. In the case of clinicians and translational scientists, it highlights pathways to biomarkers and novel interventions. Ultimately, we seek to engage anyone invested in the frontier where fundamental discovery meets solutions for health, technology, and a deeper understanding of the human mind.

Alexander N. Pisarchik
Guest Editor

applied sciences

MDPI

Editorial

Editorial for First Edition of Special Issue "Brain Functional Connectivity: Prediction, Dynamics, and Modeling"

Alexander N. Pisarchik

Center for Biomedical Technology, Universidad Politécnica de Madrid, Campus de Montegancedo, 28223 Pozuelo de Alarcón, Madrid, Spain; alexander.pisarchik@upm.es

1. Introduction

The brain is one of the most complex and mysterious systems in the known world. Understanding its functional architecture, the dynamic interplay between distinct regions, is fundamental to deciphering the neural code of cognition and behavior [1]. This connectivity can be studied in both the frequency and time domains using a robust methodological toolkit that includes coherence, correlation, and modern artificial neural networks [2]. Revealing these functional networks is more than a mapping exercise; it is essential for uncovering the mechanisms that underlie information processing and decision-making during cognitive tasks [3].

The implications of this knowledge extend far beyond the laboratory, addressing practical and profoundly challenging problems in healthcare, clinical medicine, biomedical engineering, brain–machine interfaces, and cognitive sciences [4]. From identifying biomarkers for neurological disorders to designing adaptive neural interfaces, research in functional connectivity sits at the exciting intersection of discovery and application [5].

This Special Issue provides a multifaceted snapshot of contemporary brain connectivity research through a curated collection of 13 contributions. The compilation includes two comprehensive reviews, ten original research papers (encompassing a case report), and one perspective study, each addressing distinct aspects of the field. These works collectively advance the discipline by bridging theoretical modeling, cutting-edge experimental designs, and advanced analytical techniques applied to data from EEG, fMRI, MEG, and PET. The articles highlight not only technical and methodological sophistication but also important conceptual shifts, thereby offering an integrative view of how functional connectivity is transforming our comprehension of brain dynamics in health and disease.

The contributions assembled here exemplify the richness and real-world relevance of applied neuroscience. From the laboratory to the clinic, and from theoretical models to practical interfaces, they illustrate how the study of the connected brain is increasingly embedded in solutions for health, technology, and understanding human cognition.

2. Reviews

The first review article in this issue (Contribution 1), presented by Bubliková et al., focuses on the structural connectivity of the substantia nigra, a pivotal nucleus within the brain's motor and dopaminergic circuits. Using diffusion tensor imaging and tractography, the authors provide a thorough synthesis of current knowledge concerning the substantia nigra's anatomical pathways in humans. The review outlines both classical and newly described connections, including direct pars compacta projections to the thalamus, cortical inputs, and links to limbic and hippocampal regions. This mapping reinforces the view of the substantia nigra as a functionally diversified hub, with substantial implications for understanding Parkinson's disease and related neuropsychiatric conditions [6]. By systematically addressing a gap in the literature, the article establishes a valuable framework for future investigations into dopaminergic circuitry and its role in both motor and non-motor functions.

In the second article (Contribution 2), Andreou et al. present a timely review on the shared social cognitive dysfunctions in Autism Spectrum Disorder (ASD) and schizophrenia, with a focus on the mirror neuron system (MNS) as a potential neural substrate. The authors synthesize recent neuroimaging and neurophysiological evidence, examining how altered functionality and activation within the MNS, a network of neurons responsive during both action observation and execution, may underlie deficits in imitation, empathy, and theory of mind common to both disorders. Their analysis critically evaluates whether MNS dysfunction represents a transdiagnostic mechanism linking social impairment across diagnostic boundaries, while also considering methodological challenges in measuring MNS activity in clinical populations [7]. By bridging psychiatric and neuroscience perspectives, this review offers a nuanced framework for understanding social cognition deficits and suggests pathways for future translational research aimed at biomarkers and targeted interventions [8].

3. Original Research

The ten original research papers in this Special Issue are organized into four thematic groups, each reflecting a key domain of contemporary functional connectivity research: (i) cognitive dynamics (4 papers), (ii) prediction models (2 papers), (iii) brain disorders (3 papers) and (iv) applications in sport and exercise programs (1 paper). This diversity not only illustrates the richness and methodological breadth of the field but also its expanding relevance across basic science, clinical translation, and applied human performance.

3.1. Cognitive Dynamics

A central theme across several contributions is the investigation of cognitive dynamics, specifically, how emotional and contextual factors modulate cognitive functions and decision-making processes [9]. In an electrophysiological study, Rovelli et al. (Contribution 3) demonstrate that decision contexts with higher socio-emotional salience elicit faster, emotionally guided choices. Their work underscores the value of integrating behavioral, experiential, and neural measures to characterize how individuals adaptively regulate decision-making under socially evaluative stress. Furthermore, the authors highlight the utility of dual-paradigm experimental designs in advancing both theory and application in cognitive–affective neuroscience.

Another compelling contribution to cognitive dynamics comes from Angioletti et al. (Contribution 4), who investigated functional connectivity during a negotiation process using electrophysiological correlates in dyads engaged in a shared decision-making task. Employing EEG hyperscanning, the authors analyzed both single-brain and inter-brain neural activity during a structured negotiation paradigm. Their findings reveal the dual nature of negotiation as both a cooperative endeavor and a cognitively demanding process, requiring emotional alignment and strategic adaptation between partners. By capturing the neural signatures of mutual influence and coordination, this study advances our understanding of the neurophysiological foundations of social negotiation and offers novel insights into how inter-brain dynamics underpin real-world collaborative decision-making. While the study demonstrates that negotiation elicits measurable inter-brain synchrony, a critical question remains: what is the precise functional role of this neural alignment? Is it a marker of successful communication, a mechanism for emotional contagion, a facilitator of mutual understanding, or simply an epiphenomenon of shared attention? Future research

https://doi.org/10.3390/app16020789

could dissect these possibilities by correlating specific patterns of inter-brain connectivity with behavioral outcomes such as deal quality, joint payoff, or post-negotiation trust [10]. This would move the field from demonstrating that brains synchronize to explaining why and when such synchronization is beneficial or costly.

Carvalho et al. (Contribution 5) employ advanced tensor decomposition techniques to identify individual brain fingerprints with high accuracy using resting-state fMRI data. The concept of a functional connectome (FC) fingerprint, a unique, stable neural signature identifiable across scanning sessions, has gained substantial empirical support over the past decade through studies using fMRI [11,12] and EEG [13,14]. In this work, the authors implement a Tucker decomposition framework for FC fingerprinting, enabling robust identification of an individual's functional connectome from a set of repeated fMRI sessions. Their results demonstrate the high potential of tensor-based methods to uncover individualized neural signatures, offering a mathematically elegant and computationally efficient approach to person-level brain characterization. Perhaps individualized connectome fingerprints may serve as sensitive baselines for detecting subtle pathological deviations, offering a pathway to personalized biomarkers for neurological or psychiatric disorders. The tensor methods showcased here could enhance our ability to detect early signs of Alzheimer's disease, depression, or other conditions by quantifying deviations from a person's own neural baseline [15].

The case study by Pisarchik et al. (Contribution 6) introduces a novel methodological framework for constructing both graphs and hypergraphs of functional brain connectivity from single-subject EEG data during endogenous selective visual attention. The authors generate a comprehensive set of connectivity networks using multiple complementary measures (coherence, cross-correlation, and mutual information), thereby capturing linear, nonlinear, and statistical dependencies between brain regions. This multi-metric approach provides a richer, more nuanced characterization of the functional relationships underpinning attentional control. The framework's ability to model higher-order interactions via hypergraphs is particularly notable, as it moves beyond traditional pairwise connections to capture more complex network dynamics. These methodological innovations have direct translational potential for attention monitoring systems and the clinical assessment of attention-deficit disorders [16]. The work is timely, as recent research using effective connectivity methods like Granger causality has begun to dissect the directed networks of endogenous attention [17]. The methodology presented in contribution 6 is well-suited to extend such investigations by providing tools to model both the strength and the higher-order structure of these directed and undirected functional networks. In summary, this case study provides a valuable blueprint for single-subject network neuroscience, demonstrating how advanced graph and hypergraph theory can be applied to EEG data to reveal the complex, multi-faceted nature of functional connectivity during a fundamental cognitive process.

3.2. Prediction Models

In the realm of prediction models, Mizrahi et al. (Contribution 7) present a novel neurocomputational approach that employs EEG recordings to predict individual attachment styles during performance on the Secretary Problem—a canonical optimal stopping problem in decision theory that formalizes the trade-off between exploration and exploitation [18]. Classically, the Secretary Problem prescribes a strategy of observing the first $\approx$37% of options to establish a benchmark, followed by a commitment to the first subsequent option that exceeds it, thus advocating for "informed decisiveness" over exhaustive search [19]. The authors leverage this paradigm to examine whether neural signals during decision-making can reflect stable interpersonal dispositions. Using graph-based EEG

https://doi.org/10.3390/app16020789

features and machine learning classifiers, Mizrahi et al. successfully distinguish between attachment styles, suggesting that neural signatures of exploratory and decisional phases may encode dispositional tendencies toward anxiety or avoidance in interpersonal contexts. Their findings indicate that future research should employ larger and more diverse samples to validate and refine these predictive models. If replicated in multi-site cohorts, such graph-based EEG markers could inform personalized psychological interventions by providing an objective, neural measure of attachment-related vulnerabilities [20]. This study highlights the potential of EEG-based prediction to bridge cognitive neuroscience, computational psychiatry, and personality research, offering a pathway toward more individualized assessment and intervention in both clinical and non-clinical settings.

In the next paper, Vakorin et al. (Contribution 8) introduce a novel method for estimating neurobiological brain age, a biomarker reflecting deviation from typical aging trajectories, using resting-state MEG data. Their approach moves beyond conventional univariate analyses by examining cross-frequency coupling (CFC) within functionally defined MRI networks, capturing the dynamic interplay between neuronal oscillations as a marker of neurophysiological integrity [21]. Unlike models that emphasize static baseline activity, their framework highlights neural capacity: the brain's ability to flexibly coordinate oscillatory activity across frequencies. The results suggest that such capacity-sensitive indices may offer a more nuanced and state-dependent perspective on brain aging, potentially detecting early or subtle deviations associated with cognitive resilience or decline [22]. This approach may capture aspects of neural communication efficiency and cognitive reserve that are not evident in slower-changing structural measures [23]. An important question thus arises: Are CFC-derived brain-age estimates more predictive of functional outcomes, such as memory decline, processing speed, or daily functioning, than traditional neuroimaging-based predictions? A critical nuance in brain-age research lies in distinguishing pathological from normative variability. The brain age gap—the difference between estimated brain age and chronological age—has been shown to predict risks of cognitive decline, mental health disorders, and mortality [24]. However, not all individuals with an "older" brain age exhibit cognitive impairment, and vice versa. Future models could therefore benefit from integrating behavioral, lifestyle, and genetic data to differentiate benign variations in neural aging from deviations signaling increased risk for neurodegenerative disease. Such a multimodal approach aligns with the growing emphasis on precision aging research [25].

3.3. Brain Disorders

Five contributions in this Special Issue investigate alterations in functional brain connectivity across major neuropsychiatric and neurological disorders, including Autism Spectrum Disorder (ASD) and schizophrenia (review Contribution 2), Alzheimer's disease (AD) (Contribution 9), epilepsy (Contributions 10 and 11), and cognitive depression (Contribution 12). This collective focus underscores the critical role of network-based approaches in elucidating the pathophysiological signatures of these complex conditions.

In one such study, Motta et al. (contribution 9) apply novel ^{18}F-Fluorodeoxyglucose Positron Emission Tomography (^{18}F-FDG PET) measures (ventricular uptake, cortical uptake, and their ratio) to probe metabolic dysfunction and aberrant glucose dynamics in patients with AD. These innovative indices reflect not only regional hypometabolism but also broader disturbances in cerebral energy homeostasis. The authors demonstrate that these PET-derived measures are significantly associated with key biological factors, including age, blood–brain barrier integrity, and mitochondrial dysfunction [26]. Notably, distinct metabolic patterns emerged across different *Apolipoprotein E (APOE)* genotypes, highlighting how genetic risk modulates brain energy metabolism in AD [27]. This work advances

https://doi.org/10.3390/app16020789

the use of quantitative PET biomarkers to capture the multifaceted metabolic alterations underlying AD progression. The findings by Motta et al. invite a broader discussion on the potential of metabolic network biomarkers for early detection and personalized prognosis in neurodegenerative disease. Future research integrating these PET measures with functional connectivity data from fMRI or MEG could offer a more comprehensive model linking metabolic deficits to network-level dysregulation [28]. Furthermore, longitudinal studies are needed to determine whether these biomarkers can track disease progression or response to emerging metabolic and neuroprotective therapies [29].

In the next contribution, Amoiridou et al. (Contribution 10) investigate brain network topological variations between patients with temporal lobe epilepsy (TLE) and extratemporal lobe epilepsy (ETLE). Research into the distinctive functional connectivity signatures of epilepsy subtypes is crucial for understanding their divergent neural mechanisms and clinical trajectories [30–32]. While recent studies have increasingly applied graph theory and machine learning to EEG data [33,34], the authors adopt a complementary approach using resting-state functional MRI (rs-fMRI). Their methodology is notable for its comprehensive comparison of connectivity measures: they construct individual brain graphs using three distinct metrics (undirected Pearson correlation, nonlinear undirected mutual information, and directed Granger causality) for TLE patients, ETLE patients, and healthy controls. This multi-method design allows for a robust characterization of both static and directed network properties. The graph analysis revealed that TLE patients exhibit more disassortative networks at lower density levels compared to ETLE patients, indicating differences in how hubs connect within each subtype. Furthermore, the comparison of global centralization features across groups at varying density thresholds highlighted distinct network integration and segregation profiles. These findings suggest that the brain network organization in TLE and ETLE is fundamentally shaped by the unique pathophysiology of each epilepsy type. The identification of such distinct topological signatures offers a promising path toward developing type-specific neuroimaging biomarkers for diagnosis, prognosis, and potentially for guiding targeted treatment strategies [35].

In a related network-focused study, Evans et al. (Contribution 11) investigate resting-state EEG functional connectivity between key large-scale brain networks: the dorsal attention network (DAN), the ventral attention network (VAN), and the salience network (SN), to assess their potential as electrophysiological biomarkers for depression subtypes. Their work aligns with the growing consensus that major depressive disorder (MDD) is a heterogeneous syndrome, necessitating biologically defined subtypes for more precise diagnosis and treatment [36]. The authors examined connectivity patterns across multiple frequency bands. Their analysis revealed that connectivity in the beta and gamma bands was significantly associated with the Anhedonia and Cognitive depression subtypes across and within all three networks. In contrast, alpha band connectivity showed no significant associations, and only a single significant finding emerged for the Mood or Somatic subtypes. This frequency- and subtype-specific pattern underscores the distinct neurophysiological profiles underlying different clinical presentations of depression. These results provide compelling support for the conceptualization of depression as a heterogeneous condition with identifiable neural signatures [37,38]. By identifying novel electrophysiological signatures, specifically beta/gamma hyperconnectivity in attention and salience networks for the Anhedonia and Cognitive subtypes, this study moves beyond symptom-based classification toward a more objective, network-based taxonomy. Such an approach holds significant promise for developing personalized neurobiomarkers and guiding subtype-specific therapeutic interventions [39,40].

https://doi.org/10.3390/app16020789

3.4. Applications in Sport and Exercise Programs

A pilot study by Poinsard et al. (Contribution 12) bridges functional connectivity research with applied sports science. The authors investigate the neural and physiological responses of six trained male cyclists during two distinct exercise protocols: a traditional Incremental Exercise Test (IET) and a Self-Paced $\dot{V}O_2max$ (SPV) test. Neural activity was continuously recorded via EEG, alongside comprehensive physiological monitoring of gas exchange, heart rate, stroke volume, and power output. The IET followed a conventional fixed-intensity ramp, while the SPV protocol allowed participants to self-regulate intensity using ratings of perceived exertion (RPE), a method grounded in psychophysiological models of exercise regulation [41]. The EEG analysis revealed divergent neural efficiency profiles between the protocols. During the SPV test, beta band power spectral densities increased initially but stabilized after approximately 80% of the test duration, suggesting a plateau in neural demand and effective cognitive management of effort. In contrast, the IET elicited a continuous, monotonic increase in beta activity, indicating escalating neural resource allocation and a higher likelihood of premature central fatigue [42]. These neural patterns were paralleled by differences in perceived exertion and performance sustainability. The findings provide novel evidence that self-paced exercise, guided by RPE, optimizes neural efficiency and delays the onset of fatigue compared to externally imposed, fixed-intensity protocols. This supports the psychobiological model of endurance performance, which emphasizes the role of the brain in pacing and fatigue management [43]. Consequently, this study highlights the potential of EEG-derived metrics as objective biomarkers for optimizing athletic training regimens and underscores the value of neuroergonomic approaches in sports science [44].

4. Perspective Study

The Special Issue concludes with a perspective study by Falsaperla et al. (Contribution 13), which explores the evolving paradigm of precision medicine for epileptic and developmental encephalopathies (E/DEs). These severe, genetically heterogeneous conditions are characterized by early-onset seizures and profound developmental impairments. The authors provide a comprehensive overview of the field, emphasizing the critical role of genetic insights in designing targeted therapies and advocating for a multidisciplinary clinical framework. They also candidly address the significant barriers to widespread implementation, including diagnostic delays, limited accessibility to genetic testing, and a paucity of robust clinical evidence for novel interventions. Falsaperla et al. envision a future for encephalopathy management centered on therapies that move beyond symptomatic control to directly address underlying genetic and molecular pathophysiology, offering a more effective and individualized standard of care. Building upon this perspective, a critical and complementary research frontier lies in the development of methods for the prediction and preemptive control of seizures. The future of epilepsy management will likely integrate precision medicine with advanced neuroengineering. This involves leveraging extreme event theory [45] and modern machine learning techniques [46] to analyze neural data for early seizure prediction. Such systems could enable timely intervention by sending inhibitory signals via closed-loop neuromodulation devices. Promising modalities for this intervention include Deep Brain Stimulation (DBS) [47,48] and Vagus Nerve Stimulation (VNS) [49,50], which are evolving from open-loop to responsive, adaptive systems. The convergence of genetic precision, predictive analytics, and responsive neuromodulation heralds a transformative era where encephalopathy management is not only personalized but also proactive, aiming to prevent seizures before they clinically manifest.

https://doi.org/10.3390/app16020789

5. Conclusions

This Special Issue presents a compelling cross-section of how functional connectivity research is transforming our understanding of the brain. The collected papers bridge fundamental discovery with real-world application, revealing the neural architecture of decision-making, social interaction, cognitive endurance, and clinical disorders. By weaving together methodologies from electrophysiology to artificial intelligence, the work showcased here moves beyond correlation toward prediction and mechanism. Ultimately, these contributions affirm that the study of connected brain networks is not merely an academic pursuit but a foundational tool for advancing healthcare, technology, and our fundamental understanding of human experience.

Acknowledgments: I thank the authors, reviewers, and editorial team for their invaluable work in bringing this issue to fruition. I invite the scientific community to explore these pages and join in advancing this fascinating frontier.

Conflicts of Interest: The author declares no conflicts of interest.

List of Contributions:

1. Bublíková, I.; Mareček, S.; Krajča, T.; Malá, C.; Dušek, P.; Krupička, R. Structural Connectivity of the Substantia Nigra: A Comprehensive Review of Diffusion Imaging and Tractography Studies. *Appl. Sci.* **2025**, *15*, 7902. https://doi.org/10.3390/app15147902.
2. Andreou, M.; Skrimpa, V.; Peristeri, E. Neurological Underpinnings of Socio-Cognitive Dysfunction in Schizophrenia and Autism Spectrum Disorder: Evidence from "Broken" Mirror Neurons. *Appl. Sci.* **2025**, *15*, 6629. https://doi.org/10.3390/app15126629.
3. Rovelli, K.; Daffinà, A.; Balconi, M. Metacognitive Modulation of Cognitive-Emotional Dynamics Under Social-Evaluative Stress: An Integrated Behavioural–EEG Study. *Appl. Sci.* **2025**, *14*, 10678. https://doi.org/10.3390/app151910678.
4. Angioletti, L.; Rovelli, K.; Acconito, C.; Daffinà, A.; Balconi, M. Electrophysiological Hyperscanning of Negotiation During Group-Oriented Decision-Making. *Appl. Sci.* **2025**, *15*, 6073. https://doi.org/10.3390/app15116073.
5. Carvalho, V.; Liu, M.; Harezlak, J.; Estrada Gómez, A.M.; Goñi, J. Functional Connectome Fingerprinting Through Tucker Tensor Decomposition. *Appl. Sci.* **2024**, *15*, 4821. https://doi.org/10.3390/app15094821.
6. Pisarchik, A.N.; Peña Serrano, N.; Escalante Puente de la Vega, W.; Jaimes-Reátegui, R. Hypergraph Analysis of Functional Brain Connectivity During Figurative Attention. *Appl. Sci.* **2024**, *15*, 3833. https://doi.org/10.3390/app15073833.
7. Mizrahi, D.; Laufer, I.; Zuckerman, I. Predicting Attachment Class Using Coherence Graphs: Insights from EEG Studies on the Secretary Problem. *Appl. Sci.* **2025**, *15*, 9009. https://doi.org/10.3390/app15169009.
8. Vakorin, V.A.; Liaqat, T.; Liaqat, H.; Doesburg, S.M.; Medvedev, G.; Moreno, S. Slower Ageing of Cross-Frequency Coupling Mechanisms Across Resting-State Networks Is Associated with Better Cognitive Performance in the Picture Priming Task. *Appl. Sci.* **2025**, *15*, 6880. https://doi.org/10.3390/app15126880.
9. Motta, C.; Bonomi, C.G.; Poli, M.; Mercuri, N.B.; Martorana, A.; Chiaravalloti, A. 18F-Fluorodeoxyglucose Uptake in Cerebrospinal Fluid Reflects Both Brain Glucose Demand and Impaired Blood–Brain Barrier Transport in Alzheimer's Disease. *Appl. Sci.* **2025**, *15*, 5677. https://doi.org/10.3390/app15105677.
10. Amoiridou, D.; Gkiatis, K.; Kakkos, I.; Garganis, K.; Matsopoulos, G.K. Multi-Graph Assessment of Temporal and Extratemporal Lobe Epilepsy in Resting-State fMRI. *Appl. Sci.* **2024**, *14*, 8336. https://doi.org/10.3390/app14188336.
11. Evans, I.D.; Sharpley, C.F.; Bitsika, V.; Vessey, K.A.; Williams, R.J.; Jesulola, E.; Agnew, L.L. Differences in EEG Functional Connectivity in the Dorsal and Ventral Attentional and Salience Networks Across Multiple Subtypes of Depression. *Appl. Sci.* **2025**, *15*, 1459. https://doi.org/10.3390/app15031459.

https://doi.org/10.3390/app16020789

12. Poinsard, L.; Palacin, F.; Hashemi, I.S.; Billat, V. Neural and Cardio-Respiratory Responses During Maximal Self-Paced and Controlled-Intensity Protocols at Similar Perceived Exertion Levels: A Pilot Study. *Appl. Sci.* **2024**, *14*, 10551. https://doi.org/10.3390/app142210551.
13. Falsaperla, R.; Sortino, V.; Pavone, P. Is Precision Therapy in Infantile-Onset Epileptic Encephalopathies Still Too Far to Call Upon? *Appl. Sci.* **2024**, *15*, 2371. https://doi.org/10.3390/app15052372.

References

1. Friston, K.J. Functional and effective connectivity: A review. *Brain Connect.* **2011**, *1*, 13–36. [CrossRef]
2. Hramov, A.E.; Frolov, N.S.; Maksimenko, V.A.; Kurkin, S.A.; Kazantsev, V.B.; Pisarchik, A.N. Functional networks of the brain: From connectivity restoration to dynamic integration. *Physics–Uspekhi* **2021**, *64*, 584–616. [CrossRef]
3. Kanai, R.; Bahrami, B.; Rees, G. Cortical network dynamics of perceptual decision-making in the human brain. *Front. Hum. Neurosci.* **2011**, *5*, 21. [CrossRef]
4. Hramov, A.E.; Maksimenko, V.A.; Pisarchik, A.N. Physical principles of brain-computer interfaces and their applications for rehabilitation, robotics and control of human brain states. *Phys. Rep.* **2021**, *918*, 1–133. [CrossRef]
5. Bassett, D.S.; Sporns, O. Network neuroscience. *Nat. Neurosci.* **2017**, *20*, 353–364. [CrossRef]
6. Amstutz, D.; Sousa, M.; Maradan-Gachet, M.E.; Debove, I.; Lhommée, E.; Krack, P. Psychiatric and cognitive symptoms of Parkinson's disease: A life's tale. *Nat. Neurosci.* **2025**, *181*, 265–283. [CrossRef] [PubMed]
7. Vucurovic, K.; Caillies, S.; Kaladjian, A. Neural correlates of mentalizing in individuals with clinical high risk for schizophrenia: ALE meta-analysis. *Front. Psychiatry* **2021**, *12*, 634015. [CrossRef]
8. Sasaki, A.; Suzuki, E.; Homma, K.; Mura, N.; Suzuki, K. Duration in action observation therapy: Manual dexterity, mirror neuron system activity, and subjective psychomotor effort in healthy adults. *Brain Sci.* **2025**, *15*, 457. [CrossRef]
9. Crivelli, D.; Acconito, C.; Balconi, M. Emotional and cognitive "route" in decision-making process: The relationship between executive functions. *Brain Sci.* **2024**, *14*, 734. [CrossRef] [PubMed]
10. Froese, T.; Lam Loh, C.; Putri, F. Inter-brain desynchronization in social interaction: A consequence of subjective involvement? *Front. Hum. Neurosci.* **2024**, *18*, 1359841. [CrossRef] [PubMed]
11. Finn, E.; Shen, X.; Scheinost, D.; Rosenberg, M.C.; Huang, J.; Chun, M.M.; Papademetris, X.; Constable, R.T. Functional connectome fingerprinting: Identifying individuals using patterns of brain connectivity. *Nat. Neurosci.* **2015**, *18*, 1664–1671. [CrossRef]
12. Lu, J.; Yan, T.; Yang, L.; Zhang, X.; Li, J.; Li, D.; Xiang, J.; Wang, B. Brain fingerprinting and cognitive behavior predicting using functional connectome of high inter-subject variability. *NeuroImage* **2024**, *295*, 120651. [CrossRef]
13. Maksimenko, V.A.; Runnova, A.E.; Zhuravlev, M.O.; Protasov, P.; Kulanin, R.; Khramova, M.V.; Pisarchik, A.N.; Hramov, A.E. Human personality reflects spatio-temporal and time-frequency EEG structure. *PLoS ONE* **2019**, *13*, e0197642. [CrossRef]
14. Mantwill, M.; Gell, M.; Krohn, S.; Finke, C. Brain connectivity fingerprinting and behavioural prediction rest on distinct functional systems of the human connectome. *Commun. Biol.* **2022**, *5*, 261. [CrossRef]
15. Zhao, K.; Chen, P.; Alexander-Bloch, A.; Wei, Y.; Dyrba, M.; Yang, F.; Kang, X.; Wang, D.; Fan, D.; Ye, S.; et al. A neuroimaging biomarker for Individual brain-related abnormalities In neurodegeneration (IBRAIN): A cross-sectional study. *eClinicalMedicine* **2023**, *65*, 102276. [CrossRef]
16. Sarter, M.; Gehring, W.J.; Kozak, R. More attention must be paid: The neurobiology of attentional effort. *Brain Res. Rev.* **2006**, *51*, 145–160. [CrossRef] [PubMed]
17. Escalante Puente de la Vega, W.; Pisarchik, A.N. Effective brain connectivity analysis during endogenous selective attention based on Granger causality. *Appl. Sci.* **2025**, *15*, 3833. [CrossRef]
18. Ferguson, T.S. Who solved the secretary problem? *Stat. Sci.* **1989**, *4*, 282–289. [CrossRef]
19. Bearden, J.N. A new secretary problem with rank-based selection and cardinal payoffs. *J. Math. Psychol.* **2006**, *50*, 58–59. [CrossRef]
20. Gabrieli, J.D.; Ghosh, S.S.; Whitfield-Gabrieli, S. A new secretary problem with rank-based selection and cardinal payoffs. *Neuron* **2015**, *85*, 11–26. [CrossRef] [PubMed]
21. Canolty, R.T.; Knight, R.T. The functional role of cross-frequency coupling. *Trends Cogn. Sci.* **2010**, *14*, 506–515. [CrossRef]
22. Cole, J.H.; Franke, K. Predicting age using neuroimaging: Innovative brain ageing biomarkers. *Trends Neurosci.* **2017**, *40*, 681–690. [CrossRef]
23. Tanaka, M.; Yamada, E.; Mori, F. Neurophysiological markers of early cognitive decline in older adults: A mini-review of electroencephalography studies for precursors of dementia. *Front. Aging Neurosci.* **2024**, *15*, 1486481. [CrossRef]
24. Zhang, R.; Yi, F.; Mao, H.; Huang, Z.; Wang, K; Zhang, J. Brain age gap as a predictive biomarker that links aging, lifestyle, and neuropsychiatric health. *Commun. Med.* **2025**, *5*, 441. [CrossRef] [PubMed]

25. Dartora, C.; Marseglia, A.; Mårtensson, G.; Rukh, G.; Dang, J.; Muehlboeck, J.-S.; Wahlund, L.-o.; Moreno, R.; Barroso, J.; Ferreira, D.; et al. A deep learning model for brain age prediction using minimally preprocessed T1w images as input. *Front. Aging Neurosci.* **2024**, *15*, 1303036. [CrossRef]
26. Mosconi, L.; De Santi, S.; Li, Y.; Li, J.; Zhan, J.; Tsui, W.H.; Boppana, M.; Pupi, A.; de Leon, M.J. Visual rating of medial temporal lobe metabolism in mild cognitive impairment and Alzheimer's disease using FDG-PET. *Eur. J. Nucl. Med. Mol. Imaging* **2006**, *33*, 210–221. [CrossRef] [PubMed]
27. Reiman, E.M.; Quiroz, Y.T.; Fleisher, A.S.; Chen, K.; Velez-Pardo, C.; Jimenez-Del-Rio, M.; Fagan, A.M.; Shah, A.R.; Alvarez, S.; Arbelaez, A.; et al. Brain imaging and fluid biomarker analysis in young adults at genetic risk for autosomal dominant Alzheimer's disease in the presenilin 1 E280A kindred: A case-control study. *Lancet Neurol.* **2012**, *11*, 1048–1056. [CrossRef] [PubMed]
28. Schoonhoven, D.N.; Coomans, E.M.; Millán, A.P.; van Nifterick, A.M.; Visser, D.; Ossenkoppele, R.; Tuncel, H.; van der Flier, W.M.; Golla, S.S.V.; Scheltens, P.; et al. Tau protein spreads through functionally connected neurons in Alzheimer's disease: A combined MEG/PET study. *Brain* **2023**, *146*, 4040–4054. [CrossRef]
29. Yakoub, Y.; Ashton, N.J.; Strikwerda-Brown, C.; Montoliu-Gaya, L.; Karikari, T.K.; Kac, P.R.; Gonzalez-Ortiz, F.; Gallego-Rudolf, J.; Meyer, P-F.; St-Onge, F.; et al. Longitudinal blood biomarker trajectories in preclinical Alzheimer's disease. *Alzheimers Dement.* **2023**, *19*, 5620–5631. [CrossRef]
30. Coito, A.; Genetti, M.; Pittau, F.; Iannotti, G.R.; Thomschewski, A.; Höller, Y.; Trinka, E.; Wiest, R.; Seeck, M.; Michel, C.M.; et al. Altered directed functional connectivity in temporal lobe epilepsy in the absence of interictal spikes: A high-density EEG study. *Epilepsia* **2016**, *57*, 402–411. [CrossRef]
31. Carboni, M.; De Stefano, P.; Vorderwülbecke, B.J.; Tourbier, S.; Mullier, E.; Rubega, M.; Momjian, S.; Schaller, K.; Hagmann, P.; Seeck, M.; et al. Abnormal directed connectivity of resting state networks in focal epilepsy. *NeuroImage Clin.* **2020**, *27*, 102336. [CrossRef]
32. Elkholy, M.M. Disruption of EEG resting state functional connectivity in patients with focal epilepsy. *Egypt. J. Neurol. Psychiatr. Neurosurg.* **2023**, *59*, 122. [CrossRef]
33. Gromov, N.V.; Lebedeva, A.V.; Sharkov, A.A.; Grebenyukova, A.D.; Malkov, A.E.; Gerasimova, S.A.; Smirnov, L.A.; Levanova, T.A.; Pisarchik, A.N. Automated sleep spindle analysis in epilepsy EEG using deep learning. *Technologies* **2025**, *13*, 524. [CrossRef]
34. Myers, P.; Gunnarsdottir, K.M.; Adam Li, A.; Razskazovskiy, V.; Craley, J.; Chandler, A.; Wyeth, D.; Wyeth, E.; Zaghloul, K.A.; Inati, S.K.; et al. Diagnosing epilepsy with normal interictal EEG using dynamic network models. *Ann. Neurol.* **2025**, *97*, 907–918. [CrossRef] [PubMed]
35. Machetanz, K.; Weinbrenner, E.; Wuttke, V.W.; Ethofer, S.; Helfrich, R.; Kegele, J.; Lauxmann, S.; Alber, M.; Rona, S.; Tatagiba, M.; et al. Connectome-based disentangling of epilepsy networks from insular stereoelectroencephalographic leads. *Front. Neurol.* **2025**, *15*, 1460453. [CrossRef]
36. Williams, L.M. Precision psychiatry: A neural circuit taxonomy for depression and anxiety. *Lancet Psychiatry* **2016**, *3*, 472–480. [CrossRef] [PubMed]
37. Pisarchik, A.N.; Andreev, A.V.; Kurkin, S.A.; Stoyanov, D.; Badarin, A.A.; Paunova, R.; Hramov, A.E. Topology switching during window thresholding fMRI-based functional networks of patients with major depressive disorder: Consensus network approach. *Chaos* **2023**, *33*, 093122. [CrossRef]
38. Stone, B.T.; Desrochers, P.C.; Nateghi, M.; Chitadze, L.; Yang, Y.; Cestero, G.I.; Bouzid, Z.; Chen, C.; Bull, R.; Bremner, J.D.; et al. Decoding depression: Event related potential dynamics and predictive neural signatures of depression severity. *J. Affect. Disord.* **2025**, *391*, 119893. [CrossRef] [PubMed]
39. Redlich, R., Almeida, J. R., Grotegerd, D.; Opel, N.; Kugel, H.; Heindel, W.; Arolt, V.; Phillips, M.L.; Dannlowski, U. Brain morphometric biomarkers distinguishing unipolar and bipolar depression: A voxel-based morphometry–pattern classification approach. *JAMA Psychiatry* **2014**, *71*, 1222–1230. [CrossRef]
40. Abbasi, A.A.; Jokhio, F.; Norouziyan, F.; Moradikor, N. Neurobiological and neuroimaging biomarkers: A narrative review of precision medicine for diagnosing neurodegenerative disorders. *NeuroMarkers* **2025**, *in press*. [CrossRef]
41. Wallman-Jones, A.; Perakakis, P.; Tsakiris, M.; Schmidt, M. Physical activity and interoceptive processing: Theoretical considerations for future research. *Int. J. Psychophysiol.* **2021**, *166*, 38–49. [CrossRef] [PubMed]
42. Meeusen, R.; Roelands, B. Fatigue: Is it all neurochemistry? *Eur. J. Sport Sci.* **2018**, *18*, 37–46. [CrossRef]
43. Marcora, S.M. Do we really need a central governor to explain brain regulation of exercise performance? *Eur. J. Appl. Physiol.* **2008**, *104*, 929–931. [CrossRef]
44. Mehta, R.K.; Parasuraman, R. Neuroergonomics: A review of applications to physical and cognitive work. *Front. Hum. Neurosci.* **2013**, *7*, 00889. [CrossRef]
45. Frolov, N.S.; Grubov, V.V.; Maksimenko, V.A.; Pavlov, A.N.; Sitnikova, E.; Pisarchik, A.N.; Kurths, J.; Hramov, A.E. Statistical properties and predictability of extreme epileptic events. *Sci. Rep.* **2019**, *9*, 7243. [CrossRef]

https://doi.org/10.3390/app16020789

46. Malkov, A.E.; Lebedeva, A.V.; Gerasimova, S.A.; Levanova, T.A.; Smirnov, L.A.; Sharkov, A.A.; Pisarchik, A.N. Multiparametric machine learning for predicting epileptic hyperexcitability from interitcal EEG background activity. *Commun. Nonlin. Sci. Numer. Simul.* **2026**, *152*, 109210. [CrossRef]
47. Salanova, V. Deep brain stimulation for epilepsy. *Epilepsy Behav.* **2018**, *88*, 21–24. [CrossRef] [PubMed]
48. Kaur Dhaliwal, J.; Ruiz-Perez, M.; Chari, A.; Piper, R.J.; Tisdall, M.M.; Hart, M. Deep brain stimulation for epilepsy: A systematic review and meta-analysis of randomized and non-randomized studies of thalamic targeting. *Epilepsy Res.* **2025**, *216*, 107607. [CrossRef]
49. González, H.F.J.; Yengo-Kahn, A.; Englot, D.J. Vagus nerve stimulation for the treatment of epilepsy. *Neurosurg. Clin. N. Am.* **2019**, *30*, 219–230. [CrossRef]
50. Cocoli, K.; Curley, J.; Rohatgi, P.; Abdennadher, M. Vagus nerve stimulation therapy for epilepsy: Mechanisms of action and therapeutic approaches. *Brain Sci.* **2025**, *15*, 1236. [CrossRef]

https://doi.org/10.3390/app16020789

applied sciences

Review

Structural Connectivity of the Substantia Nigra: A Comprehensive Review of Diffusion Imaging and Tractography Studies

Iva Bublíková [1], Stanislav Mareček [2], Tomáš Krajča [1], Christiane Malá [1], Petr Dušek [2] and Radim Krupička [1,*]

[1] Faculty of Biomedical Engineering, Czech Technical University in Prague, 272 01 Kladno, Czech Republic; iva.bublikova@fbmi.cvut.cz (I.B.); tomas.krajca@cvut.cz (T.K.); malachri@fbmi.cvut.cz (C.M.)
[2] Department of Neurology and Center of Clinical Neuroscience, First Faculty of Medicine, Charles University and General University Hospital in Prague, 128 21 Prague, Czech Republic; stanislav.marecek@vfn.cz (S.M.); petr.dusek@lf1.cuni.cz (P.D.)
* Correspondence: radim.krupicka@fbmi.cvut.cz

Abstract

The substantia nigra (SN) has historically been regarded as a pivotal element of the brain's motor circuits, notably within the context of the nigrostriatal pathway and Parkinson's disease. However, recent advancements in neuroimaging techniques, particularly tractography, have facilitated the delineation of its anatomical projections. These techniques have revealed the involvement of the SN in a more extensive array of functional networks encompassing cognitive, emotional, and motivational domains. This paper reviews the current knowledge on the structural connectivity of the SN in humans based on diffusion tensor imaging and tractography. It summarizes the main projection pathways, including classical and newly described connections, such as the direct SN pars compacta connections to the thalamus, cortico–neural inputs, and connections to limbic regions and the hippocampus. Furthermore, the text delves into the distinctions between the SN pars compacta and SN pars reticulata subregions, exploring their parcellation based on connectivity. The paper demonstrates that the SN is a functionally diversified nucleus, the implications of which are significant for the understanding of both motor and neuropsychiatric disorders. The present study addresses the paucity of comprehensive treatment in this area and provides a framework for further research on dopaminergic circuits.

Keywords: substantia nigra; tractography; connection

1. Introduction

Diffusion tensor imaging (DTI) is a diffusion-weighted magnetic resonance imaging (MRI) technique that quantifies the diffusivity, its direction, and anisotropy of water diffusion in voxels [1–6]. DTI can therefore be used for the reconstruction of the primary fiber direction within the white matter [7–10]. The calculation of parameters such as fractional anisotropy, axial, radial, and mean diffusivity [11–14] enables further description of microstructural properties of white matter tracts within the voxel, implying their potential damage or reorganization [1,13,15–18]. Fractional anisotropy (FA) quantifies the degree of directional dependence of water diffusion in biological tissue by measuring the variance of the diffusion tensor's eigenvalues normalized to its trace [5,12,14,19]. Mean diffusivity (MD) is defined as the arithmetic mean of the diffusion tensor's eigenvalues, providing an index of the overall magnitude of water diffusion independent of its directional bias [12,14,19].

 https://doi.org/10.3390/app15147902

Building on these principles, tractography may be used to reconstruct entire white matter tracts [2,19–21]. In deterministic tractography, the algorithm invariably follows paths of greatest diffusivity creating streamlines, theoretically reconstructing the underlying tracts [3,11,19,20]. Conversely, probabilistic tractography randomly samples the direction according to a distribution function and generates a multitude of possible trajectories [3,8,19–21]. However, tractography has limitations. It cannot determine the direction of the connection (afferent or efferent) [12,22,23], it has certain restrictions when it comes to investigating white matter (WM) areas with various fiber orientations and numerous associative fiber tracts [24,25], and can produce both false-positive and false-negative associations, complicating the interpretation of results [26,27].

The substantia nigra (SN) and the ventral tegmental area (VTA) are two key dopaminergic midbrain structures that play a crucial role in basal ganglia (BG) circuits [28–32]. The SN primarily projects to the putamen via the nigrostriatal pathway (connections from the SN pars compacta to the dorsal striatum) [28,33–36] (see Figure 1A). The VTA, in contrast, has diverse neuronal populations that project to various forebrain regions, including the nucleus accumbens and amygdala [29,37–41]. The SN has traditionally been divided into the dopaminergic part (the pars compacta (SNc)) and the GABAergic part (the pars reticulata (SNr)) [29,31,42–45]. The SNc projects to the striatum in order to modulate movement, while the SNr receives inputs from the striatum and the subthalamic nucleus, and sends outputs to the thalamus (nucleus ventralis anterior and lateralis) [28,34,46,47] (see Figure 1B). Tractography has facilitated non-invasive mapping of connections [48–52]. Recent studies have confirmed the presence of connections between the SN and various cortical areas, with the prefrontal, motor, and somatosensory cortices being of note [7,34,53]. The present findings suggest that the SN is not merely an isolated component of the basal ganglia but rather is involved in broader brain networks [7,34,36,54–56]. This is of crucial importance for understanding the pathophysiology of neurological and psychiatric disorders involving the SN [7,34].

However, despite the growing number of studies focusing on individual SN connections, a comprehensive review that systematically maps individual pathways, their anatomy, functional significance, and clinical implications is lacking. The extant literature is characterized by a preponderance of isolated descriptions of individual connections. The objective of this review is to provide an overview of the current knowledge on SN structural connectivity, with a particular emphasis on its functionally specialized projections. The text synthesizes both conventionally delineated pathways and recently identified connections, encompassing their topography, fiber types, participation in functional circuits, and pertinence to clinical neurology and neuropsychiatry. Furthermore, the differences between the SNc and SNr subregions are considered, alongside their connectivity diversity and the potential for parcellation by functional domains.

The objective of this study is twofold: firstly, to address a significant gap in the existing literature by providing an overview of the subject matter; and secondly, to offer a systematic knowledge of the structural connectivity of the SN and its functional relevance in the human brain.

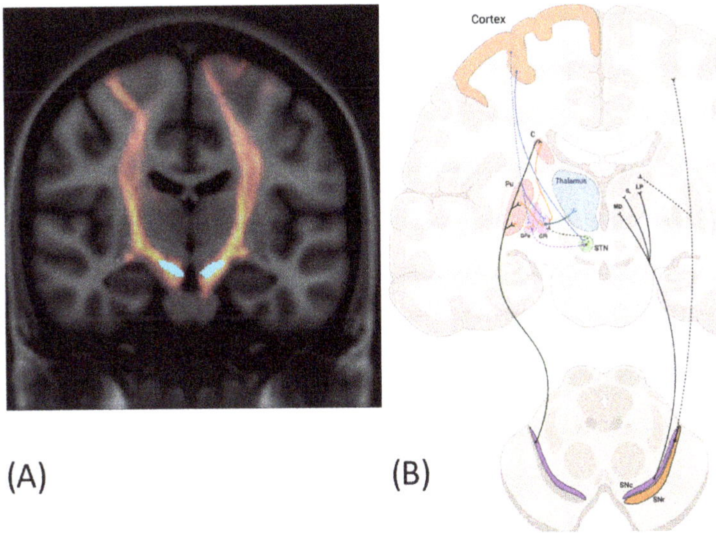

Figure 1. (**A**) Tractographic visualization of fiber pathways originating from the substantia nigra, derived from real diffusion MRI data. (**B**) Schematic representation of the nigrothalamic system. Reproduced from Cirillo et al., 2025 [28]. Efferent connections of the SNc to the dorsal striatum (caudate and putamen) (**left**) and to thalamic nuclei (MD, IL, LP) (**right**), together with nigro–cortical fibers (dashed black line). Direct, indirect and hyperdirect pathways are represented on the left side. C, caudate; Pu, putamen; MD, medio dorsal nucleus; IL, intralaminar nuclei; LP, latero posterior nucleus; GPe, external globus pallidus; GPi, internal globus pallidus; STN, subthalamic nucleus; SNc, substantia nigra pars compacta; SNr, substantia nigra pars reticulata.

2. Materials and Methods

This review was designed in accordance with the PRISMA 2020 recommendations for systematic reviews, considering its adapted application to short review studies. The objective of this study was to provide a comprehensive overview of the current state-of-the-art and novel findings concerning the utilization of DTI and tractography for the investigation of SN connectivity in humans. A systematic search for relevant literature was performed in March 2025 in the PubMed, Web of Science, and ScienceDirect databases. Furthermore, the SciSpace tool was utilized, and the outputs were subjected to a rigorous process of validation to ensure their relevance and appropriateness for inclusion. The temporal scope of the search covered articles published up to March 2025, with the aim of including recent developments in the field.

The following were included in the review: studies conducted on human subjects (including healthy volunteers and clinical populations), studies using DTI or tractography methods to investigate SN connectivity, and articles published in any language. Studies performed in animal models, post-mortem studies (ex vivo imaging), articles that did not make a relevant contribution to the topic of connectivity, and abstracts without available full text were excluded from the analysis.

A single reviewer conducted the screening and selection process for the articles. The initial phase of this process involved searching for articles based on keywords (substantia nigra, tractography, connection) and evaluating the titles and abstracts. Following this preliminary assessment, a comprehensive full-text examination was performed.

In the first step, a total of 150 records were identified, from the following databases: PubMed (n = 38), Web of Science (n = 28), ScienceDirect (n = 67) and SciSpace (n = 17).

Following the removal of duplicates (30 records), 120 articles were included in the preliminary screening.

Following an evaluation of the titles and abstracts, 75 records were excluded based on the established criteria listed above. The primary reasons for exclusion included a lack of focus on SN, an absence of DTI methods, and overall irrelevance to the subject matter.

A comprehensive full-text screening process was conducted on the remaining 45 articles, resulting in the exclusion of further 13 studies, primarily due to their emphasis on animal models, post-mortem data, or an indirect relevance to the subject matter.

The remaining 32 studies were selected for the final review. In consideration of the nature of the review, it was deemed inappropriate to apply a formal tool for the purpose of assessing the quality of the included studies. The primary objective of this study was to produce a qualitative map of the predominant trends and novel methodological approaches in the domain of SN connectivity, utilizing DTI.

3. Results

3.1. Connection with the Striatum

The tractography reliably identifies the classical pathway from the SNc to the striatum [35,57–59] (see Figure 2). In a considerable number of studies, nigrostriatal fibers have been repeatedly visualized and exhibited different diffusion properties in both healthy subjects and patients with PD [11,14,60,61].

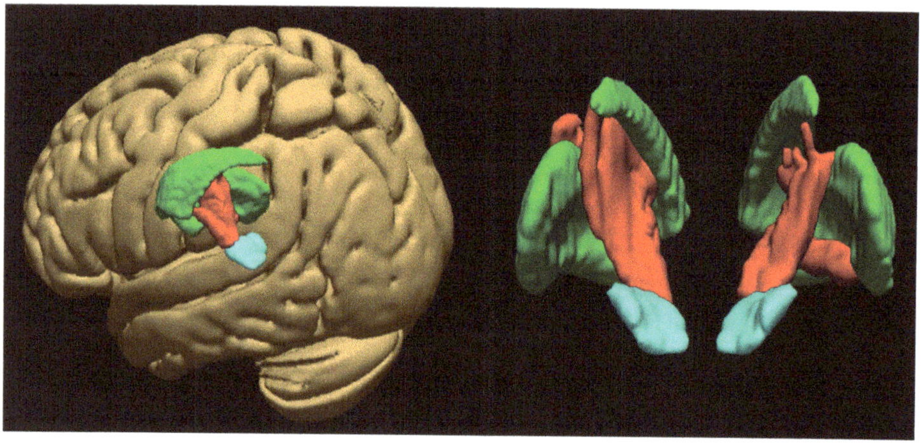

Figure 2. Nigrostriatal connectivity visualized using diffusion MRI. The substantia nigra pars compacta (SNc) are shown in cyan, the dopaminergic projection pathway in red, and the striatum in green.

In the multicenter Parkinson's Progression Markers Initiative study, 50 patients with PD and 27 controls exhibited systematic abnormalities in the diffusion parameters of the nigrostriatal tract, including reduced fractional anisotropy (FA) and changes in radial/axial diffusivity, indicative of loss of fiber integrity [62]. Furthermore, these changes were found to be associated with the severity of motor symptoms and a decline in striatal dopaminergic marker uptake on nuclear medicine methods, thereby substantiating the functional significance of the pathway damage. Tractography has been demonstrated to quantify the integrity of nigrostriatal connections [62] with the caveat that this is closely related to motor deficits. In a study with 21 PD patients, Tan et al. (2015) utilized deterministic DTI tractography to reveal a substantial decrease in FA and streamline count within the nigrostriatal pathway when compared to a control group [11]. Magnetic resonance imaging alterations are indicative of dopaminergic fiber degeneration. Furthermore, an association was observed between streamline count and motor impairment scores, thereby substantiating the clinical significance of connectivity metrics. Current 7T MRI techniques facilitate more detailed imaging of SNs and their projections. However, the increased resolution of these techniques poses a significant challenge in terms of standardization. A study by Shim and Baek (2022) revealed discrepancies in detected fiber counts between 3T and 7T systems, attributable to variations in acquisition parameters [63].

3.2. Connection with the Thalamus

The main basal ganglia output pathway originates in the SNr and the inner segment of the globus pallidus (GPi) and leads to the thalamus. These pathways are referred to as the nigrothalamic and pallidothalamic pathways, respectively [64]. D.H. Kwon et al. (2021) [64] utilized super-resolution track-density imaging (TDI) on 7T MRI to generate three-dimensional maps of the nigrothalamic and pallidothalamic projections and to identify the SNr substructures in detail. The study demonstrates that SNr projects directly to disparate thalamic nuclei. Two such pathways have been identified: the lateral pathway, which originates in the SN, leading through the reticular nucleus and internal capsule and to the ventral anterior nucleus (VA) and ventral lateral posterior nucleus (VLp); and the medial pathway, which originates in the caudal SNr, traverses the SNc, and terminates at the ventromedial nucleus (VM) of the thalamus. As demonstrated in the latter work, the superior and inferior parietal lobes provide parallel but topographically distinct outputs to the thalamus.

A recent finding revealed the existence of a direct connection between SNc dopaminergic neurons and the thalamus. The conventional model proposed by DeLong (1990) posited that the SNc exerts its influence on the thalamus solely indirectly, through the intermediary of the striatum and subsequent downstream circuits [28]. As demonstrated in the study by Cirillo et al. (2025) [28], the authors conducted a selective monitoring of connections through the SNc, thereby ascertaining that, on average, approximately 12% of all streamlines passing through the SNc proceed directly to the thalamus. This nigrothalamic pathway was reproducibly identified in all young healthy subjects in the study and formed a separate bundle not involving other structures (i.e., it was not a branch to the striatum, cortex, or cerebellum). This constitutes quantitative in vivo evidence of a direct SNc link to the thalamus in humans. This finding extends the established model of basal ganglia circuits, indicating that, in addition to the expected inhibitory pathway from the SNr, there is also modulation of the thalamus by dopamine from the SNc. This represents a significant advancement in our understanding of basal ganglia circuits. The authors illustrated the nigrothalamic system in schematic representation as previously seen in Figure 1B [28].

In addition to the direct connections of SNr and SNc to the thalamus, a recent study using generalized q-sampling imaging (GQI) tractography has allowed the complex involvement of the ansa lenticularis (AL) to be described. Li et al. (2022) [65] reconstructed four subcomponents of the AL, one of which, the globus pallidus–substantia nigra (GPSN), demonstrates a direct anatomical connection between these structures. This connection indicates the presence of additional, more precisely organized circuits that integrate basal ganglia output activity not only to the thalamus but also in the opposite direction towards the SN.

3.3. Connection with the Cortex

Until recently, the relationship between the cerebral cortex and SN was believed to be only indirect, with connections via cortex–striatum–SNr or cortex–subthalamic nucleus–

SNr circuits. However, several studies have revealed direct projections from the cortex to the SN, the so-called cortico–nigral pathway.

Cacciola et al. (2016) conducted a study that utilized probabilistic tractography to identify pathways connecting broad regions of the cerebral cortex to the SN [34]. The most prominent connections were observed to be with the prefrontal cortex and with sensorimotor regions, including the gyrus precentralis, postcentralis, and superior parietal lobule. These findings expanded the network of SN connections to include direct cortical connection, a finding that is supported by a study by H.G. Kwon and Jang (2014) [7], which also suggested that the SN has high connectivity with cortical areas. In healthy subjects, the SN demonstrated connections (>70% probability) with virtually all cortical areas tested. The authors observed the highest SN connectivity to areas of the frontal, parietal, and occipital lobes, to the primary motor and somatosensory cortex, and to the cerebellum. The presence of these pathways in humans has contributed to the clarification of certain ambiguities, such as the reasons for the alterations in SN activity during specific cognitive tasks or sensory inputs [34].

3.4. Connection with Limbic Structures

3.4.1. Mesolimbic Projection of the SNc/Ventral Tegmental Area (VTA)

The midbrain dopaminergic system is traditionally divided into the nigrostriatal pathway (SNc $\rightarrow$ dorsal striatum) and the mesolimbic pathway, which leads from the VTA to limbic structures (ventral striatum, amygdala, prefrontal cortex). However, it appears that the SNc itself, particularly its dorsal region adjacent to the VTA, also contributes to limbic innervation [28]. Consequently, SN involvement in the limbic system is predominantly through projections to the ventral striatum and limbic cortex. The ventral striatum exerts its influence on circuits associated with reward and emotion. As demonstrated by Zhang et al. (2017) [66], the orbitofrontal cortex and the anterior cingulum receive dopaminergic inputs from the SNc and the VTA. In turn, these regions send fibers (directly or indirectly) back to the aforementioned regions, modulating the activity of dopaminergic neurons.

3.4.2. Connection to the Hippocampus

In addition to connections with the striatum and cortex, the SN/VTA also appears to be structurally connected to the hippocampus via specific fibers identified by probabilistic tractography. Carmichael et al. (2021) provided a concise overview of the literature on this subject, concluding that SN/VTA projection neurons contribute only marginally to the formation of the hippocampus [47]. Nonetheless, indirect interactions have been demonstrated. For instance, the hippocampus, via the subiculum, exerts influence on the VTA/SNc, and conversely, dopamine from the midbrain modulates hippocampal function, which is important for novelty and memory consolidation [47]. A study by Elliott et al. (2022) [67] demonstrated that the degree of SN/VTA–hippocampus connectivity exhibited a correlation with performance on a task requiring motivated memory encoding. That is to say, the study revealed that learning was influenced by the expectation of reward or punishment. Conversely, SN/VTA connectivity to the striatum was not significant in this cognitive context. The results of this study indicate that individual dopaminergic mesencephalic projections are not only anatomically distinct, but also exhibit functional specialization, in this case to influence memory processes [67].

3.4.3. Functional Diversity of Limbic Connections

A study by García-Gomar et al. (2022) [68] utilized 7T probabilistic tractography to create a detailed structural brainstem connectome. The authors distinguished two parts of the substantia nigra—SN1 (corresponding to the SNr) and SN2 (SNc)—and described their structural connectivity with the brainstem nuclei, cortex and limbic regions. While the first part (SN1) exhibited predominant connections with motor areas, the second part (SN2) demonstrated direct connectivity with the amygdala, the ventral tegmental area, the periaqueductal grey, the reticular formation, and other nuclei implicated in emotion regulation, stress response, and sleep. The results provide support for the hypothesis that the SNc is not only part of the motor and reward circuits, but also an integral node of broader limbic circuits that may be relevant to mood disorders, anxiety, REM atonia, or attention disorders.

3.4.4. Projection to the Nucleus Accumbens

A significant finding of the research conducted by MacNiven et al. (2020) [69] is the identification of a specific projection from the midbrain (including the SN and VTA) to the NAc that passes through the lateral hypothalamus. This projection corresponds to the classical description of the inferior portion of the medial forebrain bundle. Utilizing diffusion parameters (FA and inverse mean diffusivity (MD)), the authors demonstrated that the diminished structural integrity of this pathway exhibited a robust correlation with heightened impulsivity in healthy subjects. This connectivity was also significantly reduced in patients with stimulant use disorder. Consequently, the results of the latter study demonstrate the pivotal function of SN/VTA–ventral striatum dopaminergic projections in the modulation of motivated behavior and impulsivity. These findings extend the classical concept of the dopaminergic system and confirm the functional diversity of the different subregions of the SN.

3.5. *Regional Subdivisions of the Substantia Nigra*

The SNc and SNr exhibit marked differences in neurochemistry, connectivity, and function. The SNc contains dopaminergic neurons whose axons form ascending projections (i.e., nigrostriatal and mesolimbic) that modulate the activity of striatal circuits [28]. In contrast, the SNr predominantly consists of GABAergic neurons that function as output cells of the basal ganglia. These neurons are analogous to GPi, in that they send inhibitory signals to the motor and associative nuclei of the thalamus and to the brainstem [64].

Tractography has enabled the differentiation of distinct functional subcomponents of the SN based on their connectivity.

Menke et al. (2010) [70] performed DTI segmentation of the SN and divided it into two parts, corresponding to the SNc and SNr. The inner segment (SNc) demonstrated predominant connections with the dorsal striatum (putamen, caudate), globus pallidus, anterior nuclei of the thalamus, and via thalamic pathways with the prefrontal cortex. In contrast, the outer segment (SNr) has been shown to be predominantly connected with the posterior and ventral thalamus, the dorsal striatum, and pathways directed to the premotor and primary motor cortex.

A complementary approach to the segmentation of dopaminergic structures is connectivity-driven parcellation, a method which Basile et al. (2020) [71] applied to the SNc and the adjacent VTA. Utilizing probabilistic tractography and clustering analysis, the researchers categorized the dopaminergic midbrain into three functionally distinct domains: the limbic, prefrontal, and sensorimotor domains. These domains exhibit topographical overlap with the dorsoventral gradient. This approach is consistent with the established functional subdivisions of dopamine neurons in primates and underscores the significant heterogeneity of the SNc/VTA in humans.

Another study by Chowdhury et al. (2013) [72] divided the SN according to connections to the striatum. This revealed that the dorsomedial SN preferentially innervates the ventral striatum (NAc), whereas the ventrolateral SN connects to the dorsal striatum (putamen and caudate).

In addition to the classical division between dorsal and ventral SN projections to the striatum, a recent study using probabilistic tractography demonstrated increased structural connectivity between the VTA and the SN and NAc in individuals who carry the fat mass and obesity-associated (FTO) risk gene. This connectivity was found to be significantly higher than in the control group and was also associated with higher levels of motor impulsivity. As demonstrated by the authors, the structural connection between the SN/VTA and the NAc partially mediated the relationship between genetic risk and impulsive behavior [73].

A study by Li et al. (2022) [65] utilizing GQI tractography described four subcomponents of the AL, including the GPSN branch, which is the connection between the globus pallidus and the SN. The present study demonstrated that this pathway runs laterally to the other outputs of the AL and is directed specifically towards the SN. This is evidenced by the presence of a special anatomical bundle connecting these two structures. The presence of the GPSN lends support to the hypothesis that the SN is not merely an exit point (SNr) or an entry node (SNc), but may also be involved in the integration of signals from other nuclei of the basal ganglia, including the globus pallidus.

As illustrated in Table 1, the key structural connections of the SN are outlined, including their anatomical course, functional significance, and related clinical implications.

Table 1. Connections of the substantia nigra.

Pathway	Connection	Significance for Neurology	Study
Nigrostriatal	SNc → dorsal striatum (putamen, caudate)	A key pathway for motor control; its degeneration leads to the motor symptoms of PD	[11,12,14,57,59–63,70]
Nigrothalamic	SNr → VA, VL, VM thalamic nuclei	Influence motor and associative cortical circuits, important for motor and cognitive function, involved in dystonia pathophysiology	[28,64,70]
SNc → thalamus	Dopaminergic connections from the SNc directly to the thalamus	Possible dopamine influence on thalamocortical circuits; potential mechanism for non-motor symptoms of PD	[28]
Cortico–Nigral	Direct projections from the cortex (prefrontal, sensorimotor areas) → SN	Cognitive and sensory modulation of dopamine neurons; SN responds to cortical inputs	[7,34,74]
SN–ventral striatum (mesolimbic pathway)	SNc/VTA → NAc and orbitofrontal cortex	Regulation of reward, emotion, and impulsiveness; key to addictions and behavioral disorders	[28,66,69,72,73]
SN–hippocampus	SN/VTA ↔ hippocampus (direct and indirect connections)	Modulation of motivation, novelty, and learning; relevance to stress; major depressive disorder (MDD)	[47,67]
SN–amygdala, limbic cortex	SN2 (part of SNc) → amygdala, cingulum, orbitofrontal region	Emotion regulation; involved in mood disorders, anxiety, REM atonia	[68]
SN–GPi/ansa lenticularis (GPSN)	Globus pallidus → SN (via specific GPSN branch)	Possible feedback effect of pallidum on SN; integration of BG outputs and feedback	[65]
SNc–subcortical circuits (reticular formation)	SNc ↔ brainstem, VTA, PAG, hypothalamus	Involved in regulation of sleep and autonomic functions; important in MDD, stress, and sleep disorders	[68,75]

4. Discussion

This paper provides a synthesis of the current knowledge on the DTI and tractography-derived structural connectivity of the SN and its relevance to the understanding of neurological and neuropsychiatric disorders. The tractographic reconstructions confirm that the SN is not merely part of the classical motor circuits but rather represents a complex integrative node that links motor, cognitive, and affective brain regions. This perspective extends conventional neuroanatomical frameworks and possesses extensive clinical ramifications, which are elaborated upon in subsequent sections.

The nigrostriatal pathway has received the most research attention of all SN connections. The degeneration of this pathway is the central pathophysiological mechanism of PD. As Zhang et al. (2015) [62] confirmed, an analysis of the available data indicates a close correlation between the diffusion parameters of the pathway in question and the motor deficits in question. A number of studies have indicated the possibility that DTI may assist in the early detection of Parkinson's disease and the subsequent monitoring of its progression [12]. The utilization of DTI markers (e.g., increased diffusivity in SN) is also being explored to distinguish idiopathic PN from atypical Parkinsonian syndromes [12]. Jang and Cho (2022) [76] measured two key parameters from diffusion tensor tractography: fractional anisotropy and tract volume for both the ipsilesional nigrostriatal tract and the corticospinal tract. The results showed that the tract volume of the ipsilesional nigrostriatal tract was strongly positively correlated with the Motricity Index score, which is a measure of motor function in patients who suffered putaminal hemorrhage. Similarly, both the fractional anisotropy and tract volume of the ipsilesional corticospinal tract had strong positive correlations with the Motricity Index score. This means that as the integrity and fiber amount (volume) of these tracts increased, the motor function of the patients improved. Wende et al. (2021) [77] used a tractography technique that extracts directional diffusion information in brain white matter using diffusion tensor imaging (DTI). Tractography reconstructs the corticospinal tract (CST), a major nerve fiber bundle responsible for motion control. This study used an automated workflow to extract the FA (fractional anisotropy) values within the reconstructed CST volume while ensuring that anatomically unrealistic fibers and noise were minimized. The study found that the mean FA in the CST was around 0.4406 and identified a lower cutoff of 0.15 as a reliable threshold for neurosurgical tractography, with infiltrated or compressed CSTs requiring consideration of an even lower cutoff (around 0.1) to accurately depict affected fibers [77]. In addition to diagnostics, structural connectivity of the SN is also studied in the context of therapy; for example, changes in diffusion parameters following the deployment of dopaminergic therapy or in the context of deep brain stimulation may reflect circuit remodeling. The capacity to quantify degenerative changes in the SN–striatum circuit furnishes objective markers for clinical assessment of patient status and treatment efficacy. This renders DTI and other advanced methods a valuable tool not only for research but potentially for clinical use.

While degeneration of the nigrostriatal pathway leads to the classical motor symptoms of PN, abnormalities in other SN pathways, particularly in SNr–thalamic wiring, may underlie the complex motor phenotypes of other disorders. For instance, in cases of dystonia, creation of a lesion (i.e., pallidotomy) or high-frequency stimulation of the GPi have been observed to result in the alleviation of symptoms, which may be attributable to the reorganization of the connections between the GPi, SNr, and thalamus. It has been hypothesized that alterations in the connectivity of these circuits may affect the transmission of inhibitory signals to the thalamus and subsequently to the cortex, with the potential to affect motor output [7,34]. The results of this study underscore the significance of tractographic imaging not only of classical pathways but also of output connections, which may serve as a target for therapeutic intervention in a range of motor disorders.

Lin et al. (2023) [78] explored the microstructural changes in the SN and its tracts to the dorsal striatum and dorsolateral prefrontal cortex in PD patients using diffusion MRI and histopathology. The research involved post-mortem analysis of PD and control donors, revealing increased MD and FA in specific SN tracts in PD compared to controls. These microstructural alterations were associated with dopaminergic degeneration and Lewy neurite pathology, suggesting that diffusion MRI can effectively capture these changes and potentially aid in understanding motor and cognitive impairments in these disorders.

In addition to motor function, the SN also affects a range of behavioral and cognitive functions, as evidenced by impulsivity disorders, addictions, and mood disorders. Impulsivity manifests in diverse forms within the context of the basal ganglia, encompassing motor impulsiveness, characterized by an inability to inhibit action, and decision-making impulsivity, marked by a propensity to seek reward irrespective of risk. A review of the extant literature confirms that dopaminergic inputs from the SN/VTA to limbic and prefrontal regions are significantly involved in modulating memory, learning, and impulsivity [66]. Research by Elliott et al. (2022) [67] demonstrated that structural connectivity between the SN/VTA and the hippocampus is associated with the capacity for memory concerning motivation (e.g., potential reward). Therefore, the SN is hypothesized to function as a central node in the modulation of motivated behavior and learning. Recent research by Liu et al. (2024) [75] utilizing 7T DTI revealed a substantial decrease in connectivity between the SN and GP in patients diagnosed with major depressive disorder (MDD). It is hypothesized that this pathway constitutes a motor component of the BG output system; however, its disruption in MDD suggests the possibility of a regulatory role in motivation, reward, and emotional state. Correlations with insomnia scores and age of onset provide further evidence for the functional importance of this pathway in non-motor symptoms of affective disorders. Dopaminergic projections from the SN/VTA to the ventral striatum have been demonstrated to play a critical role in the regulation of impulsivity and motivated behavior. The research conducted by MacNiven et al. (2020) [69] and Edwin Thanarajah et al. (2023) [73] demonstrated a robust correlation between the structural integrity of these pathways (e.g., the medial forebrain bundle) and measures of impulsiveness. The study further suggested that these pathways can be subject to developmental modulation by genetic factors, such as polymorphisms in the FTO. It has been demonstrated that individuals with higher levels of motor impulsivity, and patients diagnosed with stimulant use disorder, exhibit increased levels of connectivity within these circuits. This finding serves to substantiate their significance in the context of the neurobiology of addiction and behavioral disorders. This provides a novel basis for research, establishing a link between neurodevelopment, genetics, and imaging with a view to comprehending the brain circuits underpinning behavior.

The identification of the direct dopaminergic SNc–thalamus pathway [28] offers novel insights into the non-motor symptoms of PD, including hallucinations, REM sleep disturbances, and cognitive dysfunction. The hypothesis that the direct influence of thalamic dopamine from the SNc may modulate large-scale thalamo–cortical networks could provide a theoretical framework for explaining some non-motor symptoms in PD and other neuropsychiatric conditions. Research by Antal et al. (2014) [79] highlighted that the SN has a dual role in influencing thalamic activity through both inhibitory and excitatory outputs. This dual mechanism suggests a complex synaptic connection that allows the basal ganglia to modulate thalamic activity in a subtle manner, which is essential for various motor and cognitive functions. A study by Oishi et al. (2020) utilizing ex vivo diffusion tensor imaging and tractography identified fiber connections between the substantia nigra and thalamus within the human subthalamic area, to map these pathways [80]. The significance of this direct pathway remains to be fully elucidated; however, its existence serves to reinforce the role of the SN as a central integrator of both motor and cognitive functions, as well as sensory functions.

The evidence reviewed above illustrates that distinct components of the SN network are selectively affected in different pathological conditions, with specific disruptions correlating with clinical symptoms. For instance, degeneration of the nigrostriatal pathway is closely linked to the motor deficits of PD [12,62,70], whereas reduced connectivity between the SN and globus pallidus or limbic structures has been observed in patients with MDD and addiction [67,69,75]. Altered connectivity between the SN and the thalamus or prefrontal cortex may further contribute to cognitive dysfunction and impulsivity [69,73]. In this context, structural connectivity mapping of SN circuits may hold promise for advancing individualized diagnostic and therapeutic strategies. Subject-specific tractography can identify patterns of circuit disruption that underlie motor, affective, or cognitive impairments, potentially guiding personalized interventions. For instance, mapping the integrity of pathways such as the medial forebrain bundle or SN and thalamus projections may inform deep brain stimulation target selection or stratify patients for pharmacological therapies. Integrating diffusion imaging with genetic, behavioral, and clinical markers may further enhance the precision of diagnosis and treatment planning in disorders involving SN dysfunction [67,78].

Building on this expanded network view, the imaging of SN connectivity using advanced tractography represents a promising tool not only for research but potentially for clinical practice. In addition to its association with PD, there is an emerging body of evidence that suggests these pathways may also play a role in a wide range of neuropsychiatric conditions. Future research should aim to validate the use of connectivity markers against longitudinal data, integrating structural and functional methods, and further subdividing individual SN subregions. It is imperative that particular emphasis is placed on the standardization of methodology and the development of applications in the field of personalized medicine.

Considering the complex connectivity of the SN and its role in various pathologies, it is important to take into account additional brain regions that may act as potential nodes within SN-related networks. One such structure is the zona incerta (ZI), a subthalamic area that has attracted increasing interest due to its anatomical and functional connections with components of the SN [81,82]. Ossowska et al. (2020) [83] reported, based on both preclinical findings and therapeutic outcomes in humans, that the ZI maintains widespread anatomical connectivity with the SNr and SNc. They suggested that deep brain stimulation targeting the ZI may exert its therapeutic effects in Parkinson's disease by restoring the function of a broader SN–ZI–thalamus–striatum network. Supporting this notion, recent findings by Londei et al. (2024) revealed that neurons within the ZI exhibit functional connectivity with the SNr, thalamic nuclei, and the caudatoputamen, forming recurrent loops that may contribute to motor and non-motor regulation [84].

5. Conclusions

This paper reviews the current knowledge on SN connectivity and its importance for motor, cognitive, and emotional brain functions. Evidence from the extant literature suggests that the SN is a structurally and functionally diversified node that is connected not only to the striatum but also to the thalamus, hippocampus, limbic regions, and brainstem. In the future, the development of sophisticated brain pathway imaging techniques has the potential to facilitate the identification of biomarkers. These biomarkers may then be used to facilitate early diagnosis, to monitor treatment response, and to develop personalized therapeutic strategies.

Funding: This research was funded by the National Institute for Neurological Research (Programme EXCELES, ID Project No. LX22NPO5107)–Financed by the European Union–Next Generation EU, and was supported by the Grant Agency of the Czech Technical University in Prague, grant No. SGS25/187/OHK4/3T/17.

Conflicts of Interest: The authors declare no conflicts of interest.

Abbreviations

The following abbreviations are used in this manuscript:

AL	Ansa lenticularis
BD	Basal ganglia
C	Caudate
DTI	Diffusion tractography imaging
FA	Fractional anisotropy
GPe	External globus pallidus
GPi	Internal globus pallidus
IL	Intralaminar nuclei
LP	Latero posterior nucleus
MD	Medio dorsal nucleus
MD	Mean diffusivity
MDD	Major depressive disorder
MRI	Magnetic resonance imaging
NAc	Nucleus accumbens
PD	Parkinson's disease
Pu	Putamen
SN	Substantia nigra
SNc	Substantia nigra pars compacta
SNr	Substantia nigra pars reticulata
STN	Subthalamic nucleus
T	Tesla
VA	Ventral anterior nucleus
VLp	Ventral lateral posterior nucleus
VM	Ventromedial nucleus
VTA	Area tegmentalis ventralis
ZI	Zona incerta

References

1. Alexander, D.C.; Dyrby, T.B.; Nilsson, M.; Zhang, H. Imaging brain microstructure with diffusion MRI: Practicality and applications. *NMR Biomed.* **2019**, *32*, e3841. [CrossRef]
2. McMaster, E.M.; Newlin, N.R.; Cho, C.; Rudravaram, G.; Saunders, A.M.; Krishnan, A.R.; Remedios, L.W.; Kim, M.E.; Xu, H.; Schilling, K.G.; et al. Sensitivity of quantitative diffusion MRI tractography and microstructure to anisotropic spatial sampling. *arXiv* **2024**, arXiv:2409.18255. [CrossRef]
3. Smith, R.E.; Calamante, F.; Connelly, A. Mapping connectomes with diffusion MRI: Deterministic or probabilistic tractography? *Magn. Reson. Med.* **2020**, *83*, 787–790. [CrossRef] [PubMed]
4. Mormina, E.; Arrigo, A.; Calamuneri, A.; Granata, F.; Quartarone, A.; Ghilardi, M.F.; Inglese, M.; Di Rocco, A.; Milardi, D.; Anastasi, G.P.; et al. Diffusion tensor imaging parameters' changes of cerebellar hemispheres in Parkinson's disease. *Neuroradiology* **2015**, *57*, 327–334. [CrossRef]
5. Reid, A.T.; Camilleri, J.A.; Hoffstaedter, F.; Eickhoff, S.B. Tract-specific statistics based on diffusion-weighted probabilistic tractography. *Commun. Biol.* **2022**, *5*, 138. [CrossRef] [PubMed]
6. Qu, Y. DTI Tractography. *TNS* **2024**, *44*, 102–110. [CrossRef]
7. Kwon, H.G.; Jang, S.H. Differences in neural connectivity between the substantia nigra and ventral tegmental area in the human brain. *Front. Hum. Neurosci.* **2014**, *8*, 41. [CrossRef]

8. Carrozzi, A.; Gramegna, L.L.; Sighinolfi, G.; Zoli, M.; Mazzatenta, D.; Testa, C.; Lodi, R.; Tonon, C.; Manners, D.N. Methods of diffusion MRI tractography for localization of the anterior optic pathway: A systematic review of validated methods. *NeuroImage Clin.* **2023**, *39*, 103494. [CrossRef]
9. Shepherd, T.M.; Hoch, M.J. MRI-Visible Anatomy of the Brainstem. *Neuroimaging Clin. N. Am.* **2022**, *32*, 553–564. [CrossRef]
10. Xu, C.; Neuroth, T.; Fujiwara, T.; Liang, R.; Ma, K.-L. A Predictive Visual Analytics System for Studying Neurodegenerative Disease Based on DTI Fiber Tracts. *IEEE Trans. Visual. Comput. Graph.* **2023**, *29*, 2020–2035. [CrossRef]
11. Tan, W.-Q.; Yeoh, C.-S.; Rumpel, H.; Nadkarni, N.; Lye, W.-K.; Tan, E.-K.; Chan, L.-L. Deterministic Tractography of the Nigrostriatal-Nigropallidal Pathway in Parkinson's Disease. *Sci. Rep.* **2015**, *5*, 17283. [CrossRef] [PubMed]
12. Zhang, Y.; Burock, M.A. Diffusion Tensor Imaging in Parkinson's Disease and Parkinsonian Syndrome: A Systematic Review. *Front. Neurol.* **2020**, *11*, 531993. [CrossRef] [PubMed]
13. Figley, C.R.; Uddin, M.N.; Wong, K.; Kornelsen, J.; Puig, J.; Figley, T.D. Potential Pitfalls of Using Fractional Anisotropy, Axial Diffusivity, and Radial Diffusivity as Biomarkers of Cerebral White Matter Microstructure. *Front. Neurosci.* **2022**, *15*, 799576. [CrossRef]
14. Bergamino, M.; Keeling, E.G.; Mishra, V.R.; Stokes, A.M.; Walsh, R.R. Assessing White Matter Pathology in Early-Stage Parkinson Disease Using Diffusion MRI: A Systematic Review. *Front. Neurol.* **2020**, *11*, 314. [CrossRef]
15. Wang, J.; Zhang, F.; Zhao, C.; Zeng, Q.; He, J.; O'DOnnell, L.J.; Feng, Y. Investigation of local white matter abnormality in Parkinson's disease by using an automatic fiber tract parcellation. *Behav. Brain Res.* **2020**, *394*, 112805. [CrossRef]
16. Li, Y.; Yang, F.G.; Nguyen, C.T.; Cooper, S.R.; LaHue, S.C.; Venugopal, S.; Mukherjee, P. Independent component analysis of DTI reveals multivariate microstructural correlations of white matter in the human brain. *Hum. Brain Mapp.* **2012**, *33*, 1431–1451. [CrossRef]
17. Kuchling, J.; Brandt, A.U.; Paul, F.; Scheel, M. Diffusion tensor imaging for multilevel assessment of the visual pathway: Possibilities for personalized outcome prediction in autoimmune disorders of the central nervous system. *EPMA J.* **2017**, *8*, 279–294. [CrossRef]
18. Pieper, C.C.; Konrad, C.; Sommer, J.; Teismann, I.; Schiffbauer, H. Structural changes of central white matter tracts in Kennedy's disease—A diffusion tensor imaging and voxel-based morphometry study. *Acta Neurol. Scand.* **2013**, *127*, 323–328. [CrossRef]
19. Yang, J.Y.-M.; Yeh, C.-H.; Poupon, C.; Calamante, F. Diffusion MRI tractography for neurosurgery: The basics, current state, technical reliability and challenges. *Phys. Med. Biol.* **2021**, *66*, 15TR01. [CrossRef] [PubMed]
20. Theisen, F.; Leda, R.; Pozorski, V.; Oh, J.M.; Adluru, N.; Wong, R.; Okonkwo, O.; Dean, D.C.; Bendlin, B.B.; Johnson, S.C.; et al. Evaluation of striatonigral connectivity using probabilistic tractography in Parkinson's disease. *Neuroimage Clin.* **2017**, *16*, 557–563. [CrossRef]
21. Sarwar, T.; Ramamohanarao, K.; Zalesky, A. Mapping connectomes with diffusion MRI: Deterministic or probabilistic tractography? *Magn. Reson. Med.* **2019**, *81*, 1368–1384. [CrossRef]
22. Jbabdi, S.; Johansen-Berg, H. Tractography: Where Do We Go from Here? *Brain Connect.* **2011**, *1*, 169–183. [CrossRef] [PubMed]
23. Behrens, T.; Woolrich, M.; Jenkinson, M.; Johansen-Berg, H.; Nunes, R.; Clare, S.; Matthews, P.; Brady, J.; Smith, S. Characterization and propagation of uncertainty in diffusion-weighted MR imaging. *Magn. Reson. Med.* **2003**, *50*, 1077–1088. [CrossRef] [PubMed]
24. Behrens, T.E.J.; Berg, H.J.; Jbabdi, S.; Rushworth, M.F.S.; Woolrich, M.W. Probabilistic diffusion tractography with multiple fibre orientations: What can we gain? *NeuroImage* **2007**, *34*, 144–155. [CrossRef]
25. Jones, D.K.; Cercignani, M. Twenty-five pitfalls in the analysis of diffusion MRI data. *NMR Biomed.* **2010**, *23*, 803–820. [CrossRef]
26. Schilling, K.G.; Nath, V.; Hansen, C.; Parvathaneni, P.; Blaber, J.; Gao, Y.; Neher, P.; Aydogan, D.B.; Shi, Y.; Ocampo-Pineda, M.; et al. Limits to anatomical accuracy of diffusion tractography using modern approaches. *NeuroImage* **2019**, *185*, 1–11. [CrossRef]
27. Thomas, C.; Ye, F.Q.; Irfanoglu, M.O.; Modi, P.; Saleem, K.S.; Leopold, D.A.; Pierpaoli, C. Anatomical accuracy of brain connections derived from diffusion MRI tractography is inherently limited. *Proc. Natl. Acad. Sci. USA* **2014**, *111*, 16574–16579. [CrossRef]
28. Cirillo, G.; Caiazzo, G.; Franza, F.; Cirillo, M.; Papa, M.; Esposito, F. Evidence for direct dopaminergic connections between substantia nigra pars compacta and thalamus in young healthy humans. *Front. Neural Circuits* **2025**, *18*, 1522421. [CrossRef]
29. Beier, K.T.; Steinberg, E.E.; DeLoach, K.E.; Xie, S.; Miyamichi, K.; Schwarz, L.; Gao, X.J.; Kremer, E.J.; Malenka, R.C.; Luo, L. Circuit Architecture of VTA Dopamine Neurons Revealed by Systematic Input–Output Mapping. *Cell* **2015**, *162*, 622–634. [CrossRef]
30. Miller, D.R.; Guenther, D.T.; Maurer, A.P.; Hansen, C.A.; Zalesky, A.; Khoshbouei, H. Dopamine Transporter Is a Master Regulator of Dopaminergic Neural Network Connectivity. *J. Neurosci.* **2021**, *41*, 5453–5470. [CrossRef]
31. Groenewegen, H.J. The basal ganglia and motor control. *Neural Plast.* **2003**, *10*, 107–120. [CrossRef]
32. Ring, H.; Serra-Mestres, J. Neuropsychiatry of the basal ganglia. *J. Neurol. Neurosurg. Psychiatry* **2002**, *72*, 12–21. [CrossRef]
33. Harris, J.P.; Burrell, J.C.; Struzyna, L.A.; Chen, H.I.; Serruya, M.D.; Wolf, J.A.; Duda, J.E.; Cullen, D.K. Emerging regenerative medicine and tissue engineering strategies for Parkinson's disease. *npj Park. Dis.* **2020**, *6*, 4. [CrossRef]
34. Cacciola, A.; Milardi, D.; Anastasi, G.P.; Basile, G.A.; Ciolli, P.; Irrera, M.; Cutroneo, G.; Bruschetta, D.; Rizzo, G.; Mondello, S.; et al. A Direct Cortico-Nigral Pathway as Revealed by Constrained Spherical Deconvolution Tractography in Humans. *Front. Hum. Neurosci.* **2016**, *10*, 374. [CrossRef] [PubMed]

35. Seo, J.-P.; Koo, D.-K. Aging of the Nigrostriatal Tract in the Human Brain: A Diffusion Tensor Imaging Study. *Medicina* **2021**, *57*, 994. [CrossRef] [PubMed]
36. Basso, V.; Döbrössy, M.D.; Thompson, L.H.; Kirik, D.; Fuller, H.R.; Gates, M.A. State of the Art in Sub-Phenotyping Midbrain Dopamine Neurons. *Biology* **2024**, *13*, 690. [CrossRef] [PubMed]
37. Morales, M.; Margolis, E.B. Ventral tegmental area: Cellular heterogeneity, connectivity and behaviour. *Nat. Rev. Neurosci.* **2017**, *18*, 73–85. [CrossRef]
38. Breton, J.M.; Charbit, A.R.; Snyder, B.J.; Fong, P.T.K.; Dias, E.V.; Himmels, P.; Lock, H.; Margolis, E.B. Relative contributions and mapping of ventral tegmental area dopamine and GABA neurons by projection target in the rat. *J. Comp. Neurol.* **2019**, *527*, 916–941. [CrossRef]
39. Taylor, S.R.; Badurek, S.; Dileone, R.J.; Nashmi, R.; Minichiello, L.; Picciotto, M.R. GABAergic and glutamatergic efferents of the mouse ventral tegmental area. *J. Comp. Neurol.* **2014**, *522*, 3308–3334. [CrossRef]
40. Park, J.-Y.; Park, S.Y.; Kwon, H.; Song, Y.; Yun, B.; Lee, Y.; Cho, Y.; Joo, A.; Han, P.-L. A Group of Descending Glutamatergic Neurons Activated by Stress in Corticolimbic Regions Project to the Nucleus Accumbens. *Exp. Neurobiol.* **2018**, *27*, 387–396. [CrossRef]
41. Hou, G.; Hao, M.; Duan, J.; Han, M.-H. The Formation and Function of the VTA Dopamine System. *Int. J. Mol. Sci.* **2024**, *25*, 3875. [CrossRef] [PubMed]
42. Obeso, J.A.; Rodríguez-Oroz, M.C.; Benitez-Temino, B.; Blesa, F.J.; Guridi, J.; Marin, C.; Rodriguez, M. Functional organization of the basal ganglia: Therapeutic implications for Parkinson's disease. *Mov. Disord.* **2008**, *23*, S548–S559. [CrossRef] [PubMed]
43. Zhou, F.-M.; Lee, C.R. Intrinsic and integrative properties of substantia nigra pars reticulata neurons. *Neuroscience* **2011**, *198*, 69–94. [CrossRef] [PubMed]
44. Saitoh, K.; Isa, T.; Takakusaki, K. Nigral GABAergic inhibition upon mesencephalic dopaminergic cell groups in rats. *Eur. J. Neurosci.* **2004**, *19*, 2399–2409. [CrossRef]
45. Ma, S.; Zhong, H.; Liu, X.; Wang, L. Spatial Distribution of Neurons Expressing Single, Double, and Triple Molecular Characteristics of Glutamatergic, Dopaminergic, or GABAergic Neurons in the Mouse Ventral Tegmental Area. *J. Mol. Neurosci.* **2023**, *73*, 345–362. [CrossRef]
46. Carmichael, K.; Sullivan, B.; Lopez, E.; Sun, L.; Cai, H. Diverse midbrain dopaminergic neuron subtypes and implications for complex clinical symptoms of Parkinson's disease. *Ageing Neur. Dis.* **2021**, *1*, 4. [CrossRef]
47. Voorn, P.; Vanderschuren, L.J.M.J.; Groenewegen, H.J.; Robbins, T.W.; Pennartz, C.M.A. Putting a spin on the dorsal–ventral divide of the striatum. *Trends Neurosci.* **2004**, *27*, 468–474. [CrossRef]
48. Bingham, C.S.; Petersen, M.V.; Parent, M.; McIntyre, C.C. Evolving characterization of the human hyperdirect pathway. *Brain Struct. Funct.* **2023**, *228*, 353–365. [CrossRef]
49. Maier-Hein, K.H.; Neher, P.F.; Houde, J.-C.; Côté, M.-A.; Garyfallidis, E.; Zhong, J.; Chamberland, M.; Yeh, F.-C.; Lin, Y.-C.; Ji, Q.; et al. The challenge of mapping the human connectome based on diffusion tractography. *Nat. Commun.* **2017**, *8*, 1349. [CrossRef]
50. Lazar, M. Mapping brain anatomical connectivity using white matter tractography. *NMR Biomed.* **2010**, *23*, 821–835. [CrossRef]
51. Behrens, T.E.J.; Johansen-Berg, H.; Woolrich, M.W.; Smith, S.M.; Wheeler-Kingshott, C.A.M.; A Boulby, P.; Barker, G.J.; Sillery, E.L.; Sheehan, K.; Ciccarelli, O.; et al. Non-invasive mapping of connections between human thalamus and cortex using diffusion imaging. *Nat. Neurosci.* **2003**, *6*, 750–757. [CrossRef]
52. Leergaard, T.B.; Hilgetag, C.C.; Sporns, O. Mapping the Connectome: Multi-Level Analysis of Brain Connectivity. *Front. Neurosci.* **2012**, *6*, 14. [CrossRef] [PubMed]
53. Gerfen, C.R.; Staines, W.A.; Fibiger, H.C.; Arbuthnott, G.W. Crossed connections of the substantia nigra in the rat. *J. Comp. Neurol.* **1982**, *207*, 283–303. [CrossRef] [PubMed]
54. Cacciola, A.; Calamuneri, A.; Milardi, D.; Mormina, E.; Chillemi, G.; Marino, S.; Naro, A.; Rizzo, G.; Anastasi, G.; Quartarone, A. A Connectomic Analysis of the Human Basal Ganglia Network. *Front. Neuroanat.* **2017**, *11*, 85. [CrossRef] [PubMed]
55. Quartarone, A.; Cacciola, A.; Milardi, D.; Ghilardi, M.F.; Calamuneri, A.; Chillemi, G.; Anastasi, G.; Rothwell, J. New insights into cortico-basal-cerebellar connectome: Clinical and physiological considerations. *Brain* **2020**, *143*, 396–406. [CrossRef]
56. Kawahata, I.; Finkelstein, D.I.; Fukunaga, K. Dopamine D1–D5 Receptors in Brain Nuclei: Implications for Health and Disease. *Receptors* **2024**, *3*, 155–181. [CrossRef]
57. Burke, R.E.; O'Malley, K. Axon degeneration in Parkinson's disease. *Exp. Neurol.* **2013**, *246*, 72–83. [CrossRef]
58. Cheng, H.-C.; Ulane, C.M.; Burke, R.E. Clinical progression in Parkinson disease and the neurobiology of axons. *Ann. Neurol.* **2010**, *67*, 715–725. [CrossRef]
59. Andica, C.; Kamagata, K.; Hatano, T.; Okuzumi, A.; Saito, A.; Nakazawa, M.; Ueda, R.; Motoi, Y.; Kamiya, K.; Suzuki, M.; et al. Neurite orientation dispersion and density imaging of the nigrostriatal pathway in Parkinson's disease: Retrograde degeneration observed by tract-profile analysis. *Park. Relat. Disord.* **2018**, *51*, 55–60. [CrossRef]
60. Uhr, L.; Tsolaki, E.; Pouratian, N. Diffusion tensor imaging correlates of depressive symptoms in Parkinson disease. *J. Comp. Neurol.* **2022**, *530*, 1729–1738. [CrossRef]

61. Moghaddam, H.S.; Dolatshahi, M.; Mohebi, F.; Aarabi, M.H. Structural white matter alterations as compensatory mechanisms in Parkinson's disease: A systematic review of diffusion tensor imaging studies. *J. Neurosci. Res.* **2020**, *98*, 1398–1416. [CrossRef]
62. Zhang, Y.; Wu, I.-W.; Ba, S.B.; Coffey, C.S.; Foster, E.; Mendick, S.; Seibyl, J.; Schuff, N. Diffusion Tensor Imaging of the Nigrostriatal Fibers in Parkinson's Disease. *Mov. Disord.* **2015**, *30*, 1229–1236. [CrossRef]
63. Shim, J.-H.; Baek, H.-M. White Matter Connectivity between Structures of the Basal Ganglia using 3T and 7T. *Neuroscience* **2022**, *483*, 32–39. [CrossRef]
64. Kwon, D.-H.; Paek, S.H.; Kim, Y.-B.; Lee, H.; Cho, Z.-H. In vivo 3D Reconstruction of the Human Pallidothalamic and Nigrothalamic Pathways with Super-Resolution 7T MR Track Density Imaging and Fiber Tractography. *Front. Neuroanat.* **2021**, *15*, 739576. [CrossRef] [PubMed]
65. Li, M.; Ribas, E.C.; Zhang, Z.; Wu, X.; Wang, X.; Liu, X.; Liang, J.; Chen, G.; Li, M. Tractography of the ansa lenticularis in the human brain. *Clin. Anat.* **2022**, *35*, 269–279. [CrossRef] [PubMed]
66. Zhang, Y.; Larcher, K.M.-H.; Misic, B.; Dagher, A. Anatomical and functional organization of the human substantia nigra and its connections. *eLife* **2017**, *6*, e26653. [CrossRef]
67. Elliott, B.L.; D'Ardenne, K.; Murty, V.P.; Brewer, G.A.; McClure, S.M. Midbrain-Hippocampus Structural Connectivity Selectively Predicts Motivated Memory Encoding. *J. Neurosci.* **2022**, *42*, 9426–9434. [CrossRef] [PubMed]
68. García-Gomar, M.G.; Singh, K.; Cauzzo, S.; Bianciardi, M. In vivo structural connectome of arousal and motor brainstem nuclei by 7 Tesla and 3 Tesla MRI. *Hum. Brain Mapp.* **2022**, *43*, 4397–4421. [CrossRef]
69. MacNiven, K.H.; Leong, J.K.; Knutson, B. Medial forebrain bundle structure is linked to human impulsivity. *Sci. Adv.* **2020**, *6*, eaba4788. [CrossRef]
70. Menke, R.A.; Jbabdi, S.; Miller, K.L.; Matthews, P.M.; Zarei, M. Connectivity-based segmentation of the substantia nigra in human and its implications in Parkinson's disease. *NeuroImage* **2010**, *52*, 1175–1180. [CrossRef]
71. Basile, G.A.; Bramanti, A.; Bertino, S.; Cutroneo, G.; Bruno, A.; Tisano, A.; Paladina, G.; Milardi, D.; Anastasi, G. Structural Connectivity-Based Parcellation of the Dopaminergic Midbrain in Healthy Subjects and Schizophrenic Patients. *Medicina* **2020**, *56*, 686. [CrossRef] [PubMed]
72. Chowdhury, R.; Lambert, C.; Dolan, R.J.; Duezel, E. Parcellation of the human substantia nigra based on anatomical connectivity to the striatum. *Neuroimage* **2013**, *81*, 191–198. [CrossRef] [PubMed]
73. Thanarajah, S.E.; Hanssen, R.; Melzer, C.; Tittgemeyer, M. Increased meso-striatal connectivity mediates trait impulsivity in FTO variant carriers. *Front. Endocrinol.* **2023**, *14*, 1130203. [CrossRef] [PubMed]
74. Funk, A.T.; Hassan, A.A.O.; Brüggemann, N.; Sharma, N.; Breiter, H.C.; Blood, A.J.; Waugh, J.L. In humans, striato-pallido-thalamic projections are largely segregated by their origin in either the striosome-like or matrix-like compartments. *Front. Neurosci.* **2023**, *17*, 1178473. [CrossRef]
75. Liu, W.; Heij, J.; Liu, S.; Liebrand, L.; Caan, M.; van der Zwaag, W.; Veltman, D.J.; Lu, L.; Aghajani, M.; van Wingen, G. Structural connectivity of dopaminergic pathways in major depressive disorder: An ultra-high resolution 7-Tesla diffusion MRI study. *Eur. Neuropsychopharmacol.* **2024**, *89*, 58–70. [CrossRef]
76. Jang, S.H.; Cho, M.J. Relationship of the Nigrostriatal Tract with the Motor Function and the Corticospinal Tract in Chronic Hemiparetic Stroke Patients: A Diffusion Tensor Imaging Study. *Healthcare* **2022**, *10*, 731. [CrossRef]
77. Wende, T.; Kasper, J.; Wilhelmy, F.; Dietel, E.; Hamerla, G.; Scherlach, C.; Meixensberger, J.; Fehrenbach, M.K. Assessment of a Reliable Fractional Anisotropy Cutoff in Tractography of the Corticospinal Tract for Neurosurgical Patients. *Brain Sci.* **2021**, *11*, 650. [CrossRef]
78. Lin, C.; Knoop, L.E.; Frigerio, I.; Bol, J.G.; Rozemuller, A.J.; Berendse, H.W.; Pouwels, P.J.; van de Berg, W.D.; Jonkman, L.E. Nigral Pathology Contributes to Microstructural Integrity of Striatal and Frontal Tracts in Parkinson's Disease. *Mov. Disord.* **2023**, *38*, 1655–1667. [CrossRef]
79. Antal, M.; Beneduce, B.M.; Regehr, W.G. The Substantia Nigra Conveys Target-Dependent Excitatory and Inhibitory Outputs from the Basal Ganglia to the Thalamus. *J. Neurosci.* **2014**, *34*, 8032–8042. [CrossRef]
80. Oishi, K.; Mori, S.; Troncoso, J.C.; Lenz, F.A. Mapping tracts in the human subthalamic area by 11.7T ex vivo diffusion tensor imaging. *Brain Struct. Funct.* **2020**, *225*, 1293–1312. [CrossRef]
81. Shi, Y.; Zhang, J.; Xiu, M.; Xie, R.; Liu, Y.; Xie, J.; Shi, L. The zona incerta system: Involvement in Parkinson's disease. *Exp. Neurol.* **2024**, *382*, 114992. [CrossRef]
82. Wu, F.-L.; Chen, S.-H.; Li, J.-N.; Zhao, L.-J.; Wu, X.-M.; Hong, J.; Zhu, K.-H.; Sun, H.-X.; Shi, S.-J.; Mao, E.; et al. Projections from the Rostral Zona Incerta to the Thalamic Paraventricular Nucleus Mediate Nociceptive Neurotransmission in Mice. *Metabolites* **2023**, *13*, 226. [CrossRef]

83. Ossowska, K. Zona incerta as a therapeutic target in Parkinson's disease. *J. Neurol.* **2020**, *267*, 591–606. [CrossRef]
84. Londei, F.; Arena, G.; Ferrucci, L.; Russo, E.; Ceccarelli, F.; Genovesio, A. Connecting the dots in the zona incerta: A study of neural assemblies and motifs of inter-area coordination in mice. *iScience* **2024**, *27*, 108761. [CrossRef]

applied sciences

MDPI

Review

Neurological Underpinnings of Socio-Cognitive Dysfunction in Schizophrenia and Autism Spectrum Disorder: Evidence from "Broken" Mirror Neurons

Maria Andreou [1,*], Vasileia Skrimpa [2] and Eleni Peristeri [3]

[1] Department of Speech and Language Therapy, University of Peloponnese, 24150 Kalamata, Greece
[2] Department of English, School of Arts and Humanities, University of Cologne, 50931 Cologne, Germany; vskrimp1@uni-koeln.de
[3] Department of Theoretical and Applied Linguistics, School of English, Aristotle University of Thessaloniki, 54124 Thessaloniki, Greece; eperiste@enl.auth.gr
* Correspondence: m.andreou@go.uop.gr

Abstract: Mirror neurons (MNs), a set of neurons that are activated during the processes of observation and execution of actions, have drawn significant attention in the research of neurodegenerative and psychological disorders. Research in the field of Autism Spectrum Disorder (ASD) and schizophrenia demonstrates evidence in favour of common underlying neural mechanisms underlying the two conditions, especially with respect to *mu* rhythm suppression, a proxy for MN activation and socio-cognitive impairments. This paper aims to review the most recent studies on the neurological underpinnings of social cognition deficits and cognitive discrepancies shared by ASD and schizophrenia, as detected by measuring the functionality and activation of the mirror neuron system. The findings of the review reveal a lack of consensus with respect to the validity of the "broken mirror" theory. The review also shows that further research is warranted to shed light on the implications of mirror neuron dysfunction in neuropsychiatric conditions and assist the development of technological interventions and treatments.

Keywords: autism; schizophrenia; mirror neurons; social dysfunction

1. Introduction

Mirror neurons (MNs), a set of neurons that are activated during the processes of observing and executing actions, have drawn significant attention in the research of neurodegenerative and mental disorders. Mirror neurons are located in the inferior parietal cortex, the premotor cortex, and the primary somatosensory cortex [1], and they are hypothesised to play a significant role in the comprehension and action of imitation, since the activation patterns of MNs during the process of observing an action are very comparable to those observed when an individual is performing an action [2]. The mechanism of mirroring elements of motion and behaviour in social contexts and the ability to reproduce these patterns in an imitative manner have been attributed to the firing of this set of neurons [2].

Mirror neurons were originally described in 1992 as a set of cells located in the premotor cortex of primates (macaque monkeys) [3]. Recent research has investigated the involvement of MNs in a number of cognitive and social functions. In humans, as well as in primates, this set of neurons has been found to be linked to social cognition that includes empathy and emotion understanding, intention understanding [4], imitation and language processing [5,6]. Due to their strong relation with the ability to acquire knowledge through imitation processes [7], MNs are activated during the action of the sensorimotor system for perceiving facial expressions and gesturing [8]. Research alleges that this set of neurons can be found in brain regions that include the inferior frontal gyrus [9], the superior temporal sulcus [10] and the ventral premotor cortex [11], which are dynamically involved in the processing of facial expressions and in higher-level cognitive abilities, such as Theory of Mind and empathy [12]. The aforementioned cerebral distribution advocates the functionality of MNs in processes that go beyond simple motor imitation.

However controversial, it has been theorised that the evolution of MNs also relates to language development by enabling complex communication systems [13]. Broca's area, which plays a crucial role in language in both production and perception, has been suggested to be a critical MN region. Nevertheless, research using neuroimaging techniques has shown that MNs are most probably associated with mechanisms implicated in speech articulation and not the semantic representation of human language [14].

Brodmann area 2 (BA2) has also been identified as having a relation with the function of the MN system (MNS defined as the group of MNs) [15]. BA2 has been determined as the locus of *mu* rhythm generation [16]. Measuring *mu* rhythm suppression has been shown to be a valid method for examining the level of activation of the MNS. *Mu* rhythm suppression indicates reduction in the amplitude of *mu* waves, which are cerebral oscillations in the frequency band of 8–13 Hz observed in an electroencephalogram (EEG) in the sensorimotor cortex. This type of modulation indicates the difference between active and baseline condition in MN activation. *Mu* suppression generally emerges during the performance of motion or the observation of such actions [17]. In addition, *mu* rhythm suppression has been linked to socio-cognitive mechanisms responsible for the development of intention understanding, inference making, emotion recognition and empathy [18]. *Mu* suppression therefore seems to be the neural basis for both social cognition and sensorimotor abilities. Figure 1 below illustrates the parietal mirror neuron (MN) system located in the inferior parietal lobule, as well as the frontal MN system situated in the inferior frontal gyrus.

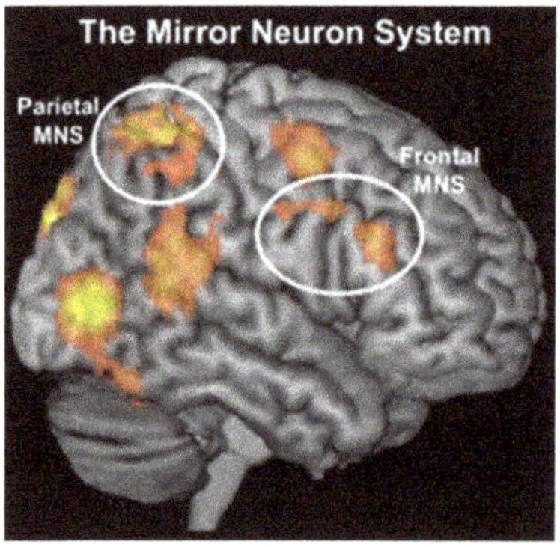

Figure 1. Schematic illustration of the mirror neuron system and its brain correlates, depicting parietal MN system in the inferior parietal lobule and frontal MN system in the inferior frontal gyrus ([19], p.186). Reproduced with permission from Liew & Aziz-Zadeh, Handbook of Neurosociology; published by Springer, Dordrecht, 2013.

Mu rhythm suppression has been a fruitful area of study in the examination of the neurobiological underpinnings of Autism Spectrum Disorder (ASD) and schizophrenia. ASD and schizophrenia are two distinct conditions in terms of clinical profiling and diagnostic traits, which nevertheless demonstrate particular commonalities, such as deficits in social cognition and difficulties with social navigation. These impairments have been associated with the function of the MNS and reduced *mu* rhythm suppression that individuals with ASD and schizophrenia exhibit while performing in social observation, imitation and empathy tasks [20]. In addition, both conditions present altered connectivity in specific cerebral regions, such as the superior temporal sulcus and sensorimotor cortex. These regions are related to *mu* rhythm activity, and contribute to discrepancies identified in both ASD and schizophrenia when processing and responding to social stimuli. The aim of this review is to provide insight into both the distinct and shared impairments in social cognition that characterise the two conditions, as reflected in the aberrant activation of *mu* rhythm suppression. The examination of *mu* rhythm suppression in ASD and schizophrenia seems to be crucial for comprehending the neurobiological mechanisms underlying social cognition impairments in these disorders.

Although previous reviews have examined the function of MNs in both ASD and schizophrenia [16,21], the present review will focus on investigating the phenomenon of *mu* rhythm suppression in both conditions through studies that have employed the electroencephalography (EEG) technique. This review will visit upon findings of studies within the past 10 years, while selection criteria involve the administration of EEG tests to measure *mu* rhythm suppression in both conditions with relevance to social dysfunction and impairments in social cognition. To secure the transparent selection of data, this review adheres to the PRISMA (Preferred Reporting Items for Systematic Reviews and Meta-Analyses) 2020 checklist (www.prisma-statement.org, accessed on 15 January 2025). Table 1 below demonstrates the selection and rejection criteria for the studies reviewed in the current paper according to PRISMA guidelines. This review hypothesises that both ASD and schizophrenic individuals demonstrate abnormal *mu* suppression activation and similarities in the hypofunction of MNs.

Table 1. Inclusion and exclusion criteria.

Category	Inclusion Criteria	Exclusion Criteria
Population and clinical data	Participants with ASD or schizophrenia	Studies involving disorders other than ASD or schizophrenia. Animal or in vitro studies
Study type	Empirical peer-reviewed research articles	Opinion papers or non-peer-reviewed work, unpublished dissertations, conference abstracts
Methodology	EEG as the main technique	Studies using only pharmacological or behavioural interventions without neurophysiological measurements of the MNS. Studies using other than EEG neuroimaging tests.
Publication date	Published from 2010 onward	Older publications
Language	Published in English	Publications in other languages than English
Outcomes	Reports on MNS activity and *mu* rhythm suppression in ASD and schizophrenia	Studies not isolating MNS-related processes (e.g., only general social cognition without measuring mirroring activity)

2. Autism Spectrum Disorder (ASD) vs. Schizophrenia: General Overview

2.1. Autism Spectrum Disorder

Autism Spectrum Disorder (ASD) is characterised by marked phenotypic heterogeneity in terms of manifestation of symptoms, level of severity and traits [22]. It is defined as a pervasive neurodevelopmental disorder that demonstrates impairments in social interaction, restricted or/and repetitive behavioural patterns, limited interests, and deficits in communication, both verbal and non-verbal [23]. The autistic spectrum reflects a set of characteristics that correspond to a range between high- and low-functioning individuals [24]. The classification of DSM-5 (Diagnostic and Statistical Manual of Mental Disorders) for ASD has replaced the term "Pervasive Developmental Disorders" with a single category that comprises previous mentions of diagnostic classes (i.e., Asperger's Syndrome, Pervasive Developmental Disorder not otherwise specified (PDD-NOS), Rett Syndrome, and Childhood Disintegrative Disorder). It also identifies three severity levels that rely on the extent of support required for each individual in the spectrum [25].

The autistic spectrum is very often linked to motor sensory deficits and problems with coordination, as well as mental disabilities, attentional impairments and anxiety [26]. However, despite the commonality of traits, individuals with ASD demonstrate individual variation in their general skills and severity of symptomatology. The aetiology of the disorder is multifaceted, encompassing epigenetic factors that stem both from environmental and biological origins or genetic predisposition. Nevertheless, it is still unclear which of those factors mostly contribute to the emergence of ASD, and to what extent. In addition, comorbidity between ASD and other disorders, such as Attention Deficit Hyperactivity Disorder (ADHD), Developmental Language Disorder (DLD), learning disabilities or/and neuropsychological conditions, is very often evinced in ASD, making it difficult to detect symptoms that are specific to the autistic spectrum [26].

Research on functional connectivity in ASD demonstrates atypical brain connectivity and deviations from neurotypical populations in terms of brain structure, such as differences in connectivity in the prefrontal cortex, amygdala and cerebellum [27]. Despite impairments in the neurobiological basis of autism and the neural circuits that are associated with social understanding, some individuals with ASD demonstrate strong abilities in several cognitive areas such as mathematics, art or music [28]. ASD is more frequently diagnosed in male (vs. female) individuals with a ratio of 4:1, and also more frequently observed in individuals who have siblings with the disorder. Research reveals a large increase in the reported diagnosed cases of ASD within the past 30 years, which is attributed to the existence of standardised diagnostic tools that can more accurately detect autistic traits and separate the diagnostic criteria for ASD from symptoms ascribed to other conditions [29].

2.2. Schizophrenia

Schizophrenia is one of the most common mental and neuropsychiatric disorders that affects in approximation a minimum of 0.7% of the global population [30,31]. Characteristics of the disorder encompass a variety of symptoms, i.e., hallucinations in the sense of seeing non-existing things or people or hearing voices, delusions and unrealistic thoughts, disoriented thinking and language, a sense of disconnection and disassociation from interests, feelings, people, personal care and emotions [26]. The disorder has an impact on the affected individuals in day-to-day life, negatively influencing their functioning, communication, and social interaction skills, and can result into severe disability, while in many cases, it can also be fatal [32].

The onset of the disease emerges early in adulthood, or late adolescence, and it tends to prevail more often in male individual than females. Its aetiology lies in a combination of factors that involve genetic, biological and environmental origins [33]. Research in the field of genetics and genome-wide association studies (GWAS) has identified genes associated with neural connectivity, synaptic function, calcium signalling and neurotransmitter systems [34]. Prenatal and perinatal factors as well as abusive experiences constitute environmental factors that have also been found to be involved in the occurrence of schizophrenia [35,36]. In terms of the pathophysiology of the disorder, neuroimaging studies show that individuals with schizophrenia exhibit cerebral abnormalities [37,38] that encompass alterations in connectivity between brain regions whose grey matter volume is pathologically reduced, such as the temporal lobes, the prefrontal cortex and the hippocampus. In addition, irregularities in terms of dopamine and glutamatergic neurotransmission [39] have been hypothesised to have an impact on the positive and negative symptoms of the disorder.

The symptomatologic classification of schizophrenia reflects the heterogeneity that defines it, and it involves positive, negative and cognitive symptoms. In terms of positive symptoms, they normally appear when the individual experiences a psychotic episode, and involve delusional situations, hallucinations, inappropriate behaviours, and distorted thoughts. Negative symptoms fall into five distinctive categories, namely anhedonia, apathy, avolition, asociality alogia and blunted affect [40]. Cognitive symptoms can be detected in about 70% of the individuals diagnosed with schizophrenia, and in most cases, they are present before the onset of the disease [41]. This category of symptoms comprises impairments in verbal and non-verbal memory, attention, reasoning, visual processing especially when experiencing visual illusions, and general executive functioning [42]. Social cognition is also impaired in individuals with schizophrenia: these individuals have problems with emotion recognition, phacial emotion perception, intention attribution, mentalising, perception of non-verbal social cues, non-literate language and general social contexts [43,44].

2.3. Commonalities Between ASD and Schizophrenia

When comparing the two conditions, ASD and schizophrenia share significant similarities with respect to deficits in social cognition, despite the fact that they are both separately diagnosed neuropsychiatric disorders. Individuals with either of these conditions demonstrate impairments in mentalising and Theory of Mind, which refers to the ability to comprehend and infer the mental states of others and of oneself, to take other people's perspectives, and interpret others' emotions, beliefs and intentions [45,46]. Theory of Mind deficits result in multiple challenges in everyday social navigation. In ASD, individuals display problems with reasoning and interpreting social cues and behavioural patterns that are impelled by internal mental states and emotions [47,48]. In individuals with schizophrenia, on the other hand, social and cognitive symptoms are manifested as the negative symptoms of the disorder, and involve a lack of motivation for social integration, problems with emotion perception and expression, and impaired Theory of Mind and sense of empathy [49].

Neuroimaging studies have demonstrated impairments in the MNS in both ASD and schizophrenia [50,51], further supporting the hypothesis that socio-cognitive deficits stem from common neural mechanisms across the two disorders. Shared neural dysfunctions, for instance, irregular function in cerebral areas such as the amygdala, prefrontal cortex and superior temporal sulcus, indicate shared dysfunctional social cognition. The results of recent studies underscore the notable commonality of neuro- and socio-cognitive impairments in ASD and schizophrenia, placing emphasis on the individuals' emotion recognition and irregular processing speed [52].

3. The Mechanisms Behind Broken MNs in ASD and Schizophrenia

3.1. Mu Rhythm Suppression in ASD

Attenuations in the function of the MNS in autism have been included in the theory of "broken mirror neurons", which has been researched over recent decades to explore possible links between atypical MN functionality and ASD symptomatology. Cognitive deficits in individuals with autism, and impairments in communication and social integration may be traced back to these links [53]. Dysfunctions in MNs have received attention in the field of autism research; nevertheless, they have also received criticism, mostly due to the mixed findings. It has been claimed that deficits in the sensorimotor system in individuals with ASD is a more plausible reasoning for explaining problems in mentalizing, imitation, and action interpretation. In addition, it has been proposed that the activation of MNS is fundamental for social cognition and reasoning [54,55] as well as the understanding of mental and emotional states [56]. These abilities have been well researched in ASD, which is often characterized by considerable discrepancies in internal state reasoning and Theory of Mind [57].

Measurements of suppression of the *mu* wave EEG have been prominent in exploring MN function in individuals. In clinical populations, however, and in ASD in particular, *mu* rhythm suppression has not been the most effective method. This may be attributed to various reasons ranging from sample sizes to issues with electrode selection for data collection. Neuroimaging studies provide insights into abnormalities in ASD in cerebral areas that are crucially implicated in social cognition and processing of social stimuli, for instance, the prefrontal cortex and the superior temporal sulcus [50].

One of the most prominent components that is hypothesised to trigger the activation of MNs is the observation of an action. Action observation involves, among others, the processing of facial expressions further related to ToM skills and emotion recognition [58]. Deschrijver et al. [59] critically examined the motor interference effect in ASD, which corresponds to more errors and slower reaction times when an individual visually perceives a movement that contradicts the movement of intention. A group of 23 adults with high-functioning autism and 23 age-, handedness- and sex-matched healthy controls participated in an EEG paradigm. The study focused on the P3 (a positive deflection in EEG recordings that occurs approximately 300 ms after stimulus onset, and is associated with attention and stimulus evaluation), the N190 (a negative ERP component occurring around 190 ms after stimulus onset, commonly linked to early visual processing and face recognition), and the RP ("Readiness Potential", a slow negative shift in EEG activity preceding voluntary movement, indicating motor preparation) that were congruent to the conditions of the stimulus. The participants were instructed to visually follow the gestural movements of a videotaped hand. Directly after the end of the observation, the participants were instructed to perform a pre-instructed gesture. The experiment involved a congruent condition, under which the required action matched the shown hand movement; an incongruent condition, where the observed gesture did not match the one for execution; and a baseline condition (hand in resting state). The aim of the study was to examine whether impairments in the processing of self- and other-action performances could provide a justification for imitation attenuations in high-functioning ASD. The results demonstrated that, in contrast with the predictions of the researchers of the study, there was no reported difference between the high-functioning autistic group and the control group in the P3 ERP component. In particular, both groups presented larger numbers of the P3 component when observing the congruent hand movement condition, so distinguishing between congruent and incongruent hand movements with high processing levels was intact. On the other hand, the clinical group demonstrated a larger value of RP for the congruent condition in comparison to the baseline, similarly to the incongruent condition, which implies elevated brain activation for motor preparation in both congruent and incongruent observations, while the baseline condition remained neutral. Comparing the compatible and non-compatible trials, no particular difference was reported. Crucially, the outcomes of the study show that the neural mechanisms underlying the motor interference effect in high-functioning ASD manifest discrepancies in the early stages of motor preparation rather than in higher-level cognitive processes, e.g., interpreting social cues, which contradicts the theory of "broken" MNs in ASD.

Another study that measures *mu* rhythm suppression in ASD through the observation of gestures is that of Fan et al. [60]. The objective of their study was to test the integrity of the MN hypothesis in ASD by administering an EEG experiment that measured *mu* rhythm suppression over the sensorimotor cortex while the participants observed and later imitated gestural hand movements. The study involved two groups of participants, 20 male adolescents and young adults, and 20 age-matched neurotypical controls. The two groups were required to complete an eye-tracking and an EEG test that included four conditions: *baseline* condition (observation of a static object on a screen), *hand* condition (observation of a video-recorded gesturing action), *dot* condition (observation of a video with a white dot), and *execution* condition (manipulation of an object in the same manner as in the *hand* condition). The study also controlled for fixation duration and normalised fixation number on the stimuli to account for potential differences in visual attention. The results strongly support the inefficacy of the theory of dysfunction in the MNS in ASD. Specifically, *mu* suppression over the sensorimotor cortex in the observation and imitation conditions did not present differences between the two groups. Essentially, *mu* rhythm suppression was stronger in both groups during the observation of the hand rather than the moving dot. Contrasting previous research [61], no correlation between *mu* rhythm suppression and imitation was detected. Although the ASD participants scored lower than neurotypicals in the imitation condition, the *mu* rhythm suppression was not different from controls. The outcomes highlighted the heterogeneity of ASD in terms of traits, and concluded that the MNS activation appears to be intact to a certain extent.

A more recent study by Sotoodeh et al. [62] examined the way *mu* suppression is modulated in ASD by visual attention during the process of perceiving biological motion. The researchers recruited 20 participants diagnosed with high-functioning ASD, and 20 age- and sex-matched healthy controls. An EEG paradigm was conducted to measure the suppression of *mu* rhythm over the sensorimotor cortex while participants were required to watch point-light displays presenting biological motions, e.g., walking, cartwheeling, free-throwing and underarm throwing, as well as non-biological motion, e.g., moving dots. Eye-tracking technology was used to monitor visual attention and ensure that participants were fixating on the stimuli. The EEG findings showed that visual attention in both groups had a significant impact on *mu* suppression; nevertheless, *mu* rhythm suppression in the ASD group was reduced compared to that of the control group during biological motion observation. This finding can be interpreted as a cue for neural deviations in biological motion processing in ASD, and that visual attention is not the only determining factor for impaired *mu* rhythm suppression. Furthermore, a positive correlation was detected between level of *mu* suppression and social abilities in the group with ASD, thus revealing an association between social cognition and *mu* suppression.

The study by Cole et al. [63] aimed to investigate MN activation in individuals with ASD in the process of making inferences about other people's intentions. In total, 43 participants took part in the study, distributed in two groups in compliance with the level of autistic traits. The 20 neurotypical controls that participated in the study reported no history of mental or neurological illness. The study used both transcranial magnetic stimulation (TMS) and EEG experiments to measure MNS activity, giving emphasis on the connection between autistic traits and neural responses during mentalizing tasks. The participants were instructed to watch short clip videos with mentalizing and non-mentalising actions, and place a judgement afterward with respect to the intention of the action performed. The study employed three screening tasks: an EEG task that was conducted during the projection of the videos, an eye-tracking task and a TMS-induced EMG (electromyography) task. In terms of *mu* rhythm suppression, this was found to be at a lower level in the right hemisphere among participants with higher level of autistic traits when performing in the mentalizing task. Nevertheless, *mu* rhythm suppression on the right hemisphere did not present significant correlation with how well the participants performed in the task. Notably, there was a positive correlation between lower MN activation in the left hemisphere and poor performance in the mentalising task, with no present effect occurring from the autistic traits. TMS-induced MEP did not present any significant variation between ASD individuals and controls. Similarly, no visual processing differences were reported for the eye-tracking task. The authors concluded that attenuated MNs and impaired ToM in ASD are the reasons behind the irregularities of MN activation in the right hemisphere.

Dumas et al. [64] revisited the MN hypothesis aiming by placing emphasis on two main aspects of the existing EEG literature: the functional segregation of *mu* rhythm into two discrete sub-bands (8–10 Hz and 10–12/13 Hz), and the focus on central electrodes (C3/C4) in previous studies in ASD. Ten high-functioning individuals with ASD and 30 age-matched healthy controls participated in the study. The two groups were tested on a task involving three conditions, namely simple observation of gestures, free imitation of gestures, and imitation of a pre-recorded video. The outcomes of the study revealed a distinct pattern of *mu* suppression in the ASD group compared to the TD controls. Specifically, when the *mu* rhythm was analysed as a homogeneous phenomenon covering the 8–13 Hz frequency range, no significant differences were found between the groups. However, when the two sub-bands were distinguished, a different *mu* response to observation appeared for subjects with ASD in the upper sub-band (10–12/13 Hz) over the sensorimotor cortex, while the lower sub-band (8–10 Hz) responded similarly in both groups. Source reconstructions demonstrated that this effect was related to a joint *mu*-suppression deficit over the occipito-parietal regions and an increase over the frontal regions in the ASD group. The researchers argues that *mu* suppression deficits in individuals with ASD do not support a global irregular functionality of the MNS but are rather specific to the neural mechanisms underlying the observation of intentions.

3.2. *Mu Rhythm Suppression in Schizophrenia*

Mirror neurons have received quite some criticism in research in recent years due to controversial findings of studies and the soundness of the hypotheses concerning the exact function and significance of the MNS [65,66]. Nevertheless, they have also drawn significant attention, as they have been shown to be actively involved in complex cognitive processes, such as empathy, mentalising, perspective taking, language and imitation [67]. As mentioned before, the MNS has mostly been researched in ASD, mainly due to the hypothesis of the "broken mirror neurons" which negatively affect language, Theory of Mind and social navigation [68]. Nevertheless, as shown in the previous section, the results do not demonstrate a clear consensus as to the role of the MNS in social and cognitive discrepancies in autism [69].

Research on the connection between the functionality of the MNS and schizophrenic symptoms has gained ground over the last decade. Schizophrenia is often characterised as the disorder of the "social brain" [70], which underlines the attenuations in social cognition that schizophrenic patients exhibit. Similarly to ASD, the MNS system has also been investigated in schizophrenia using a variety of neuroimaging methods, each of which measures different aspects of the functions of the MNS (ex. fMRI measuring fluctuations in blood oxygenation, MEG measuring alpha band suppression and gamma band amplification, TMS, EEG measuring *mu* rhythm suppression, and electromyography to measure rapid face muscle activation during observation of facial expressions).

With respect to *mu* rhythm suppression as a proxy for MNS activation, research demonstrates abnormalities in the MNS in schizophrenia, which are mainly reflected in varying levels of *mu* suppression. Variation in *mu* rhythm suppression has in turn been linked to deficits in Theory of Mind, empathy, emotion recognition and processing of social cues [71]. Several studies have explored *mu* rhythm suppression in patients with schizophrenia; however, the outcomes are heterogeneous.

Horan et al. [72] conducted an EEG study to explore impairments in the functioning of the MNS in patients with schizophrenia as indexed by *mu* suppression. The participants of their study included 32 outpatients diagnosed with schizophrenia, and 26 controls with no clinical diagnosis The two groups took part in an EEG paradigm, which involved six conditions of an either observed or executed action. These conditions contained varying levels of social interaction. In addition, validated empathy questionnaires were collected from the participants. The results revealed that both groups exhibited a crucial linear increase in *mu* rhythm suppression with respect to those conditions that contained a higher demand for socialisation; however, neither interactions nor group differences were found to be significant. In the questionnaires, patients described their empathy and mentalising ability as being low, nevertheless, as evinced from the *mu* suppression measurements, their MNS functioning appeared to be relatively intact. The researchers concluded that there was an impairment in emotion recognition processing in schizophrenia, with no evidence in favour of inability of experiencing them. However, it should be mentioned that the tasks in Horan et al.'s [72] study involved non-complex social and emotional conditions, which may have biased the interpretation of the results.

Other studies, on the other hand, have reported reduced *mu* rhythm suppression in patients with schizophrenia. Mitra et al. [73] investigated the efficacy of the MNS operation in individuals with schizophrenia. The researchers' objective was to detect whether the occurrence of psychotic episodes of the illness was attributed to attenuations of the MNS system by measuring *mu* rhythm suppression. *Mu* suppression was measured through an EEG task, in which the participants were presented with a clip containing short sequences involving white background, biological motion with social inferences, and white visual noise. The study recruited 30 participants: 15 drug-naïve or drug-free patients with schizophrenia at the early stages of the illness, and 15 controls with no record of neuropsychiatric disorder, matched on age, sex and educational level. According to the results, the level of *mu* rhythm suppression over the sensorimotor cortex (SMC) among patients with schizophrenia and healthy controls presented significant deviations, which as interpreted as an indicator for the MNS dysfunction. In addition, the findings highlighted a significant negative correlation between *mu* suppression over the right sensorimotor cortex (rSMC) and the thought disturbance cluster on PANSS (Positive and Negative Syndrome Scale) [74]. Thought disorder refers to cognitive discrepancies that attenuate communication, language and thought [75]. Mitra et al.'s [73] results link irregularities in the function of the MNS to cognitive deficits in patients with schizophrenia.

Differences in research outcomes across studies may be attributed to differences in study design, population samples, and/or the tasks administered to measure the suppression of the *mu* wave. Minichino et al. [76] conducted an analysis of relevant studies in order to investigate the activation of MNs in schizophrenia and ASD by measuring deviations in *mu* rhythm suppression during the observation of biological motion. For their analysis, they selected three groups of participants: individuals with high-functioning ASD, individuals with schizophrenia in the early stages of psychosis (EP) that had active negative symptoms, and a group of controls with no clinical diagnoses. The authors examined neural commonalities between the disorders that might stem from the negative symptoms of schizophrenia. The studies selected had all used EEG during the observation of biological motion. The analysis involved four conditions: baseline/ball condition [77], moving hand condition [77], social interactive condition [78] and a biological motion/point light display animation [79]. The results indicated that reactions to the observation of biological movement resulted in crucially reduced *mu* rhythm suppression in both clinical groups (i.e., individuals with ASD and individuals with early psychosis experiencing negative symptoms) as opposed to the healthy control group. For the EP group, the findings showed that individuals that exhibited more severe negative symptoms demonstrated less *mu* suppression. As mentioned before, negative symptoms in schizophrenia correspond to impairments in social cognition and executive functioning. Minichino et al.'s [76] study has highlighted the similarities between ASD and schizophrenia with respect to dysfunctional neural mechanisms associated with impaired socio-cognitive performance.

The relationship between *mu* rhythm suppression and degree of severity of symptoms in schizophrenia has also been a fruitful area of research. For instance, Singh et al. [80] tested 20 first-episode psychosis patients with schizophrenia for Event Related Desynchronisation (ERD) of *mu* waves, comparing their performance to 12 healthy controls. Their findings demonstrated that reduced *mu* suppression was associated with more severe negative symptoms and poorer social functioning in first-episode psychosis patients, further implying that deficits with respect to *mu* suppression may be related to the social and cognitive impairments observed in the early stages of schizophrenia.

Similarly, the objective of the study by McCormick et al. [81] was to explore the connection among MN function, psychosis and empathic processes in patients with schizophrenia, by measuring *mu* rhythm suppression and its correlation with empathy and psychotic episodes. A group of 32 participants were recruited for this study: 8 patients with active schizophrenic psychosis, 8 patients in the residual phase of the disease, and 16 controls with no reported neurological or psychiatric disorders. An EEG paradigm was administered that measured *mu* rhythm suppression over the sensorimotor cortex while the participants were observing their own hand movements and hand movements of other individuals. Furthermore, levels of empathy were evaluated by the implementation of the Interpersonal Reactivity Index (IRI), which involves a classification of empathy and perspective taking (Perspective Taking subscale, Empathic Concern subscale, Personal Distress subscale and Fantasy subscale). The study revealed that the levels of *mu* rhythm suppression in the schizophrenic patients that experience active psychotic episodes were greater over the left sensorimotor cortex, when compared to the other groups. Healthy controls and patients with schizophrenia in the residual phase exhibited no considerable differences in the levels of *mu* rhythm suppression. McCormick et al. [81] attributed the higher levels observed in the first group to the occurrence of psychotic episodes. In particular, the findings have shown that the more severe the psychotic symptoms, the greater the left-sided *mu* rhythm suppression. In addition, according to the findings from the IRI, the degree of personal distress was significantly stronger in patients with schizophrenia than controls. The outcomes of the study suggest that impaired social cognition and empathy can be attributed to abnormal function of the MNS in schizophrenic patients, and can be further negatively impacted by increased levels of distress.

A study by Wynn et al. [82] has investigated the optimal dose of oxytocin (OT) for boosting the processing of social cues and contexts in patients with schizophrenia. The main aim of the study was to measure *mu* rhythm suppression as a proxy for the function of the MNS, and pupil dilation, which served as an index for emotion recognition and processing. The researchers recruited 47 individuals diagnosed with schizophrenia that were randomly instructed to receive one of eight doses of OT (8, 12, 24, 36, 48, 60, 72, or 84 IU), or the placebo substance. The experiment involved two tasks associated with social cues: an EEG task that measure *mu* rhythm suppression in reaction to biological motion, and a pupillometry task that demonstrated the level of the dilation of the pupil when observing faces that present emotion. The outcomes of the study were that OT doses of 36 and 48 IU substantially boosted *mu* suppression, as opposed to the placebo. Crucially, the study demonstrated no effect of the OT on the face affect pupillometry test, which can be attributed to the well-established impairment that individuals with schizophrenia present in face emotion processing and recognition [83]. Table 2 provides an overview of the studies used in this review, the methodological approach and their main findings in both conditions.

3.3. *Methodological Limitations and Strengths of the Included Studies*

The selected studies that have investigated the MNS function in ASD and schizophrenia present a range of methodological strengths and limitations, particularly concerning the EEG protocols, sample sizes, and statistical rigor. Deschrijver et al. [59] employed high-density EEG with appropriate artifact correction to localize motor interference effects in high-functioning ASD, but their modest sample size ($n \approx 20$ per group) limits generalisability of findings and statistical power. Fan et al.'s [60] study lacked detailed reporting on EEG preprocessing parameters, raising concerns about the reproducibility of the findings and signal quality, however, the particular study has been notable for combining behavioural and EEG measures to argue against the MNS dysfunction in ASD. Sotoodeh et al.'s [62] methodological design represents a methodological advance by integrating eye-tracking with EEG to control for visual attention, which has been a critical confound in MNS research, however, the relatively large dataset would benefit from more transparent reporting of effect sizes and corrections for multiple comparisons. Finally, Cole et al. [63] demonstrated methodological rigor through the integration of EEG and TMS, offering multimodal evidence for intact MNS activity in ASD. Nevertheless, the small sample sizes and the complexity of dual-modality analyses introduce further potential noise and interpretation challenges.

With respect to the studies that have investigated the MNS in schizophrenia, Horan et al. [72] applied a well-established *mu* rhythm suppression paradigm, and included a healthy control group enhancing interpretability. Crucially, their moderate sample size and lack of detailed artifact rejection procedures may limit replicability. Mitra et al. [73] contributed valuable cross-cultural data using EEG to examine *mu* rhythm suppression activity in schizophrenia, but their relatively small and clinically heterogeneous sample, along with limited statistical correction for multiple comparisons, weakens the robustness of their findings. Minichino et al. [76] demonstrated methodological strength by comparing both early psychosis (EP) and ASD participants, yet the EEG analysis focused narrowly on sensor-level *mu* rhythm suppression without source localization, potentially reducing neuroanatomical specificity. McCormick et al. [81] incorporated both fMRI and EEG measures, offering a multimodal perspective on empathy and the MNS function in schizophrenia; nevertheless, their modest sample and the limited temporal resolution of fMRI reduced the study's capacity to link neural activity directly to mirroring processes. Finally, Wynn et al. [82] employed a pharmacological intervention design using oxytocin, and incorporated neurophysiological measures to examine social processing, showcasing innovative methodology. Nonetheless, EEG outcomes were secondary to the pharmacological aims, and the study lacked *mu* rhythm-specific analyses, making their study less directly informative about the MNS function. Collectively, while these studies provide valuable insights into *mu* rhythm suppression activity and its link to socio-cognitive deficits in ASD and schizophrenia, future work would benefit from larger cohorts, standardized EEG processing pipelines, and robust statistical corrections to improve replicability and validity.

Table 2. Overview of the selected research studies for this review.

Authors	Participants	Method	Findings
[59]	46 participants 23 ASD and 23 controls	EEG	No significant variation between ASD and controls in P3 ERP component. Larger number RP (readiness potential) Laplacian both in congruent and incongruent trials in ASD. No effect of intended action on early visual processing detected.
[64]	40 participants 10 ASD and 30 controls	EEG	When *mu* frequency was distinguished into two sub-bands, group difference was observed in the upper sub-band (10–12/13 Hz) of the sensorimotor cortex in ASD in the condition of gestural observation; no significant variation in lower sub-band (8–10 Hz) among the two groups. No globally dysfunctional MN system in ASD.
[62]	40 participants 20 with high-functioning autism and 20 controls	EEG, Eye-Tracker	Visual attention in both groups had a significant impact on *mu* suppression. The *mu* rhythm suppression of the ASD group was reduced during biological motion observation.
[63]	43 participants with autistic traits	EEG, Eye-Tracker, TMS-EMG	Lower level of *mu* suppression in the right hemisphere in ASD during mentalising task. Positive correlation of lower performance in mentalising task with poorer activation of mirror neurons in left hemisphere but not linked to level of autistic traits. Autistic traits predictive factor for *mu* suppression in the 8–10 Hz for mentalising task and poorer mirror neuron firing in right hemisphere. During non-mentalising task, no low-level *mu* suppression detected.
[60]	40 participants 20 male ASD and 20 controls	EEG	No significant variation among groups in *mu* suppression occurring from EEG monitoring of observation and imitation of a gestural action. Stronger *mu* suppression during gestural action observation than dot observation in ASD. No imitation of the observed action while MN system activation intact in ASD. Relation between attenuated communication capacities and reduced *mu* rhythm.
[72]	32 outpatients with schizophrenia and 26 controls	EEG	Linear increase in *mu* rhythm suppression in conditions with higher social demand for both groups. Schizophrenia patients self-reported low empathic and mentalising ability, however, the modulation of MN system functioning appeared relatively intact.
[82]	47 participants with schizophrenia	EEG	Attenuated *mu* rhythm suppression was substantially boosted by OT doses of 36 and 48 IU. No effect of the OT on face affect pupillometry test.

Table 2. *Cont.*

Authors	Participants	Method	Findings
[81]	32 participants with schizophrenia: 8 actively psychotic patients, 8 patients in the residual phase of illness, 16 healthy controls	EEG	Greater *mu* suppression over the left sensorimotor cortex in patients with active psychotic schizophrenia compared to both the residual phase SSD patients and healthy controls. Severity of symptoms correlated with the level of *mu* suppression.
[73]	15 drug-naïve or drug-free patients with schizophrenia and 15 healthy controls	EEG	Significant differences in the level of *mu* rhythm suppression over the SMC among patients with schizophrenia and controls. Negative correlation between *mu* suppression over the rSMC and thought disturbance cluster on PANSS.
[76]	Studies involving schizophrenia patients with active negative symptoms, high-functioning ASD participants and healthy controls	EEG	Reduced *mu* rhythm suppression in both ASD group and EP group. Negative correlation between reduced *mu* suppression and degree of severity of negative symptoms in EP group.

4. Conclusions

Research in the field of MNs has advanced in the past years. However, the findings with respect to ASD and schizophrenia do not reach a consensus with respect to the functionality of this set of neurons and/or the validity of the theories around it. Mirror neurons are a class of neurons that fire both during the performance of an action or action observation, and have considerable impact on the individuals' social cognition skills, including imitation, empathy, intention, emotion recognition and Theory of Mind. In both ASD and schizophrenia, research has shown that the MNS may be dysfunctional, indicating abnormal activation and impaired perception of motion activity, which has in turn detrimental effects on the social cognitive skills of the individuals. On the other hand, the two disorders differ markedly in terms of the neural abnormalities they demonstrate, despite the shared social cognitive deficits that are often linked to MN dysfunction. In individuals with ASD, MN dysfunction is frequently associated with reduced *mu* rhythm suppression during both action observation and execution tasks, suggesting impaired resonance mechanisms that may hinder the development of imitation and Theory of Mind [63]. By contrast, schizophrenia patients may exhibit more variable MN responses; while some studies show preserved *mu* suppression, others report reduced MN activity specifically in relation to negative symptoms and disrupted intention attribution [72,76]. This dissociation may reflect differences in the timing and nature of the neurodevelopmental disruptions that occur in the two conditions: in ASD, MNS anomalies often emerge early and are relatively stable, whereas in schizophrenia, they may develop later and interact with broader network dysfunctions, such as impaired salience attribution and executive functioning. Moreover, compensatory mechanisms such as attentional engagement, may differentially influence MN responses in each disorder [62].

The MN hypothesis, which posits that dysfunctions in the MNS underlie the social and communicative deficits observed in ASD and schizophrenia, has garnered significant attention but remains highly controversial. Initial support for the hypothesis stemmed from findings of reduced *mu* suppression in individuals with ASD, which was interpreted as a manifestation of impaired MNS activity [77]. However, subsequent research has challenged this interpretation, suggesting that such findings may be confounded by atypical attentional engagement rather than specific MNS dysfunctions [84]. Moreover, alternative models posit that deficits in ASD and schizophrenia may instead reflect broader disruptions in large-scale neural networks, such as the default mode or salience networks, which are crucial for social cognition [85,86]. In schizophrenia, for instance, impaired social cognition has been linked to dysfunctional connectivity in frontotemporal and limbic circuits, rather than isolated MN abnormalities [87]. Research also suggests that observed MNS abnormalities may instead be secondary to deficits in visual attention. With respect to ASD, studies have indicated that reduced *mu* suppression during biological motion perception correlates with decreased visual attention, as measured by eye-tracking metrics. This suggests that attentional factors, rather than intrinsic MNS dysfunction, may underlie observed neural differences in ASD [62]. Similarly, in schizophrenia, impairments in social attention and deficits in processing salient stimuli have been found to influence neural responses during action observation tasks [88], raising questions about the specificity of the MNS dysfunction. This counter-evidence underscores the necessity of cautiously interpreting MNS -related findings, and further advocates for a more integrative, network-based approach to understanding social deficits in neurodevelopmental and psychiatric disorders.

Table 3 illustrates the supporting and opposing evidence of the "broken" MN hypothesis as explored by the studies reviewed in this paper:

Table 3. Comparative evidence supporting and opposing the broken mirror theory in ASD and schizophrenia.

Position	Study	Population	Key Findings	Interpretation
Supporting Evidence	[61]	ASD	Reduced *mu* suppression during action observation	Suggests deficient sensorimotor mirroring contributing to social deficits
	[73]	Schizophrenia	Attenuated *mu* rhythm suppression during motor tasks	Indicates impaired MNS functioning in schizophrenia
	[76]	Early psychosis and ASD	Reduced biological motion-induced *mu* suppression linked to negative symptoms	Supports MNS impairment as a shared mechanism in both conditions
Opposing Evidence	[60]	ASD	Typical *mu* suppression and preserved imitation abilities	Challenges the universality of MNS deficits in ASD
	[72]	Schizophrenia	Normal *mu* suppression in response to biological motion	Implies intact mirroring mechanisms in at least some schizophrenia subgroups
	[62]	ASD	*mu* suppression modulated by visual attention	Suggests attentional factors may account for prior findings of MNS dysfunction

The studies selected for this paper have investigated *mu* rhythm suppression in schizophrenia and ASD as a proxy for discrepancies in the function and activation of the MNS, implementing EEG paradigms. In the case of ASD, numerous research studies have presented evidence for reduced *mu* rhythm suppression during tasks that require observation of biological actions. The theory of broken MNs has been approached through a critical lens in this review, including studies that report no links between *mu* rhythm suppression and task performance, whereas others present findings that support this hypothesis and attribute the core deficits in social navigation and communication in ASD to a dysfunctional MNS.

A thematic comparison of MNS abnormalities in ASD and schizophrenia on the basis of the studies that have been reviewed, reveals both overlapping features and condition-specific differences. In ASD, Oberman and Ramachandran [61] proposed widespread MNS dysfunction as a core mechanism behind social and communicative impairments. However, more recent investigations offer a more nuanced picture. Fan et al. [60] and Cole et al. [63] found intact MNS activity under certain task conditions, suggesting that task effects, such as task demands and cognitive engagement, may influence MNS responses. Furthermore, Sotoodeh et al. [62] demonstrated that visual attention plays a key modulatory role in *mu* rhythm suppression in autistic individuals, indicating that attentional deficits may underlie MNS anomalies. Deschrijver et al. [59] and Cole et al. [63] also emphasized that while automatic motor resonance may be atypical in ASD, higher-order inferential processes remain relatively preserved.

In schizophrenia, the picture is similarly complex. Horan et al. [72] observed preserved *mu* suppression in individuals with schizophrenia, challenging the assumption of a pervasive MNS dysfunction. Yet, Mitra et al. [73] and McCormick et al. [81] reported deficits in *mu* rhythm modulation and empathy-related activation, suggesting that MNS anomalies may emerge more prominently in relation to negative symptoms or reduced social functioning in schizophrenia. Minichino et al. [76] further demonstrated that biological motion-induced *mu* rhythm suppression was diminished in early psychosis patients with concurrent ASD traits, thus highlighting potential transdiagnostic features of MNS dysfunction. Wynn et al. [82], while not directly measuring *mu* rhythm suppression, contributed to the debate by showing that oxytocin modulates social brain activity, which may interact with MNS-related pathways. Overall, both disorders show altered MNS-related activity, but ASD appears to be more influenced by attentional and contextual modulation, while MNS changes seem to be more variable in schizophrenia and may depend on symptom profile and disease phase. The overall findings support a model in which MNS abnormalities in ASD and schizophrenia are neither uniform nor wholly distinct, but shaped by broader cognitive and neurophysiological contexts.

The controversies evident in the literature might stem from a variety of factors, such as the marked heterogeneity in terms of symptomatology and traits in ASD, as well as the small sample sizes and the different methodological designs across studies, or/and the targeting of limited cerebral areas. Similarly, the studies that have examined *mu* suppression in schizophrenia also present controversies in their outcomes. Reduced *mu* suppression in schizophrenia has been linked to negative symptoms, severity of symptoms and also thought disorder. In addition, divergent findings in the case of schizophrenia range from outcomes that depict intact *mu* suppression and MN activation, to severe MNS dysfuntions. Divergence in the results may be attributed to differences in study design, task selection, and size and representability of the samples. In total, the present review has demonstrated impairments in the MNS in both ASD and schizophrenia. Correlations between severity of symptoms or frequency of traits and level of *mu* rhythm suppression have been detected. In the case of schizophrenic patients, reduced levels of *mu* rhythm suppression were found to correlate with the severity of negative symptoms, whereas in ASD, reduced levels of *mu* rhythm suppression correlated with intense deficits in social cognition.

Crucially, the overall findings of the selected studies demonstrate robust deficits in socio-cognitive understanding and Theory of Mind, i.e., the ability to mentalise, infer and understand emotions while observing actions in both clinical conditions. Further research in the field is of essential importance in order to reach consensus for the neural operations that determine the function of the MNS in both schizophrenia and ASD.

Future Research Implications and Clinical Applications

This line of research is fruitful for the development of specific biomarkers in both clinical conditions that can assist diagnostic processes as well as provide information about the overlapping neurodevelopmental origins of those conditions. Further research on *mu* rhythm suppression on schizophrenia and ASD will offer a more wholesome understanding of the factors that shape the aetiology of the disorders and will aid in the development and design of appropriate interventions. Priority should be given on disentangling core deficits from secondary effects related to attention, motivation, and symptom heterogeneity. Advances in neuroimaging and electrophysiological techniques, particularly when combined with eye-tracking and individualized symptom profiling, may help clarify whether the MNS anomalies reflect a distinct neurobiological marker or are contingent upon broader cognitive disruptions. Longitudinal and transdiagnostic studies are also essential to determine the developmental trajectories and shared versus divergent MNS mechanisms across the two conditions. Clinically, improved understanding of the MNS function may inform targeted interventions aimed at enhancing social cognition, such as neurofeedback protocols based on *mu* rhythm modulation or/and adjunctive treatments using neuromodulators, like oxytocin. However, translating these findings into practice will require rigorously controlled trials with standardised methodologies to ensure replicability and therapeutic efficacy across diverse clinical populations.

Mirror neurons and *mu* rhythm suppression contribute to our understanding of the neural basis of communication and language, stipulating valuable insights for natural language processing. It is crucial to find the link and bridge the gap between artificial intelligence (AI) systems and neuroscientific evidence, in order to develop high-order models that mimic human communication, language and interaction in neurotypical and neurodivergent populations, such as ASD and schizophrenia. A deep understanding of the brain functions that define complex cognitive systems in individuals with neurodegenerative and neuropsychiatric disorders will enhance the advancement of innovative technological applications of AI systems on language impairments, and also improve language and communicational interventions and treatments [89,90].

Author Contributions: Conceptualization, M.A., V.S. and E.P.; investigation, M.A., V.S. and E.P.; writing—original draft preparation, M.A., V.S. and E.P.; writing—review and editing, M.A., V.S. and E.P.; supervision, M.A. All authors have read and agreed to the published version of the manuscript.

Funding: This research received no external funding.

Conflicts of Interest: The authors declare no conflicts of interest.

References

1. Molenberghs, P.; Cunnington, R.; Mattingley, J.B. Is the mirror neuron system involved in imitation? A short review and meta-analysis. *Neurosci. Biobehav. Rev.* **2009**, *33*, 975–980. [CrossRef] [PubMed]
2. Rizzolatti, G.; Craighero, L. The mirror-neuron system. *Annu. Rev. Neurosci.* **2004**, *27*, 169–192. [CrossRef]
3. Bonini, L.; Rotunno, C.; Arcuri, E.; Gallese, V. Mirror neurons 30 years later: Implications and applications. *Trends Cogn. Sci.* **2022**, *26*, 767–781. [CrossRef]

4. Bonini, L.; Rozzi, S.; Serventi, F.U.; Simone, L.; Ferrari, P.F.; Fogassi, L. Ventral premotor and inferior parietal cortices make distinct contribution to action organization and intention understanding. *Cereb. Cortex* **2010**, *20*, 1372–1385. [CrossRef]
5. Gallese, V.; Keysers, C.; Rizzolatti, G. A unifying view of the basis of social cognition. *Trends Cogn. Sci.* **2004**, *8*, 396–403. [CrossRef] [PubMed]
6. Gallese, V.; Goldman, A. Mirror neurons and the simulation theory of mind-reading. *Trends Cogn. Sci.* **1998**, *2*, 493–501. [CrossRef]
7. de Jong, T.; Van Gog, T.; Jenks, K.; Manlove, S.; Van Hell, J.; Jolles, J.; Van Merrienboer, J.; Van Leeuwen, T.; Boschloo, A. *Explorations in Learning and the Brain: On the Potential of Cognitive Neuroscience for Educational Science*; Springer Science & Business Media: New York, NY, USA, 2009.
8. Leslie, K.R.; Johnson-Frey, S.H.; Grafton, S.T. Functional imaging of face and hand imitation: Towards a motor theory of empathy. *Neuroimage* **2004**, *21*, 601–607. [CrossRef]
9. Kilner, J.M.; Neal, A.; Weiskopf, N.; Friston, K.J.; Frith, C.D. Evidence of mirror neurons in human inferior frontal gyrus. *J. Neurosci.* **2009**, *29*, 10153–10159. [CrossRef] [PubMed]
10. de la Rosa, S.; Schillinger, F.L.; Bülthoff, H.H.; Schultz, J.; Uludag, K. fMRI adaptation between action observation and action execution reveals cortical areas with mirror neuron properties in human BA 44/45. *Front. Hum. Neurosci.* **2016**, *10*, 78. [CrossRef]
11. Gazzola, V.; Keysers, C. The observation and execution of actions share motor and somatosensory voxels in all tested subjects: Single-subject analyses of unsmoothed fMRI data. *Cereb. Cortex* **2009**, *19*, 1239–1255. [CrossRef]
12. Singer, T. The neuronal basis and ontogeny of empathy and mind reading: Review of literature and implications for future research. *Neurosci. Biobehav. Rev.* **2006**, *30*, 855–863. [CrossRef] [PubMed]
13. Théoret, H.; Pascual-Leone, A. Language acquisition: Do as you hear. *Curr. Biol.* **2002**, *12*, R736–R737. [CrossRef]
14. Cerri, G.; Cabinio, M.; Blasi, V.; Borroni, P.; Iadanza, A.; Fava, E.; Fornia, L.; Verpozzi, V.; Riva, M.; Casarotti, A.; et al. The mirror neuron system and the strange case of Broca's area. *Hum. Brain Mapp.* **2015**, *36*, 1010–1027. [CrossRef] [PubMed]
15. Damiani, D.; Nascimento, A.M.; Pereira, L.K. Cortical brain functions–the brodmann legacy in the 21st century. *Arq. Bras. Neurocir. Braz. Neurosurg.* **2020**, *39*, 261–270. [CrossRef]
16. Valizadeh, A.; Mbwogge, M.; Rasouli Yazdi, A.; Hedayati Amlashi, N.; Haadi, A.; Shayestefar, M.; Moassefi, M. The mirror mechanism in schizophrenia: A systematic review and qualitative meta-analysis. *Front. Psychiatry* **2022**, *13*, 884828. [CrossRef] [PubMed]
17. Oberman, L.M.; McCleery, J.P.; Ramachandran, V.S.; Pineda, J.A. EEG evidence for mirror neuron activity during the observation of human and robot actions: Toward an analysis of the human qualities of interactive robots. *Neurocomputing* **2007**, *70*, 2194–2203. [CrossRef]
18. Genzer, S.; Ong, D.C.; Zaki, J.; Perry, A. *Mu* rhythm suppression over sensorimotor regions is associated with greater empathic accuracy. *Soc. Cogn. Affect. Neurosci.* **2022**, *17*, 788–801. [CrossRef]
19. Liew, S.L.; Aziz-Zadeh, L. The human mirror neuron system, social control, and language. In *Handbook of Neurosociology*; Springer: Dordrecht, The Netherlands, 2013; pp. 183–205.
20. Chisholm, K.; Lin, A.; Armando, M. Schizophrenia spectrum disorders and autism spectrum disorder. In *Psychiatric Symptoms and Comorbidities in Autism Spectrum Disorder*; Springer: Berlin/Heidelberg, Germany, 2016; pp. 51–66.
21. Gallese, V.; Rochat, M.J.; Berchio, C. The mirror mechanism and its potential role in autism spectrum disorder. *Dev. Med. Child Neurol.* **2013**, *55*, 15–22. [CrossRef]
22. Jeste, S.S.; Geschwind, D.H. Disentangling the heterogeneity of autism spectrum disorder through genetic findings. *Nat. Rev. Neurol.* **2014**, *10*, 74–81. [CrossRef]
23. Sicile-Kira, C. *Autism Spectrum Disorder: The Complete Guide to Understanding Autism*; Tarcher Perigee: New York, NY, USA, 2014.
24. Attwood, T. Asperger's Syndrome. *Tizard Learn. Disabil. Rev.* **2006**, *11*, 3–11. [CrossRef]
25. American Psychiatric Association. *Diagnostic and Statistical Manual of Mental Disorders (DSM-5®)*; American Psychiatric Publishing: Washington, DC, USA, 2013.
26. Goldstein, G.; Johnson, C.R.; Minshew, N.J. Attentional processes in autism. *J. Autism Dev. Disord.* **2001**, *31*, 433–440. [CrossRef] [PubMed]
27. Parellada, M.; Penzol, M.J.; Pina, L.; Moreno, C.; González-Vioque, E.; Zalsman, G.; Arango, C. The neurobiology of autism spectrum disorders. *Eur. Psychiatry* **2014**, *29*, 11–19. [CrossRef]
28. Nicpon, M.F.; Doobay, A.F.; Assouline, S.G. Parent, teacher, and self-perceptions of psychosocial functioning in intellectually gifted children and adolescents with autism spectrum disorder. *J. Autism Dev. Disord.* **2010**, *40*, 1028–1038. [CrossRef]
29. Fombonne, E. The rising prevalence of autism. *J. Child Psychol. Psychiatry* **2018**, *59*, 717–720. [CrossRef] [PubMed]
30. Javitt, D.C. Balancing therapeutic safety and efficacy to improve clinical and economic outcomes in schizophrenia: A clinical overview. *Am. J. Manag. Care* **2014**, *20* (Suppl. S8), S160–S165. [PubMed]
31. Charlson, F.J.; Ferrari, A.J.; Santomauro, D.F.; Diminic, S.; Stockings, E.; Scott, J.G.; McGrath, J.J.; Whiteford, H.A. Global epidemiology and burden of schizophrenia: Findings from the global burden of disease study 2016. *Schizophr. Bull.* **2018**, *44*, 1195–1203. [CrossRef]

32. McGrath, J.; Saha, S.; Chant, D.; Welham, J. Schizophrenia: A concise overview of incidence, prevalence, and mortality. *Epidemiol. Rev.* **2008**, *30*, 67–76. [CrossRef]
33. Adamu, M.J.; Qiang, L.; Nyatega, C.O.; Younis, A.; Kawuwa, H.B.; Jabire, A.H.; Saminu, S. Unraveling the pathophysiology of schizophrenia: Insights from structural magnetic resonance imaging studies. *Front. Psychiatry* **2023**, *14*, 1188603. [CrossRef]
34. Li, G.; Dai, J.; Liu, H.; Lin, Y.; Liu, Q.; Zheng, K.; Li, S.; Chen, S.; Ye, Y. Association study between genetic variants and the risk of schizophrenia in the Chinese population based on GWAS-implicated 6p21. 3–23.1 human genome region: A case-control study. *BMC Psychiatry* **2021**, *21*, 483. [CrossRef]
35. Jenkins, T.A. Perinatal complications and schizophrenia: Involvement of the immune system. *Front. Neurosci.* **2013**, *7*, 110. [CrossRef]
36. Vila-Badia, R.; Del Cacho, N.; Butjosa, A.; Arumí, C.S.; Santjusto, M.E.; Abella, M.; Cuevas-Esteban, J.; Pardo, M.; Muñoz-Samons, D.; Usall, J. Prevalence and types of childhood trauma in first episode psychosis patients. Relation with clinical onset variables. *J. Psychiatr. Res.* **2022**, *146*, 102–108. [CrossRef] [PubMed]
37. Nyatega, C.O.; Qiang, L.; Adamu, M.J.; Younis, A.; Kawuwa, H.B. Altered dynamic functional connectivity of cuneus in schizophrenia patients: A resting-state fMRI study. *Appl. Sci.* **2021**, *11*, 11392. [CrossRef]
38. Ippolito, G.; Bertaccini, R.; Tarasi, L.; Di Gregorio, F.; Trajkovic, J.; Battaglia, S.; Romei, V. The role of alpha oscillations among the Main neuropsychiatric disorders in the adult and developing human brain: Evidence from the last 10 years of research. *Biomedicine* **2022**, *10*, 3189. [CrossRef]
39. Reddy-Thootkur, M.; Kraguljac, N.V.; Lahti, A.C. The role of glutamate and GABA in cognitive dysfunction in schizophrenia and mood disorders–A systematic review of magnetic resonance spectroscopy studies. *Schizophr. Res.* **2022**, *249*, 74–84. [CrossRef]
40. Correll, C.U.; Schooler, N.R. Negative symptoms in schizophrenia: A review and clinical guide for recognition, assessment, and treatment. *Neuropsychiatr. Dis. Treat.* **2020**, *16*, 519–534. [CrossRef]
41. Bozikas, V.P.; Andreou, C. Longitudinal studies of cognition in first episode psychosis: A systematic review of the literature. *Aust. N. Z. J. Psychiatry* **2011**, *45*, 93–108. [CrossRef]
42. Valentina, C.; De Carolis, A.; Trovini, G.; Dehning, J.; Di Pietro, S.; Curto, M.; Donato, N.; De Pisa, E.; Girardi, P.; Comparelli, A. Neurocognition in schizophrenia: From prodrome to multi-episode illness. *Psychiatry Res.* **2014**, *220*, 129–134.
43. Green, M.F.; Horan, W.P.; Lee, J. Nonsocial and social cognition in schizophrenia: Current evidence and future directions. *World Psychiatry* **2019**, *18*, 146–161. [CrossRef] [PubMed]
44. Savla, G.N.; Vella, L.; Armstrong, C.C.; Penn, D.L.; Twamley, E.W. Deficits in domains of social cognition in schizophrenia: A meta-analysis of the empirical evidence. *Schizophr. Bull.* **2013**, *39*, 979–992. [CrossRef]
45. Matthews, N.L.; Goldberg, W.A. Theory of mind in children with and without autism spectrum disorder: Associations with the sibling constellation. *Autism* **2018**, *22*, 311–321. [CrossRef]
46. Samson, F.; Mottron, L.; Jemel, B.; Belin, P.; Ciocca, V. Can spectro-temporal complexity explain the autistic pattern of performance on auditory tasks? *J. Autism Dev. Disord.* **2006**, *36*, 65–76. [CrossRef] [PubMed]
47. Shamsi, F.; Hosseini, S.; Tahamtan, M.; Bayat, M. The Impaired Theory of Mind in Autism Spectrum Disorders and the Possible Remediative Role of Transcranial Direct Current Stimulation. *J. Adv. Med. Sci. Appl. Technol.* **2017**, *3*, 175–178. [CrossRef]
48. Begeer, S.; Malle, B.F.; Nieuwland, M.S.; Keysar, B. Using theory of mind to represent and take part in social interactions: Comparing individuals with high-functioning autism and typically developing controls. *Eur. J. Dev. Psychol.* **2010**, *7*, 104–122. [CrossRef]
49. Owen, M.J.; O'Donovan, M.C. The genetics of cognition in schizophrenia. In *Genomic Psychiatry*; Academic Press: Cambridge, MA, USA, 2024.
50. Sugranyes, G.; Kyriakopoulos, M.; Corrigall, R.; Taylor, E.; Frangou, S. Autism spectrum disorders and schizophrenia: Meta-analysis of the neural correlates of social cognition. *PLoS ONE* **2011**, *6*, e25322. [CrossRef] [PubMed]
51. King, B.H.; Lord, C. Is schizophrenia on the autism spectrum? *Brain Res.* **2011**, *1380*, 34–41. [CrossRef]
52. Eack, S.M.; Bahorik, A.L.; McKnight, S.A.; Hogarty, S.S.; Greenwald, D.P.; Newhill, C.E.; Phillips, M.L.; Keshavan, M.S.; Minshew, N.J. Commonalities in social and non-social cognitive impairments in adults with autism spectrum disorder and schizophrenia. *Schizophr. Res.* **2013**, *148*, 24–28. [CrossRef]
53. Ramachandran, V.S.; Oberman, L.M. Broken mirrors: A theory of autism. *Sci. Am.* **2006**, *295*, 62–69. [CrossRef]
54. Pacherie, E.; Dokic, J. From mirror neurons to joint actions. *Cogn. Syst. Res.* **2006**, *7*, 101–112. [CrossRef]
55. Praszkier, R. Empathy, mirror neurons and SYNC. *Mind Soc.* **2016**, *15*, 1–25. [CrossRef]
56. Debes, R. Which empathy? Limitations in the mirrored "understanding" of emotion. *Synthese* **2010**, *175*, 219–239. [CrossRef]
57. Altschuler, M.; Sideridis, G.; Kala, S.; Warshawsky, M.; Gilbert, R.; Carroll, D.; Burger-Caplan, R.; Faja, S. Measuring Individual Differences in Cognitive, Affective, and Spontaneous Theory of Mind Among School-Aged Children with Autism Spectrum Disorder. *J. Autism Dev. Disord.* **2018**, *48*, 3945–3957. [CrossRef] [PubMed]
58. Iacoboni, M.; Woods, R.P.; Brass, M.; Bekkering, H.; Mazziotta, J.C.; Rizzolatti, G. Cortical mechanisms of human imitation. *Science* **1999**, *286*, 2526–2528. [CrossRef] [PubMed]

59. Deschrijver, E.; Wiersema, J.R.; Brass, M. Disentangling neural sources of the motor interference effect in high functioning autism: An EEG-study. *J. Autism Dev. Disord.* **2017**, *47*, 690–700. [CrossRef] [PubMed]
60. Fan, Y.T.; Decety, J.; Yang, C.Y.; Liu, J.L.; Cheng, Y. Unbroken mirror neurons in autism spectrum disorders. *J. Child Psychol. Psychiatry* **2010**, *51*, 981–988. [CrossRef]
61. Oberman, L.M.; Ramachandran, V.S. The simulating social mind: The role of the mirror neuron system and simulation in the social and communicative deficits of autism spectrum disorders. *Psychol. Bull.* **2007**, *133*, 310. [CrossRef]
62. Sotoodeh, M.S.; Chien, S.H.L.; Hadjikhani, N. Visual attention modulates *mu* suppression during biological motion perception in autistic individuals. *Eur. J. Neurosci.* **2024**, *60*, 6668–6685. [CrossRef]
63. Cole, E.J.; Barraclough, N.E.; Enticott, P.G. Investigating mirror system (MS) activity in adults with ASD when inferring others' intentions using both TMS and EEG. *J. Autism Dev. Disord.* **2018**, *48*, 2350–2367. [CrossRef]
64. Dumas, G.; Soussignan, R.; Hugueville, L.; Martinerie, J.; Nadel, J. Revisiting *mu* suppression in autism spectrum disorder. *Brain Res.* **2014**, *1585*, 108–119. [CrossRef]
65. Spaulding, S. Mirror neurons and social cognition. *Mind Lang.* **2013**, *28*, 233–257. [CrossRef]
66. Hickok, G. Eight problems for the mirror neuron theory of action understanding in monkeys and humans. *J. Cogn. Neurosci.* **2009**, *21*, 1229–1243. [CrossRef]
67. Rizzolatti, G.; Sinigaglia, C. The functional role of the parieto-frontal mirror circuit: Interpretations and misinterpretations. *Nat. Rev. Neurosci.* **2010**, *11*, 264–274. [CrossRef] [PubMed]
68. Andreou, M.; Skrimpa, V. Theory of mind deficits and neurophysiological operations in autism spectrum disorders: A review. *Brain Sci.* **2020**, *10*, 393. [CrossRef] [PubMed]
69. Hamilton, A.F.D.C. Reflecting on the mirror neuron system in autism: A systematic review of current theories. *Dev. Cogn. Neurosci.* **2013**, *3*, 91–105. [CrossRef]
70. Jonathan, B. The social brain hypothesis of schizophrenia. *World Psychiatry* **2006**, *5*, 77.
71. Green, M.F.; Horan, W.P. Social cognition in schizophrenia. *Curr. Dir. Psychol. Sci.* **2010**, *19*, 243–248. [CrossRef]
72. Horan, W.P.; Pineda, J.A.; Wynn, J.K.; Iacoboni, M.; Green, M.F. Some markers of mirroring appear intact in schizophrenia: Evidence from *mu* suppression. *Cogn. Affect. Behav. Neurosci.* **2014**, *14*, 1049–1060. [CrossRef]
73. Mitra, S.; Nizamie, S.H.; Goyal, N.; Tikka, S.K. *Mu*-wave activity in schizophrenia: Evidence of a dysfunctional mirror neuron system from an indian study. *Indian J. Psychol Med.* **2014**, *36*, 276–281. [CrossRef]
74. Kay, S.R.; Fiszbein, A.; Opler, L.A. The positive and negative syndrome scale (PANSS) for schizophrenia. *Schizophr. Bull.* **1987**, *13*, 261–276. [CrossRef] [PubMed]
75. Hart, M.; Lewine, R.R. Rethinking thought disorder. *Schizophr. Bull.* **2017**, *43*, 514–522. [CrossRef]
76. Minichino, A.; Singh, F.; Pineda, J.; Friederich, E.; Cadenhead, K.S. Biological Motion induced *mu* suppression is reduced in Early Psychosis (EP) patients with active negative symptoms and Autism Spectrum Disorders (ASD). *Psychiatry Res.* **2016**, *238*, 374–377. [CrossRef]
77. Oberman, L.M.; Hubbard, E.M.; McCleery, J.P.; Altschuler, E.L.; Ramachandran, V.S.; Pineda, J.A. EEG evidence for mirror neuron dysfunction in autism spectrum disorders. *Cogn. Brain Res.* **2005**, *24*, 190–198. [CrossRef] [PubMed]
78. Oberman, L.M.; Pineda, J.A.; Ramachandran, V.S. The human mirror neuron system: A link between action observation and social skills. *Soc. Cogn. Affect. Neurosci.* **2007**, *2*, 62–66. [CrossRef] [PubMed]
79. Ulloa, E.R.; Pineda, J.A. Recognition of point-light biological motion: *Mu* rhythms and mirror neuron activity. *Behav. Brain Res.* **2007**, *183*, 188–194. [CrossRef]
80. Singh, F.; Pineda, J.; Cadenhead, K.S. Association of impaired EEG *mu* wave suppression, negative symptoms and social functioning in biological motion processing in first episode of psychosis. *Schizophr. Res.* **2011**, *130*, 182–186. [CrossRef]
81. McCormick, L.M.; Brumm, M.C.; Beadle, J.N.; Paradiso, S.; Yamada, T.; Andreasen, N. Mirror neuron function, psychosis, and empathy in schizophrenia. *Psychiatry Res. Neuroimaging* **2012**, *201*, 233–239. [CrossRef] [PubMed]
82. Wynn, J.K.; Green, M.F.; Hellemann, G.; Reavis, E.A.; Marder, S.R. A dose-finding study of oxytocin using neurophysiological measures of social processing. *Neuropsychopharmacology* **2019**, *44*, 289–294. [CrossRef]
83. Marwick, K.; Hall, J. Social cognition in schizophrenia: A review of face processing. *Br. Med. Bull.* **2008**, *88*, 43–58. [CrossRef]
84. Hyatt, C.J.; Wexler, B.E.; Pittman, B.; Nicholson, A.; Pearlson, G.D.; Corbera, S.; Bell, M.D.; Pelphrey, K.; Calhoun, V.D.; Assaf, M. Atypical dynamic functional network connectivity state engagement during social–emotional processing in schizophrenia and autism. *Cereb. Cortex* **2022**, *32*, 3406–3422. [CrossRef]
85. Uddin, L.Q.; Supekar, K.; Lynch, C.J.; Khouzam, A.; Phillips, J.; Feinstein, C.; Ryali, S.; Menon, V. Salience network–based classification and prediction of symptom severity in children with autism. *JAMA Psychiatry* **2013**, *70*, 869–879. [CrossRef]
86. Yang, G.J.; Murray, J.D.; Wang, X.J.; Glahn, D.C.; Pearlson, G.D.; Repovs, G.; Krystal, J.H.; Anticevic, A. Functional hierarchy underlies preferential connectivity disturbances in schizophrenia. *Proc. Natl. Acad. Sci. USA* **2016**, *113*, E219–E228. [CrossRef]
87. Mehta, U.M.; Thirthalli, J.; Aneelraj, D.; Jadhav, P.; Gangadhar, B.N.; Keshavan, M.S. Mirror neuron dysfunction in schizophrenia and its functional implications: A systematic review. *Schizophr. Res.* **2014**, *160*, 9–19. [CrossRef] [PubMed]

88. Martínez, A.; Gaspar, P.A.; Bermudez, D.H.; Aburto-Ponce, M.B.; Beggel, O.; Javitt, D.C. Disrupted third visual pathway function in schizophrenia: Evidence from real and implied motion processing. *NeuroImage Clin.* **2024**, *41*, 103570. [CrossRef] [PubMed]
89. Alkhatib, A.J.; Darabseh, R.S. Mirror Neuron Cells, Autism, Language Deficit, and the Potential Role of Artificial Intelligence. *J. Ecohumanism* **2024**, *3*, 12552–12559. [CrossRef]
90. Dickerson, K.; Gerhardstein, P.; Moser, A. The role of the human mirror neuron system in supporting communication in a digital world. *Front. Psychol.* **2017**, *8*, 698. [CrossRef]

applied sciences

MDPI

Article

Metacognitive Modulation of Cognitive-Emotional Dynamics Under Social-Evaluative Stress: An Integrated Behavioural–EEG Study

Katia Rovelli [1,2], Angelica Daffinà [1,2,*] and Michela Balconi [1,2]

[1] International Research Center for Cognitive Applied Neuroscience (IrcCAN), Università Cattolica del Sacro Cuore, 20123 Milan, Italy

[2] Research Unit in Affective and Social Neuroscience, Department of Psychology, Università Cattolica del Sacro Cuore, 20123 Milan, Italy

* Correspondence: angelica.daffina@unicatt.it; Tel.: +39-2-7234-5929

Abstract

Background/Objectives: Decision-making under socially evaluative stress engages a dynamic interplay between cognitive control, emotional appraisal, and motivational systems. Contemporary models of multi-level co-regulation posit that these systems operate in reciprocal modulation, redistributing processing resources to prioritise either rapid socio-emotional alignment or deliberate evaluation depending on situational demands. Methods: Adopting a neurofunctional approach, a novel dual-task protocol combining the MetaCognition–Stress Convergence Paradigm (MSCP) and the Social Stress Test Neuro-Evaluation (SST-NeuroEval), a simulated social–evaluative speech task calibrated across progressive emotional intensities, was implemented. Twenty professionals from an HR consultancy firm participated in the study, with concurrent recording of frontal-temporoparietal electroencephalography (EEG) and bespoke psychometric indices: the MetaStress-Insight Index and the TimeSense Scale. Results: Findings revealed that decision contexts with higher socio-emotional salience elicited faster, emotionally guided choices (mean RT difference emotional vs. cognitive: −220 ms, $p = 0.026$), accompanied by oscillatory signatures (frontal delta: $F(1,19) = 13.30$, $p = 0.002$; gamma: $F(3,57) = 14.93$, $p \leq 0.001$) consistent with intensified socio-emotional integration and contextual reconstruction. Under evaluative stress, oscillatory activity shifted across phases, reflecting the transition from anticipatory regulation to reactive engagement, in line with models of phase-dependent stress adaptation. Across paradigms, convergences emerged between decision orientation, subjective stress, and oscillatory patterns, supporting the view that cognitive–emotional regulation operates as a coordinated, multi-level system. Conclusions: These results underscore the importance of integrating behavioural, experiential, and neural indices to characterise how individuals adaptively regulate decision-making under socially evaluative stress and highlight the potential of dual-paradigm designs for advancing theory and application in cognitive–affective neuroscience.

Keywords: decision-making; metacognition; social–evaluative stress; cognitive–emotional integration; EEG oscillations; frontal–temporo-parietal networks; stress regulation

1. Introduction

Decision-making in contexts of heightened socio-emotional salience constitutes a multi-layered process in which cognitive evaluation and affective appraisal are continuously interwoven. Classical models have conceptualised this interplay within the framework of dual-process theories, distinguishing between an intuitive, heuristic-based system (System 1) operating rapidly and automatically, and a deliberative, rule-based system (System 2) characterised by slower, more effortful reasoning [1,2]. While influential, this dichotomy has increasingly been viewed as overly static and reductive. Contemporary perspectives argue that decision-making is better understood as an emergent property of dynamic, context-dependent interactions between multiple neurocognitive subsystems [3–5]. Within this revised view, affective and cognitive contributions are not simply competitive but are mutually modulatory. Emotional signals can bias attentional allocation, evidence accumulation, and threshold setting in decision formation, while cognitive control processes can recalibrate or suppress affective influences when such modulation supports goal attainment. Crucially, this interplay unfolds on a continuum shaped by the temporal structure, social meaning, and motivational relevance of the decision context [3,6–8].

Socially evaluative stress represents a paradigmatic condition in which this dynamic coupling is tested to its limits. Evaluation by others—explicitly communicated or implicitly inferred—constitutes a high-relevance socio-emotional cue that can potentiate limbic–prefrontal reactivity and reorganise decision priorities [9,10].

Under such conditions, empirical evidence suggests a bias towards emotionally resonant, high-salience options, potentially reflecting an adaptive mechanism for maintaining social attunement or mitigating perceived interpersonal threat. This shift may occur at the expense of extended deliberation, as the cost–benefit calculus is recalibrated in favour of immediacy and affective congruence [11].

Recent integrative models conceptualise this phenomenon within the framework of multi-level co-regulation, wherein cognitive control, emotional appraisal, and motivational systems jointly contribute to adaptive functioning under pressure. Rather than operating in isolation, these systems dynamically redistribute processing resources in real time, prioritising either rapid socio-emotional alignment or more deliberate evaluation depending on the contextual affordances and the individual's regulatory profile. Neurocognitively, this involves flexible coordination across prefrontal executive regions, anterior cingulate cortex, limbic structures (including amygdala and ventral striatum), and posterior association cortices implicated in social cognition and contextual integration [11–14].

From this perspective, investigating decision-making under socio-emotional and evaluative stress requires not only behavioural measurement but also precise mapping of the underlying neural dynamics. Oscillatory brain activity offers a particularly sensitive index of these processes, capable of capturing the temporal interplay between rapid affective appraisals and controlled cognitive elaboration as they unfold within the decision window.

Delta (1–4 Hz) oscillations have been associated with large-scale integrative processes, motivated attention, and the mobilisation of neural resources in contexts of heightened motivational or affective relevance. In socially or emotionally charged decision contexts, frontal delta is often implicated in the rapid integration of affectively salient cues into ongoing cognitive evaluations, while temporo-parietal delta has been linked to contextual reconstruction and the binding of complex situational features into coherent mental representations [15–18].

Theta (4–8 Hz) activity, particularly in the midline frontal regions, has been linked to conflict monitoring, working memory maintenance, and the allocation of cognitive control [19–21]. Frontal Theta supports the detection of discrepancies between competing response tendencies and the upregulation of control mechanisms to resolve them. Posterior and temporo-parietal Theta, in contrast, is associated with episodic retrieval, semantic integration, and the assimilation of contextual information into decision-making processes [15,19,22]. In dynamic decision contexts, the interplay between frontal control-related Theta and posterior context-related Theta is considered a hallmark of adaptive regulatory functioning [17].

Alpha (8–12 Hz) oscillations, traditionally interpreted as an index of cortical inhibition [23–25], have more recently been reconceptualised as markers of functional gating in both sensory and higher-order cortical systems [26–28]. Alpha desynchronisation in task-relevant regions reflects the release of inhibition and the facilitation of information processing, whereas synchronisation in task-irrelevant areas may support the suppression of distracting inputs. In the context of decision-making under socio-emotional load, alpha modulations are often observed as a reflection of attentional focusing and the selective gating of affectively or contextually salient information [29,30].

Beta (13–30 Hz) oscillations have been linked to top-down maintenance of cognitive sets, sensorimotor integration, and the stabilisation of current task-relevant representations [31–33]. In decision-making paradigms, beta activity may index the maintenance of a chosen response framework and the suppression of alternative action plans. In socio-emotional contexts, beta modulations can also be associated with the suppression or reinforcement of affectively driven response tendencies, reflecting the balance between motor preparation and regulatory control [29].

Gamma (30–100 Hz) oscillations are widely recognised as markers of feature binding and rapid integrative processing across distributed cortical sites [31,34,35]. In decision-making contexts, gamma activity supports the synthesis of perceptual, mnemonic, and affective components into a unified decisional construct. Frontal gamma has been associated with the binding of abstract rules and social–emotional contingencies, whereas temporo-parietal gamma is implicated in perspective-taking, social simulation, and the projection of decision consequences.

To sum up, oscillatory brain activity plays a central role in coordinating and integrating information across distributed neural systems. Neurophysiological measures—particularly EEG—offer a direct, temporally precise window into the neural mechanisms underpinning decision-making and stress regulation. Region-specific modulations (e.g., frontal vs. temporo-parietal) and frequency-specific signatures (e.g., delta for motivated attention, theta for control allocation, gamma for integrative binding) map the allocation of processing resources in ways not accessible through behavioural or self-report data alone.

By providing a temporal framework for neural communication, oscillations enable the flexible alignment of cognitive and emotional processes with moment-to-moment demands. In the context of socio-emotional and evaluative challenges, they offer a sensitive index of how processing resources are dynamically redistributed across functional networks. Different frequency bands are associated with distinct, yet interacting, operations, and their spatial modulation reflects the neural strategies engaged to balance rapid affective appraisal with more deliberate cognitive evaluation.

Understanding these oscillatory signatures is therefore essential for a comprehensive account of how decision-making unfolds under socio-emotional and evaluative stress. However, despite their potential to bridge multiple levels of analysis, investigations in this field have often been constrained by methodological compartmentalization. Research streams have tended to focus either on behavioural performance (e.g., choice frequencies, reaction times—RTs), electrophysiological activity (EEG), or subjective experience (e.g., perceived stress, emotional valence) with limited attempts to integrate these layers into a coherent explanatory framework. While each approach has yielded valuable insights, their isolated application risks underestimating the inherently multi-layered nature of adaptive regulation in complex social–emotional contexts. Behavioural indices capture the overt expression of decision-making tendencies—such as the propensity towards emotionally driven versus cognitively analytical choices—and their modulation under varying affective loads. Yet behavioural data alone cannot fully illuminate the regulatory mechanisms that guide these tendencies, particularly when decisions unfold in the presence of social evaluation, a condition known to amplify affective weighting and alter decisional thresholds.

Subjective self-report measures complement behavioural indices by providing access to internal experiential states, including perceived stress intensity, metacognitive awareness of stress, and affective stability. However, these measures are inherently retrospective and self-referential, subject to biases in introspection and reporting, and may lag behind the rapid, dynamic changes that occur during real-time decision-making under pressure.

Crucially, these three domains—behavioural, neurophysiological and subjective—are not parallel but isolated streams; rather, they form a mutually constraining system in which each level influences, and is influenced by, the others. Integrating these perspectives aligns with multi-level co-regulation frameworks [18,36,37], which conceptualise cognitive–emotional functioning as the product of coordinated adjustments across multiple representational and control systems.

The challenge, therefore, is methodological: to design paradigms that elicit meaningful variability across all three levels, enabling their convergence to be observed and quantified. The present study addresses this challenge through the conceptual coupling of two complementary paradigms—the MetaCognition–Stress Convergence Paradigm (MSCP) and the Social Stress Test—NeuroEval (SST-NeuroEval)—each of which captures different but interconnected facets of decision–stress interaction.

Conceptually, the two paradigms are mutually informative, in that each targets the reciprocal coupling between socio-emotional appraisal and cognitive control processes, but through orthogonal manipulations. The MSCP focuses on how varying decisional demands shape stress-related cognitive–emotional dynamics, whereas the SST-NeuroEval examines how escalating evaluative stress modulates regulatory engagement and cognitive–emotional allocation over time. Their integration within a single experimental framework allows for the joint exploration of content-related effects, such as those determined by the socio-emotional load of decision scenarios, and temporal effects, as revealed by the unfolding of evaluative stress from anticipation to direct exposure.

This dual-paradigm design advances existing approaches by simultaneously manipulating decisional content and evaluative stress trajectories, enabling convergence across behavioural, subjective, and oscillatory indices that isolated paradigms cannot capture.

From a theoretical perspective, we hypothesise that decision scenarios characterised by higher socio-emotional salience will favour rapid, emotionally guided choices over more deliberate cognitive–analytical strategies. Likewise, the transition from anticipatory to reactive phases of evaluative stress is expected to amplify subjective stress and increase reliance on emotional over controlled responses. When the two paradigms are considered jointly, these tendencies are predicted to manifest as stable cognitive–emotional regulatory profiles, whereby individuals with a predisposition toward emotionally driven decisions will exhibit reduced cognitive control and heightened socio-emotional engagement under evaluative load.

Accordingly, the present study aims to systematically examine how decision-making orientation interacts with socio—evaluative stress across behavioural, subjective, and oscillatory indices, providing an integrated account of cognitive—emotional regulation under conditions of ecological stress.

To test these predictions, we operationalized regulatory dynamics across multiple levels. Behaviourally, we examined decision latencies as an index of cognitive versus emotional decision strategies. Subjective stress was assessed through self-reported measures capturing the perceived intensity of evaluative pressure. At the neurophysiological level, oscillatory patterns were analysed as markers of cognitive–emotional integration: heightened socio-emotional engagement was expected to be reflected by increased frontal delta and gamma power and enhanced temporo-parietal theta activity, whereas the escalation of evaluative stress was anticipated to evoke elevated frontal delta and theta activity during early orienting and conflict-monitoring stages, accompanied by transient modulations in gamma power at peaks of evaluative demand.

2. Materials and Methods

2.1. Sample

The experimental population consisted of twenty adult participants employed within the same private organisation operating in the field of strategic consulting and organisational development (north of Italy). Participants (balanced in gender) were recruited through purposive internal sampling to ensure homogeneity of the professional context in line with the aims of the study. The mean age was 31.3 years (SD = 9.10; range: 21–53 years), and the average length of formal education was 17.5 years. All individuals reported being in good general health and demonstrated intact sensory functions, as verified during a preliminary anamnesis-based screening. Eligibility was restricted to full-time employees of the organisation who were able to understand and comply with the experimental instructions and had no uncorrected sensory deficits. Exclusion criteria comprised documented neurological or psychiatric disorders, use of psychotropic medication in the three months preceding data collection, and clinical evidence of burnout or severe chronic stress.

All participants provided written informed consent without compensation. The study was approved by the Ethics Committee of the Department of Psychology (approval code: 2021 PhDTD), Catholic University of The Sacred Heart, Milan, Italy, and was conducted in adherence to the guidelines outlined in the Declaration of Helsinki (2013) and according to the General Data Protection Regulation—Reg. UE 2016/679 and its ethical guidelines.

2.2. Experimental Procedure

The experimental protocol was conducted in a controlled environment, naturally lit and free from distracting stimuli, within a dedicated room at the company premises. Participants were welcomed individually and seated at a computer workstation positioned approximately 70cm from the display, where the experimental tasks were presented. Following a brief introduction to the general aims and procedures, the neurophysiological recording apparatus was applied.

Before the experimental phase commenced, a 120 s neurophysiological baseline was recorded (60 s with eyes closed and 60 s with eyes open) to establish a stable reference free from cognitive or emotional stimulation.

Prior to the commencement of the tasks, participants completed a battery of validated psychometric instruments, including the Perceived Stress Scale (PSS-10), the Emotion Regulation of Others and Self scale (EROS), the General Decision-Making Style inventory (GDMS), and the Balanced Empathy Emotional Scale (BEES). These measures provided individual profiles of emotion regulation, perceived stress, decision-making style, and empathic balance.

The experimental phase consisted of two functionally integrated components.

The *MSCP* served as the profiling phase for decision-making tendencies and metacognitive orientation [8]. Participants were presented with four realistic decision-making scenarios—two set in an organisational context and two in an emergency medical context—each requiring the selection between two alternatives, one predominantly emotion-driven and the other analytical-rational. Each decision was followed by a series of five-point Likert-scale items probing the underlying motivations and the degree of emotional and cognitive engagement. A maximum of twenty seconds was allowed for each decision to encourage spontaneous responding. The decision-making patterns identified in the MSCP provided an individualised index of cognitive–emotional decision style, which was subsequently used to interpret the neurophysiological and behavioural responses recorded in the second task.

The *SST-NeuroEval* was an adaptation of the Trier Social Stress Test [38] designed to elicit progressive evaluative pressure [18,34,37,39–41]. Participants completed five oral trials of increasing difficulty, each preceded by a video sequence depicting a virtual evaluation panel whose facial expressions became progressively more negative and unsupportive. Participants were instructed to prepare and deliver a short speech relevant to a given prompt, maintaining eye contact with the virtual panel for the full duration of the trial and aiming to convince the panel of their suitability. The structure was calibrated to provoke moderate stress responses both during the preparation and delivery phases. In this way, the individual differences observed in the MSCP were tested under conditions of high social–evaluative stress, enabling analysis of the extent to which decision-making style and metacognitive awareness modulated the observed physiological and behavioural responses.

Upon completion of the SST-NeuroEval, a purpose-designed metacognitive and metaperceptual questionnaire (VAS-Prep and VAS-Esp) was administered to assess subjective stress awareness, temporal perception, perceived emotional control, and the perceived effectiveness of self-regulatory strategies during the preparation and performance phases of the stress task.

The entire experimental session lasted approximately fifty minutes per participant, including arrival, instrumentation, baseline acquisition, task performance, psychometric assessment, and final debriefing.

2.3. Behavioural Data Acquisition and Processing

Behavioural data were digitally recorded throughout both experimental phase—the MetaCognition–Stress Convergence Paradigm (MSCP) and the SST-NeuroEva—via automated logging by the task administration platform. In the MSCP, each response was pre-classified as "emotional" or "cognitive" according to predefined coding criteria. For each participant, and within each decision-making domain (organisational and medical emergency), the arithmetic means of the scores assigned to emotional responses and, separately, to cognitive responses was calculated. These values constituted the EmoScore_Org and EmoScore_Med indices for the emotional component, and the CogScore_Org and CogScore_Med indices for the cognitive component. In parallel, mean RTs for each response type and domain—RT_Emo_Org, RT_Cog_Org, RT_Emo_Med, RT_Cog_Med—were computed as the arithmetic mean of the completion times (i.e., RTs) for all scenarios within the same category.

In the SST-NeuroEval, performance evaluation integrated quantitative and qualitative indices. Stress Regulation (StressReg) was estimated based on the promptness of speech initiation: for each oral trial, the latency between the beginning of the exposure phase and the actual start of performance was converted into a score on a five-point scale, with higher values indicating greater promptness. The final StressReg score was computed as the mean of the values obtained across the five trials. Stress Tolerance (StressTol) was calculated by relating the overall mean preparation time to the preparation times for the final two trials, and then averaging the two resulting ratios to obtain a single composite index.

Conceptually, the two indices capture complementary facets of adaptive functioning under evaluative stress. StressReg primarily reflects an individual's ability to promptly engage with performance demands, indexing regulatory efficiency and readiness to act under social pressure. In contrast, StressTol provides a measure of persistence and stability across repeated exposure to evaluative stress, capturing how individuals sustain cognitive and emotional resources when facing cumulative demands. Empirical findings from related paradigms [18,34,37,39–41] suggest that higher StressReg scores are typically associated with more effective top-down control and faster mobilisation of attentional and motivational resources, while higher StressTol scores indicate resilience mechanisms supporting consistent performance despite escalating social–evaluative challenges.

Subjective measures, collected via visual analogue scales (0–100), were obtained separately for the preparation phase (VAS-Prep) and the exposure phase (VAS-Exp). In addition, two indices were derived from the metacognitive–metaperceptual questionnaire: Stress Metacognition (MetaStress), calculated as the mean of responses to items concerning perceived ability to identify and manage external, cognitive, and physiological sources of stress; and Temporal Awareness (MetaTime), obtained as the mean of responses to items assessing perception and management of time across the two phases of the task.

2.4. Electrophysiological Data Acquisition and Processing

EEG activity was recorded using an integrated, non-invasive wearable system. Recordings spanned the entire experimental session (baseline, MSCP, SST-NeuroEval), amounting to approximately 35 min of continuous acquisition per participant. Cortical activity was acquired through the Muse™ Headband (version 2; InteraXon Inc., Toronto, QC, Canada), a four-channel dry-electrode EEG device positioned according to the international 10–20 system, with active electrodes located over frontal (AF7, AF8) and temporo-parietal (TP9, TP10) regions, and reference electrodes placed at Fpz. Signals were sampled at 256 Hz, with a 50 Hz notch filter applied to reduce mains interference. Data were transmitted in real time via Bluetooth to a mobile device running the *Mind Monitor* application, which performed a fast Fourier transform (Hamming window, 10% length, 0.5 Hz resolution) to extract the spectral power density for the canonical EEG frequency bands: delta (1–4 Hz), theta (4–8 Hz), alpha (7.5–13 Hz), beta (13–30 Hz), and gamma (30–44 Hz). Power spectral density values were expressed as the logarithm of the absolute spectral power, typically ranging from -1 to $+1$.

Spectral power was selected as the primary feature due to its validated sensitivity to stress-related modulation and compatibility with the device's resolution, while acknowledging that future analyses may integrate additional indices (e.g., instantaneous frequency, connectivity metrics) to enhance granularity.

To control for interindividual variability, resting baseline was used to normalise task-related data. Movement-related artefacts were monitored using the integrated accelerometer and, when detected, affected segments were excluded from the analysis. Additional visual inspection ensured removal of segments with eye blinks or muscular artefacts. The use of the Muse device facilitated ecological data collection but its limited spatial resolution constitutes a methodological constraint, preventing advanced source localization. EEG processing considered regional averages for the frontal and temporo-parietal sites, allowing for the evaluation of region-specific modulations in spectral power across task conditions.

For subsequent analyses, EEG spectral power values were averaged within each experimental phase and condition. This allowed the derivation of condition-specific neurophysiological profiles for each participant, which were then correlated with behavioural and psychometric measures.

2.5. Data Analyses

Statistical analyses were performed using Jamovi software (version 2.6.26). For EEG data, repeated-measures analyses of variance (ANOVAs) were conducted separately for each of the five canonical frequency bands—delta (1–4 Hz), theta (4–8 Hz), alpha

(8–13 Hz), beta (13–30 Hz), and gamma (30–44 Hz). In the MSCP, the within-subject factors included Region of Interest (ROI: frontal vs. temporo-parietal) and scenario type (4 decision-making scenarios: 2 organisational—S1, S2—2 medical emergency—S3, S4). In the SST-NeuroEval, separate repeated-measures ANOVAs were run for the preparation and exposure phases, each including ROI (frontal vs. temporo-parietal) and trial number (5 preparation—Prep1–5—or 5 exposure trials—Exp1–5) as within-subject factors.

When Mauchly's test indicated a violation of sphericity, degrees of freedom were adjusted using the Greenhouse-Geisser epsilon. In cases of statistically significant main effects or interactions, simple effects were explored using Bonferroni-corrected pairwise comparisons. Effect sizes were reported as partial eta squared (η^2), with the significance threshold set at $\alpha = 0.05$.

Behavioural indices derived from both the MSCP (EmoScore_Org, EmoScore_Med, CogScore_Org, CogScore_Med, RT_Emo_Org, RT_Cog_Org, RT_Emo_Med, RT_Cog_Med) and SST-NeuroEval (StressReg, StressTol, VAS-Prep, VAS-Exp, MetaStress, MetaTime) were analysed in parallel using repeated-measures ANOVAs where applicable.

Additionally, exploratory Pearson's correlation analyses were performed to examine associations between behavioural outcomes, psychometric profiles, and EEG measures. Correlation models included the subscales and total scores of the administered self-report instruments (PSS-10, EROS, GDMS, BEES) alongside the behavioural indices from both experimental tasks. All significance values are reported as two-tailed, with p-values below 0.05 considered statistically. Bonferroni corrections were systematically applied to all pairwise comparisons and correlation analyses to control for multiple testing.

3. Results

3.1. *Behavioural Results*

3.1.1. MSCP Behavioural Results

In the MSCP, a main effect was found for decision type ($F(1,19) = 9.72$, $p = 0.006$, $\eta^2 = 0.338$), with a higher emotional than cognitive choices across domains. RTs showed a main effect of decision type ($F(1,19) = 5.84$, $p = 0.026$, $\eta^2 = 0.235$), with faster RTs for emotional than cognitive choices. The decision type × domain interaction was significant ($F(1,19) = 6.21$, $p = 0.022$, $\eta^2 = 0.246$). Post hoc Bonferroni-corrected comparisons indicated that, in both the organisational domain ($p = 0.012$) and the medical domain ($p = 0.038$), mean RTs for emotional decisions were significantly shorter than those for cognitive decisions (Figure 1a).

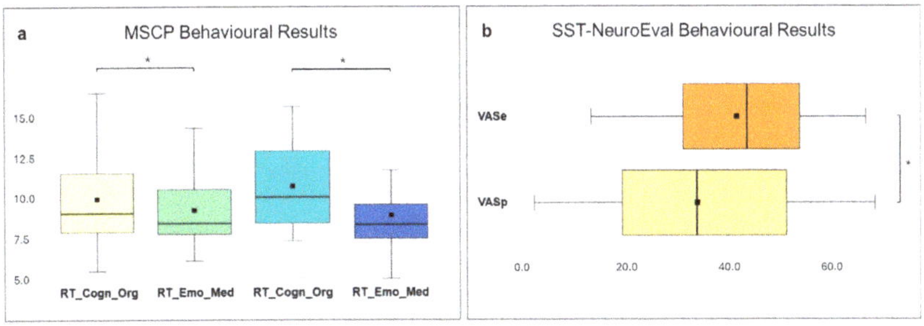

Figure 1. (**a**) Distribution of mean reaction times (RTs) in the MetaCognition–Stress Convergence Paradigm (MSCP) across decision type (cognitive vs. emotional) and domain (organisational vs. medical). (**b**) Distribution of Visual Analogue Scale (VAS) scores assessing perceived stress levels, plotted separately for the two compared conditions. Bars represent ±1 standard error and stars (*) mark statistically significant effects.

3.1.2. SST-NeuroEval Behavioural Results

In the SST-NeuroEval, subjective stress ratings (VAS-Prep vs. VAS-Exp) showed a main effect of phase (F(1,19) = 11.38, p = 0.003, η^2 = 0.375), with Bonferroni-corrected comparisons indicating higher VAS-Exp than VAS-Prep scores (p = 0.003) (Figure 1b).

3.2. EEG Results

3.2.1. MSCP EEG Results

Delta band. A main effect of ROI was observed (F(1,19) = 13.30, p = 0.002, η^2 = 0.042), with higher temporo-parietal than frontal delta power. A main effect of scenario was also found (F(2.14,40.68) = 23.20, $p \leq 0.001$, η^2 = 0.194). Post hoc Bonferroni-corrected comparisons indicated higher values in S1 than in S2 ($p \leq 0.001$) and S3 ($p \leq 0.001$), and in S4 than in S2 ($p \leq 0.001$). The ROI × scenario interaction was significant (F(3,57) = 21.40, $p \leq 0.001$, η^2 = 0.245). In the frontal ROI, S1 was higher than S2 ($p \leq 0.001$); S3 ($p \leq 0.001$) and S4 were higher than S2 ($p \leq 0.001$). In the temporo-parietal ROI, S1 was higher than S2 (p = 0.004) and S3 ($p \leq 0.001$); S3 and S4 was higher than S2 (p = 0.003; $p \leq 0.001$) (Figure 2).

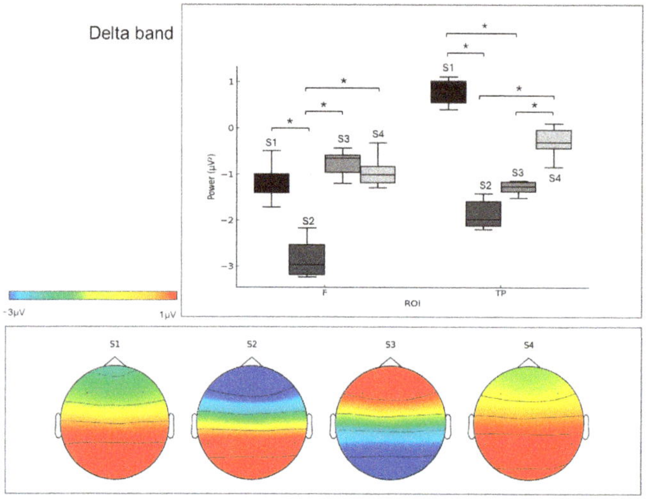

Figure 2. Top: Interaction between region of interest (ROI; frontal, temporo-parietal) and scenario (S1–S4) for delta band power (µV^2) in the MetaCognition–Stress Convergence Paradigm (MSCP). Data points represent mean values ± standard error, highlighting scenario-dependent modulations. Bottom: Scalp topographies for the delta band derived from Muse EEG recordings. Power values were computed using Welch's method on artefact-cleaned data, averaged across trials and expressed in µV^2. Electrode positions (TP9, AF7, AF8, TP10) were projected to the standard 10–20 system and visualised with mne.viz.plot_topomap (MNE-Python 3.12) using spherical spline interpolation to create a continuous scalp representation for illustrative purposes. Bars represent ±1 standard error and stars (*) mark statistically significant effects.

Theta band. A main effect of ROI was found (F(1,19) = 46.60, $p \leq 0.001$, η^2 = 0.041), with higher temporo-parietal than frontal theta power. A main effect of scenario was also observed (F(3,57) = 87.60, $p \leq 0.001$, η^2 = 0.519). Post hoc tests showed higher values in S1 than S3 ($p \leq 0.001$) and S4 ($p \leq 0.001$), in S2 than S3 ($p \leq 0.001$) and S4 ($p \leq 0.001$), and in S3 than S4 ($p \leq 0.001$).

The ROI × scenario interaction was significant (F(3,57) = 15.10, $p \leq 0.001$, η^2 = 0.118). In the frontal ROI, theta power was higher in S1 than in S3 ($p \leq 0.001$) and S4 ($p \leq 0.001$), and in S2 than in S3 ($p \leq 0.001$) and S4 ($p \leq 0.001$). In the temporo-parietal ROI, theta power was higher in S3 than in S4 ($p \leq 0.001$). Additionally, theta power was significantly lower in the temporo-parietal ROI than in the frontal ROI for S3 ($p \leq 0.001$), and higher in the temporo-parietal ROI than in the frontal ROI for S4 ($p \leq 0.001$).

Gamma band. A main effect of scenario was detected ($F(3,57) = 14.93$, $p \leq 0.001$, $\eta^2 = 0.163$), with post hoc comparisons showing higher values in S1 than in S2 ($p \leq 0.001$) and S3 ($p \leq 0.001$), and in S4 than in S2 ($p \leq 0.001$).

The ROI × scenario interaction was significant ($F(2.12,40.37) = 13.70$, $p \leq 0.001$, $\eta^2 = 0.194$). In the frontal ROI, gamma power was higher in S4 than in S2 ($p \leq 0.001$) and in S1 than in S3 ($p \leq 0.001$). In the temporo-parietal ROI, gamma power was higher in S1 than in S2 ($p \leq 0.001$), in S3 ($p \leq 0.001$), and in S4 ($p = 0.022$). Additionally, gamma power was significantly higher in the temporo-parietal ROI than in the frontal ROI for S4 ($p = 0.001$) and significantly higher in the frontal ROI than in the temporo-parietal ROI for S1 ($p \leq 0.001$).

No other significant results were found (all $p < 0.05$).

3.2.2. SST-NeuroEval EEG Results

Delta band. A main effect of preparation was observed ($F(4,76) = 68.23$, $p \leq 0.001$, $\eta^2 = 0.597$). Post hoc Bonferroni-corrected comparisons indicated higher values in Prep2 than in Prep1 ($p \leq 0.001$), Prep3 ($p \leq 0.001$), and Prep5 ($p \leq 0.001$); Prep3 than Prep1 ($p \leq 0.001$); Prep4 than Prep1 ($p \leq 0.001$), Prep3 ($p \leq 0.001$), and Prep5 ($p \leq 0.001$).

A main effect of exposure was also found ($F(4,76) = 113.77$, $p \leq 0.001$, $\eta^2 = 0.624$). Post hoc comparisons showed higher values in Esp1 than in Esp3 ($p \leq 0.001$), Esp4 ($p \leq 0.001$), and Esp5 ($p \leq 0.001$); Esp2 than Esp1 ($p \leq 0.001$), Esp3 ($p \leq 0.001$), Esp4 ($p \leq 0.001$), and Esp5 ($p \leq 0.001$); Esp3 than Esp5 ($p \leq 0.001$); and Esp4 than Esp5 ($p \leq 0.001$).

Theta band. A main effect of exposure was found ($F(2.42,45.99) = 44.31$, $p \leq 0.001$, $\eta^2 = 0.459$). Post hoc comparisons indicated higher values in Esp2 than in Esp1 ($p \leq 0.001$) and Esp4 ($p \leq 0.001$); Esp3 than Esp1 ($p \leq 0.001$) and Esp4 ($p \leq 0.001$); and Esp5 than Esp1 ($p \leq 0.001$), Esp2 ($p \leq 0.001$), and Esp4 ($p \leq 0.001$).

Gamma band. A main effect of preparation was detected ($F(2.60,49.41) = 23.54$, $p \leq 0.001$, $\eta^2 = 0.347$). Post hoc comparisons revealed higher values in Prep1 than in Prep4 ($p \leq 0.001$), Prep2 than in Prep4 ($p \leq 0.001$), Prep3 than in Prep4 ($p \leq 0.001$), and Prep5 than in Prep4 ($p \leq 0.001$). A main effect of exposure was also observed ($F(2.81,53.38) = 46.60$, $p \leq 0.001$, $\eta^2 = 0.507$). Post hoc comparisons showed higher values in Esp4 than in Esp1 ($p \leq 0.001$), Esp2 ($p \leq 0.001$), Esp3 ($p \leq 0.001$), and Esp5 ($p \leq 0.001$); and in Esp5 than in Esp1 ($p = 0.003$) and Esp2 ($p = 0.016$) (Figure 3).

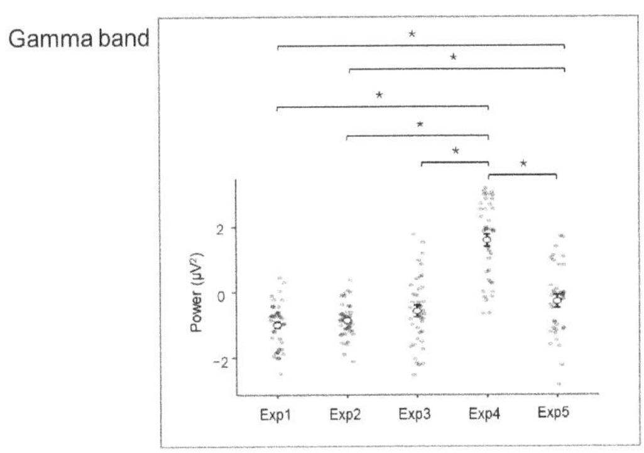

Figure 3. Gamma band power (μV^2) across exposure phases (Exp1–Exp5) in the Social Stress Test—NeuroEval (SST-NeuroEval). Each point represents an individual participant; black circles indicate group means ± standard error. Bars represent ±1 standard error and stars (*) mark statistically significant effects.

No other significant results were found (all $p < 0.05$).

3.3. Correlations Results

EmoScore_Org correlated positively with VAS-Exp (r = 0.448, $p = 0.048$). CogScore_Org correlated positively with MetaTime (r = 0.463, $p = 0.040$). RTs for emotional responses in the organisational domain correlated negatively with MetaStress (r = −0.518, $p = 0.019$). RTs for emotional responses in the medical domain correlated negatively with VAS-Exp (r = −0.462, $p = 0.040$).

PSS-10 scores correlated positively with VAS-Prep (r = 0.528, $p = 0.017$) and VAS-Exp (r = 0.560, $p = 0.010$). StressReg correlated negatively with EROS-Intrinsic_Affect_Worsening (r = −0.588, $p = 0.006$). StressTol correlated negatively with EROS-Intrinsic_Affect_Worsening (r = −0.493, $p = 0.027$). GDMS rational style correlated positively with MetaStress (r = 0.585, $p = 0.007$). GDMS dependent style correlated positively with VAS-Prep (r = 0.606, $p = 0.005$). GDMS avoidant style correlated positively with VAS-Prep (r = 0.680, $p < 0.001$).

4. Discussion

The present findings provide converging evidence that individual differences in decision-making orientation and stress regulation are systematically reflected across behavioural, subjective, and neurophysiological indices. By integrating a metacognitive decision-making paradigm with a socially evaluative stress task, the study offers a multilevel perspective on how cognitive–emotional processes unfold under controlled yet ecologically valid conditions. This discussion addresses the behavioural, EEG, and correlational patterns observed, situating them within current theoretical frameworks and identifying their potential implications for applied settings.

4.1. Behavioural Adaptation and Decision-Making Under Stress

Behavioural results from the MSCP indicated a clear predominance of emotional over cognitive–analytical choices, accompanied by significantly shorter RTs for emotional responses. This pattern suggests preferential activation of decision-making processes grounded in the rapid retrieval of affectively salient representations, consistent with dual-process models of decision-making [1,2]. Within such models, immediate access to socio-emotional heuristics is supported by limbic–frontal networks, which can substantially enhance response speed, even at the expense of extended analytical deliberation.

These observations align with recent literature emphasising that decision-making under temporal pressure or social–evaluative stress emerges from a dynamic interplay between bottom-up affective processes and top-down cognitive control mechanisms [3–5]. Under stress, neural pathways involved in the rapid appraisal of socially and biologically salient cues—particularly those connecting the amygdala, orbitofrontal cortex, and anterior cingulate cortex—may assume a predominant role, favouring decisions aimed at immediate threat reduction or optimisation of emotional resonance with the social context [11,42–44].

In the SST-NeuroEval, the systematic increase in perceived stress from the preparation to the exposure phase confirms the paradigm's effectiveness in eliciting a socially evaluative load with high ecological validity. This increase is consistent with contemporary models of social–evaluative stress [11,34,45,46], which posit that confrontation with an evaluative audience—whether real or virtual—triggers co-activation of limbic–hypothalamic and fronto-parietal circuits, producing measurable effects on both physiological responses and cognitive self-regulation processes.

Taken together, these behavioural findings support the view that evaluative pressure and task demands can substantially influence the trajectory of cognitive and emotional strategy selection. They align with integrative models conceptualising decision-making under stress as a multilevel co-regulation process between affective and cognitive systems, dynamically modulating decision priorities and balancing accuracy–speed trade-offs [18,36,47].

Within this framework, the observed predominance of rapid emotional responses may be interpreted not as a simple reduction in analytical capacity, but rather as an adaptive strategy aimed at preserving behavioural responsiveness under conditions of social and temporal pressure.

4.2. EEG Correlates of Cognitive–Emotional Processing in Decision-Making (MSCP)

The EEG analysis conducted within the MSCP revealed an oscillatory architecture markedly modulated by the variables scenario and ROI, with distinct profiles in the Delta, Theta, and Gamma frequency bands. The most relevant differences emerged between scenarios with high socio-emotional salience (S1, S4), scenarios with neutral valence and lower affective urgency (S3), and intermediate scenarios characterised by a combination of pragmatic constraints and interpersonal relevance (S2).

Delta activity exhibited significant frontal increases in S1 and S4 compared to the other conditions, with marked differences relative to S2 and S3. In S3, an inverse topographical pattern was observed, with predominance in temporo-parietal regions over frontal regions, indicating a functional reconfiguration as a function of scenario characteristics. In the temporo-parietal ROI, delta power was significantly elevated in S1, S4, and S2 compared to S3.

This pattern is consistent with models linking frontal delta to motivated attention, selective orienting, and rapid decision synthesis in contexts of high affective pressure [15–17,48], while temporo-parietal delta is associated with contextual encoding and semantic–narrative reconstruction. The shift from frontal predominance (S1, S4) to temporo-parietal predominance (S3) reflects a flexible resource allocation mechanism [41] whereby oscillatory input is redistributed according to the interplay between affective urgency and contextual complexity.

Frontal theta power was significantly higher in S1 and S2 than in S3 and S4, whereas temporo-parietal Theta showed a marked increase in S3, exceeding that observed in S4 and, in some comparisons, in S1 and S2.

Frontal theta has been consistently associated with conflict monitoring, executive control, and the concurrent maintenance of multiple competing decision representations [21]—processes that are central in scenarios with strong normative conflict (S1) and pragmatic–interpersonal tension (S2). The temporo-parietal increase in S3 is consistent with the role of posterior Theta in mnemonic retrieval, contextual encoding, and semantic integration [24,49]. The inversion of ROI predominance between S3 and S4 illustrates a dynamic reorganisation of oscillatory networks in response to decision pressure and the semantic nature of the scenario.

Frontal gamma was significantly increased in S1 and S4 compared to S2 and S3, whereas temporo-parietal gamma showed higher values in S1, S4, and S2 relative to S3. The ROI-specific contrast was particularly evident: frontal predominance occurred in conditions with greater socio-emotional integrative demands, while temporo-parietal predominance was more pronounced in scenarios requiring extended representational construction.

These findings are consistent with models assigning gamma a pivotal role in multimodal feature binding and high-frequency interregional synchronisation [31,50]. Frontal Gamma enhancement in S1 and S4 indicates rapid integration of perceptual, moral, and emotional inputs into a unified decision, whereas temporo-parietal Gamma enhancement reflects engagement of mechanisms for interpersonal simulation and projection of decision consequences.

To sum up, the ROI × scenario distribution in Delta, Theta, and Gamma demonstrates that the MSCP does not elicit a uniform increase in oscillatory activity but rather a system of selective and topographically targeted modulations. These configurations align with multi-level co-regulation models [41,51–53], whereby frontal cortical networks—responsible for decision control and synthesis—and temporo-parietal networks—specialised in contextual integration and semantic representation—operate within a dynamic resource reallocation regime, adjusting in real time to the interplay between decision urgency, emotional salience, and contextual complexity.

4.3. EEG Correlates of Stress Regulation (SST-NeuroEval)

The EEG analysis of the Social Stress Test—NeuroEval (SST-NeuroEval), similarly, revealed marked modulations in delta, theta, and gamma frequency bands as a function of the experimental phase (preparation vs. exposure) and trial progression. The observed patterns delineate the transition from anticipatory cognitive–emotional regulation processes to reactive behavioural engagement under evaluative stress.

Delta activity displayed a distinctive temporal profile across both preparation and exposure phases. During preparation, power increased from Prep2 and was maintained at Prep4, relative to the initial (Prep1) and terminal (Prep5) stages. This suggests that Delta mobilisation is not immediate but emerges after an initial alignment of cognitive–emotional resources to the impending evaluative situation, peaking at an intermediate stage before declining as direct exposure approaches.

During exposure, Delta power was elevated in the first two trials (Esp1, Esp2) and progressively decreased thereafter. This pattern indicates strong early engagement of attentional orienting mechanisms and low-frequency processing systems that register and appraise socially relevant cues at the moment of maximal evaluative impact. The subsequent decrease may reflect habituation or partial reallocation of resources towards higher-frequency bands, in response to reduced novelty or progressive adaptation to social demands.

From a theoretical perspective, these findings align with models linking Delta to motivated attention and to widespread cortical synchronisation under social stress conditions [54,55]. In preparation, the intermediate delta increase may represent peak activation of fronto-limbic and fronto-parietal circuits involved in performance readiness; in exposure, the initial Delta peak likely reflects maximal alertness and environmental registration, followed by oscillatory rebalancing as the situation becomes more familiar.

Theta showed significant increases in specific exposure trials (Esp2, Esp3, Esp5) compared to phases of lower engagement. The temporal profile suggests oscillatory peaks coinciding with moments of heightened demand for cognitive control and emotional regulation. This modulation is consistent with the role of frontal theta in conflict monitoring and management of cognitive–affective load [20,56], and may also reflect context updating processes when the social stressor retains dynamic and unpredictable features [17,57,58].

Gamma activity exhibited a biphasic pattern. In preparation, power was reduced in Prep4 relative to all other stages, suggesting a transient disengagement of integrative mechanisms immediately prior to direct exposure. During exposure, gamma power markedly increased in Esp4, exceeding all other phases.

These findings accord with the established role of gamma in rapid multimodal integration and perceptual–decisional synthesis [35]. The pre-exposure reduction may index a strategic suspension of integrative processing, followed by a potentiated reactivation during peak evaluative engagement.

Overall, the SST-NeuroEval EEG data indicate that delta, theta, and gamma bands are modulated in a phase-specific manner, delineating an oscillatory cycle that parallels the shift from anticipatory regulation to reactive engagement under stress.

4.4. Integrated Behavioural–Neurophysiological Relationships

The integration of behavioural, psychometric, and EEG results provides converging evidence that individual differences in decision-making tendencies and stress regulation capacities manifest coherently across multiple measurement levels. Correlational analyses revealed systematic associations linking task-derived indices, self-reported measures, and oscillatory activity patterns, highlighting the functional interplay between cognitive–emotional style and stress responsivity.

From a behavioural perspective, a higher emotional decisions in organisational scenarios was associated with elevated subjective stress ratings during the exposure phase of the SST-NeuroEval. This suggests that a preferential orientation towards affect-laden decision strategies may amplify perceived stress when confronted with evaluative social demands. Conversely, higher cognitive–analytical decision scores were related to greater temporal awareness of stress, pointing to a more deliberate monitoring of stress dynamics across task phases.

RTs further supported this relationship: faster latencies for emotional decisions were negatively correlated with stress metacognition measures, indicating that rapid affective responses may be accompanied by reduced explicit appraisal of stress states. These findings align with dual-process models in which fast, heuristic-based decision-making relies less on explicit metacognitive monitoring and more on prepotent emotional evaluations [11,59].

Self-report measures corroborated these patterns. Higher PSS-10 scores were consistently associated with increased subjective stress ratings in both preparation and exposure phases, reinforcing the ecological validity of the SST-NeuroEval as a stress-induction paradigm. Stress regulation and tolerance indices were inversely related to self-reported deterioration in intrinsic affect (EROS), in line with models linking effective emotion regulation under stress to preserved affective stability [60].

Oscillatory activity patterns also converged with these behavioural and psychometric profiles. Participants with higher rational decision-making tendencies (GDMS) displayed stronger associations with stress metacognition scores, suggesting a shared reliance on controlled, reflective processing. In contrast, higher dependent and avoidant decision styles correlated with increased subjective stress in preparation, implying a heightened anticipatory affective load.

Together, these multi-domain relationships underscore the value of integrating behavioural indices, subjective reports, and neurophysiological signatures to capture the dynamic coupling between decision orientation and stress regulation. Such integration not only enhances theoretical understanding of cognitive–emotional interaction under evaluative pressure but also supports the development of personalised intervention models that target both decision-making style and stress resilience.

5. Conclusions: Theoretical and Applied Implications, Limitations, and Future Directions

In conclusion, from a theoretical perspective, these results refine current models of decision-making under stress by demonstrating that the integration of task-specific decision contexts and social–evaluative demands produces distinct oscillatory "signatures" that can be reliably mapped onto decision orientation and stress responsivity. The observed convergence across measurement levels underscores the necessity of approaching cognitive–emotional regulation as a multi-domain construct that cannot be adequately characterised by a single level of analysis.

Applied implications emerge for contexts where decision-making is embedded within socially evaluative or high-stakes environments, such as organisational leadership, emergency response, or clinical decision-making under time pressure. Beyond these, potential applications extend to educational and training settings, where stress regulation and decision style profiling may inform targeted curricula and resilience-building programmes. In organisational domains, these findings could inform resilience-oriented training for leaders and teams exposed to evaluative pressure. In clinical practice, the integration of behavioural, subjective, and EEG indices may support diagnostic refinement in stress-related disorders and guide personalised interventions targeting cognitive–emotional regulation.

Profiling individuals based on their decision orientation, stress metacognition, and oscillatory markers may inform personalised training protocols aimed at optimising decision quality and resilience under stress.

However, certain limitations must be acknowledged. First, the study employed a relatively small and homogeneous preliminary sample, which constrains the generalisability of the findings and may have limited the detection of subtler interaction effects. Furthermore, although the present study was not designed for systematic gender comparisons, future larger-scale investigations may explore potential gender-related effects in stress responsivity and cognitive–emotional regulation. Furthermore, only spectral power was considered, while other digital features could provide additional granularity in future studies.

Second, while the EEG Muse Headband allowed for ecologically flexible acquisition, its reduced spatial resolution precludes fine-grained source localisation. Finally, the correlational nature of the multi-domain analyses precludes strong causal inferences regarding the directionality of the observed associations.

Future research should extend this work by employing larger and more diverse samples, integrating higher-density EEG or multimodal neuroimaging to refine spatial characterisation of the observed oscillatory patterns, and adopting counterbalanced or adaptive task sequencing to disentangle order-related effects. Longitudinal or intervention-based designs could also determine whether targeted modulation of decision-making strategies and stress regulation capacities produces measurable shifts in oscillatory dynamics and behavioural outcomes.

In conclusion, the present study advances the understanding of cognitive–emotional regulation under stress by demonstrating a consistent multi-level correspondence between decision-making orientation, subjective stress appraisal, and oscillatory brain activity. This integrated framework not only enriches theoretical models of decision-making under evaluative pressure but also offers practical pathways for developing tailored strategies to enhance resilience, adaptability, and performance in socially demanding contexts.

Author Contributions: Conceptualization, K.R. and M.B.; methodology, K.R. and M.B.; software, K.R. and A.D.; validation, M.B.; formal analysis, K.R.; investigation, K.R. and A.D.; resources, M.B.; data curation, K.R. and A.D.; writing—original draft preparation, K.R. and A.D.; writing—review and editing, M.B.; visualisation, K.R.; supervision, M.B.; project administration, M.B.; funding acquisition, M.B. All authors have read and agreed to the published version of the manuscript.

Funding: This research received no external funding.

Institutional Review Board Statement: This experimental work was conducted according to the Declaration of Helsinki and approved by the Ethics Committee of the Department of Psychology, Catholic University of the Sacred Heart, Milan, Italy (approval code: 2021 PhDTD; approval date: 1 December 2021).

Informed Consent Statement: Informed consent was obtained from all subjects involved in the study.

Data Availability Statement: The data presented in this study are available on request from the corresponding author due to ethical reasons for sensitive personal data protection (requests will be evaluated according to the GDPR—Reg. UE 2016/679 and its ethical guidelines).

Acknowledgments: We would like to acknowledge Lorini Camilla for her support in data collection.

Conflicts of Interest: The authors declare no conflicts of interest.

References

1. Evans, J.S.B.T.; Stanovich, K.E. Dual-Process Theories of Higher Cognition. *Perspect. Psychol. Sci.* **2013**, *8*, 223–241. [CrossRef]
2. Kahneman, D. *Thinking, Fast and Slow*; Farrar, Straus and Giroux: New York, NY, USA, 2011; 499p.
3. Astle, D.E.; Bassett, D.S.; Viding, E. Understanding Divergence: Placing Developmental Neuroscience in Its Dynamic Context. *Neurosci. Biobehav. Rev.* **2024**, *157*, 105539. [CrossRef]
4. Dricu, M.; Frühholz, S. A Neurocognitive Model of Perceptual Decision-Making on Emotional Signals. *Hum. Brain Mapp.* **2020**, *41*, 1532–1556. [CrossRef]
5. Levine, D. Neuroscience of Emotion, Cognition, and Decision Making: A Review. *Med. Res. Arch.* **2022**, *10*. [CrossRef]
6. Harlé, K.M.; Shenoy, P.; Paulus, M.P. The Influence of Emotions on Cognitive Control: Feelings and Beliefs—Where Do They Meet? *Front. Hum. Neurosci.* **2013**, *7*, 508. [CrossRef] [PubMed]
7. Dignath, D.; Eder, A.B.; Steinhauser, M.; Kiesel, A. Conflict Monitoring and the Affective-Signaling Hypothesis—An Integrative Review. *Psychon. Bull. Rev.* **2020**, *27*, 193–216. [CrossRef]
8. Balconi, M.; Rovelli, K.; Angioletti, L. Social Influence in Persuasion and Negotiation: A Hyperscanning EEG and Autonomic Measures Study. *Front. Neurosci.* **2025**, *19*, 1604389. [CrossRef] [PubMed]
9. Minguillon, J.; Lopez-Gordo, M.A.; Pelayo, F. Stress Assessment by Prefrontal Relative Gamma. *Front. Comput. Neurosci.* **2016**, *10*, 101. [CrossRef] [PubMed]
10. Navarro-Nolasco, D.A.; Chi-Castañeda, D.; López-Meraz, M.L.; Beltran-Parrazal, L.; Morgado-Valle, C. The Medial Prefrontal Cortex as a Proposed Regulatory Structure in the Relationship between Anxiety and Perceived Social Support: A Review. *BMC Psychol.* **2025**, *13*, 152. [CrossRef]
11. Yu, R. Stress Potentiates Decision Biases: A Stress Induced Deliberation-to-Intuition (SIDI) Model. *Neurobiol. Stress* **2016**, *3*, 83–95. [CrossRef]
12. Yang, Y.; Zhang, X.; Peng, Y.; Bai, J.; Lei, X. A Dynamic Causal Model on Self-Regulation of Aversive Emotion. *Brain Inform.* **2020**, *7*, 20. [CrossRef]
13. Kenwood, M.M.; Kalin, N.H.; Barbas, H. The Prefrontal Cortex, Pathological Anxiety, and Anxiety Disorders. *Neuropsychopharmacology* **2022**, *47*, 260–275. [CrossRef]
14. Gangopadhyay, P.; Chawla, M.; Monte, O.D.; Chang, S.W.C. Prefrontal–Amygdala Circuits in Social Decision-Making. *Nat. Neurosci.* **2021**, *24*, 5–18. [CrossRef] [PubMed]
15. Balconi, M.; Angioletti, L. Inter-Brain Entrainment (IBE) during Interoception. A Multimodal EEG-fNIRS Coherence-Based Hyperscanning Approach. *Neurosci. Lett.* **2024**, *831*, 137789. [CrossRef] [PubMed]
16. Harmony, T. The Functional Significance of Delta Oscillations in Cognitive Processing. *Front. Integr. Neurosci.* **2013**, *7*, 83. [CrossRef] [PubMed]
17. Merino, E.; Raya-Salom, D.; Teruel-Martí, V.; Adell, A.; Cervera-Ferri, A.; Martínez-Ricós, J. Effects of Acute Stress on the Oscillatory Activity of the Hippocampus–Amygdala–Prefrontal Cortex Network. *Neuroscience* **2021**, *476*, 72–89. [CrossRef]
18. Balconi, M.; Rovelli, K. Does Emotional Valence Affect Cognitive Performance and Neurophysiological Response during Decision Making? A Preliminary Study. *Front. Neurosci.* **2024**, *18*, 1408526. [CrossRef]
19. Tan, E.; Troller-Renfree, S.V.; Morales, S.; Buzzell, G.A.; McSweeney, M.; Antúnez, M.; Fox, N.A. Theta Activity and Cognitive Functioning: Integrating Evidence from Resting-State and Task-Related Developmental Electroencephalography (EEG) Research. *Dev. Cogn. Neurosci.* **2024**, *67*, 101404. [CrossRef]
20. Chikhi, S.; Matton, N.; Blanchet, S. EEG Power Spectral Measures of Cognitive Workload: A Meta-analysis. *Psychophysiology* **2022**, *59*, e14009. [CrossRef]
21. Cavanagh, J.F.; Frank, M.J. Frontal Theta as a Mechanism for Cognitive Control. *Trends Cogn. Sci.* **2014**, *18*, 414–421. [CrossRef]
22. Balconi, M.; Vanutelli, M.E. Functional EEG Connectivity during Competition. *BMC Neurosci.* **2018**, *19*, 63. [CrossRef]
23. Klimesch, W. Alpha-Band Oscillations, Attention, and Controlled Access to Stored Information. *Trends Cogn. Sci.* **2012**, *16*, 606–617. [CrossRef]
24. Klimesch, W. EEG Alpha and Theta Oscillations Reflect Cognitive and Memory Performance: A Review and Analysis. *Brain Res. Rev.* **1999**, *29*, 169–195. [CrossRef]
25. Klimesch, W.; Sauseng, P.; Hanslmayr, S. EEG Alpha Oscillations: The Inhibition–Timing Hypothesis. *Brain Res. Rev.* **2007**, *53*, 63–88. [CrossRef] [PubMed]
26. Tognoli, E.; Kelso, J.A.S. The Coordination Dynamics of Social Neuromarkers. *Front. Hum. Neurosci.* **2015**, *9*, 563. [CrossRef] [PubMed]
27. Gallina, J.; Marsicano, G.; Romei, V.; Bertini, C. Electrophysiological and Behavioral Effects of Alpha-Band Sensory Entrainment: Neural Mechanisms and Clinical Applications. *Biomedicines* **2023**, *11*, 1399. [CrossRef] [PubMed]

28. Gratton, G. Brain Reflections: A Circuit-based Framework for Understanding Information Processing and Cognitive Control. *Psychophysiology* **2018**, *55*, e13038. [CrossRef] [PubMed]
29. Balconi, M.; Rovelli, K.; Angioletti, L.; Allegretta, R.A. Working Memory Workload When Making Complex Decisions: A Behavioral and EEG Study. *Sensors* **2024**, *24*, 5754. [CrossRef]
30. Balconi, M.; Crivelli, D.; Cassioli, F. "We Will Let You Know": An Assessment of Digital vs. Face-to-Face Job Interviews via EEG Connectivity Analysis. *Information* **2022**, *13*, 312. [CrossRef]
31. Rovelli, K.; Balconi, M. Belief Shift Status in Moral Reasoning: Evidence from EEG Hyperscanning and Real-Time Dialogic Framing. *Neuroscience* **2025**, *582*, 63–71. [CrossRef]
32. Wróbel, A. Beta Activity: A Carrier for Visual Attention. *Acta Neurobiol. Exp.* **2000**, *60*, 247–260. [CrossRef] [PubMed]
33. Townsend, B.; Legere, J.K.; Mohrenschildt, M.V.; Shedden, J.M. Beta-Band Power Is an Index of Multisensory Weighting during Self-Motion Perception. *Neuroimage Rep.* **2022**, *2*, 100102. [CrossRef] [PubMed]
34. Crivelli, D.; Rovelli, K.; Balconi, M. Different Types of Social Feedback Can Enhance or Reduce Performance, and Induce or Alleviate Psychosocial Stress: An Exploratory Study of the Underlying Neurophysiological Mechanisms. *Adapt. Hum. Behav. Physiol.* **2025**, *11*, 3. [CrossRef]
35. Rieder, M.K.; Rahm, B.; Williams, J.D.; Kaiser, J. Human Gamma-Band Activity and Behavior. *Int. J. Psychophysiol.* **2011**, *79*, 39–48. [CrossRef]
36. Crivelli, D.; Allegretta, R.A.; Balconi, M. The "Status Quo Bias" in Response to External Feedback in Decision-Makers. *Adapt. Hum. Behav. Physiol.* **2023**, *9*, 426–441. [CrossRef]
37. Balconi, M.; Angioletti, L.; Rovelli, K. Sensorimotor Simulation's Influence on Stress: EEG and Autonomic Responses in Digital Interviews. *Brain Sci.* **2024**, *14*, 608. [CrossRef]
38. Allen, A.P.; Kennedy, P.J.; Dockray, S.; Cryan, J.F.; Dinan, T.G.; Clarke, G. The Trier Social Stress Test: Principles and Practice. *Neurobiol. Stress* **2017**, *6*, 113–126. [CrossRef]
39. Balconi, M.; Angioletti, L.; Rovelli, K. Neurophysiological Response to Social Feedback in Stressful Situations. *Eur. J. Neurosci.* **2024**, *60*, 6030–6045. [CrossRef]
40. Angioletti, L.; Rovelli, K.; Balconi, M. Be Ready to Manage Stress "Before" and "After" a Critical Event. What the EEG and Autonomic Correlates Tell Us. *Brain Cogn.* **2025**, *183*, 106244. [CrossRef]
41. Allegretta, R.A.; Rovelli, K.; Balconi, M. The Role of Emotion Regulation and Awareness in Psychosocial Stress: An EEG-Psychometric Correlational Study. *Healthcare* **2024**, *12*, 1491. [CrossRef]
42. Yi, W.; Li, X.; Chen, W.; Yan, L.; Xin, F.; Buchanan, T.W.; Wu, J. An Inverted U-Shaped Relationship between Chronic Stress and the Motivation to Expend Effort for Reward. *Neurobiol. Stress* **2025**, *36*, 100724. [CrossRef]
43. Starcke, K.; Brand, M. Decision Making under Stress: A Selective Review. *Neurosci. Biobehav. Rev.* **2012**, *36*, 1228–1248. [CrossRef]
44. Porcelli, A.J.; Delgado, M.R. Acute Stress Modulates Risk Taking in Financial Decision Making. *Psychol. Sci.* **2009**, *20*, 278–283. [CrossRef] [PubMed]
45. von Dawans, B.; Kirschbaum, C.; Heinrichs, M. The Trier Social Stress Test for Groups (TSST-G): A New Research Tool for Controlled Simultaneous Social Stress Exposure in a Group Format. *Psychoneuroendocrinology* **2011**, *36*, 514–522. [CrossRef] [PubMed]
46. van der Mee, D.; Duivestein, Q.; Gevonden, M.; Westerink, J.; de Geus, E. The Short Sing-a-Song Stress Test: A Practical and Valid Test of Autonomic Responses Induced by Social-Evaluative Stress. *Auton. Neurosci.* **2020**, *224*, 102612. [CrossRef] [PubMed]
47. Šimić, G.; Tkalčić, M.; Vukić, V.; Mulc, D.; Španić, E.; Šagud, M.; Olucha-Bordonau, F.E.; Vukšić, M.; Hof, P.R. Understanding Emotions: Origins and Roles of the Amygdala. *Biomolecules* **2021**, *11*, 823. [CrossRef]
48. Knyazev, G.G. EEG Delta Oscillations as a Correlate of Basic Homeostatic and Motivational Processes. *Neurosci. Biobehav. Rev.* **2012**, *36*, 677–695. [CrossRef]
49. Balconi, M.; Angioletti, L.; Cassioli, F. Hyperscanning EEG Paradigm Applied to Remote vs. Face-To-Face Learning in Managerial Contexts: Which Is Better? *Brain Sci.* **2023**, *13*, 356. [CrossRef]
50. Michalareas, G.; Vezoli, J.; van Pelt, S.; Schoffelen, J.-M.; Kennedy, H.; Fries, P. Alpha-Beta and Gamma Rhythms Subserve Feedback and Feedforward Influences among Human Visual Cortical Areas. *Neuron* **2016**, *89*, 384–397. [CrossRef]
51. Veer, I.M.; Oei, N.Y.; Spinhoven, P.; van Buchem, M.A.; Elzinga, B.M.; Rombouts, S.A. Endogenous Cortisol Is Associated with Functional Connectivity between the Amygdala and Medial Prefrontal Cortex. *Psychoneuroendocrinology* **2012**, *37*, 1039–1047. [CrossRef]
52. Von Gunten, C.D.; Volpert-Esmond, H.I.; Bartholow, B.D. Temporal Dynamics of Reactive Cognitive Control as Revealed by Event-related Brain Potentials. *Psychophysiology* **2018**, *55*, e13007. [CrossRef] [PubMed]
53. Vinberg, K.; Rosén, J.; Kastrati, G.; Ahs, F. Whole Brain Correlates of Individual Differences in Skin Conductance Responses during Discriminative Fear Conditioning to Social Cues. *eLife* **2022**, *11*, e69686. [CrossRef] [PubMed]
54. Balconi, M.; Lucchiari, C. EEG Correlates (Event-Related Desynchronization) of Emotional Face Elaboration: A Temporal Analysis. *Neurosci. Lett.* **2006**, *392*, 118–123. [CrossRef] [PubMed]

55. Cavanagh, J.F. Cortical Delta Activity Reflects Reward Prediction Error and Related Behavioral Adjustments, but at Different Times. *NeuroImage* **2015**, *110*, 205–216. [CrossRef]
56. Lundqvist, M.; Herman, P.; Lansner, A. Theta and Gamma Power Increases and Alpha/Beta Power Decreases with Memory Load in an Attractor Network Model. *J. Cogn. Neurosci.* **2011**, *23*, 3008–3020. [CrossRef]
57. Vanhollebeke, G.; De Smet, S.; De Raedt, R.; Baeken, C.; van Mierlo, P.; Vanderhasselt, M.-A. The Neural Correlates of Psychosocial Stress: A Systematic Review and Meta-Analysis of Spectral Analysis EEG Studies. *Neurobiol. Stress* **2022**, *18*, 100452. [CrossRef]
58. Barzegar, M.; Jahromi, G.P.; Meftahi, G.H.; Hatef, B. The Complexity of Electroencephalographic Signal Decreases during the Social Stress. *J. Med. Signals Sens.* **2023**, *13*, 57–64. [CrossRef]
59. Zucchelli, M.M.; Trotti, N.M.A.A.; Pavan, A.; Piccardi, L.; Nori, R. The Dual Process Model: The Effect of Cognitive Load on the Ascription of Intentionality. *Front. Psychol.* **2025**, *16*, 1451590. [CrossRef]
60. Gross, J.J. Emotion Regulation: Current Status and Future Prospects. *Psychol. Inq.* **2015**, *26*, 1–26. [CrossRef]

applied sciences

MDPI

Article

Electrophysiological Hyperscanning of Negotiation During Group-Oriented Decision-Making

Laura Angioletti [1,2], Katia Rovelli [1,2], Carlotta Acconito [1,2], Angelica Daffinà [1,2,*] and Michela Balconi [1,2]

1 International Research Center for Cognitive Applied Neuroscience (IrcCAN), Università Cattolica del Sacro Cuore, 20123 Milan, Italy

2 Research Unit in Affective and Social Neuroscience, Department of Psychology, Università Cattolica del Sacro Cuore, 20123 Milan, Italy

* Correspondence: angelica.daffina@unicatt.it

Abstract: Background: This study investigates the electrophysiological (EEG) correlates underlying negotiation dynamics in dyads engaged in a shared decision-making process. Methods: Using EEG hyperscanning, we examined single-brain and inter-brain neural activity in 26 participants (13 dyads) during a structured negotiation task. The participants, selected for their group-oriented decision-making preference, discussed a realistic group decisional scenario while their EEG activity was recorded. EEG frequency bands (delta, theta, alpha, beta, and gamma) were analyzed and Euclidean Distances were computed for measuring dissimilarity at the inter-brain neural level. Results: At the single-brain level, the results show increased delta and theta power in frontal regions, reflecting emotional engagement and goal-directed control, alongside heightened beta and gamma activity in parieto-occipital areas, linked to cognitive integration and decision-monitoring during the negotiation process. At the inter-brain neural level, we observed significant dissimilarity in frontal delta activity compared to temporo-central and parieto-occipital one, suggesting that negotiation involves independent cognitive regulation within the members of the dyads rather than complete neural synchrony. Conclusions: These findings highlight the dual role of negotiation as both a cooperative and cognitively demanding process, requiring emotional alignment and strategic adaptation. This study advances our understanding of the neurophysiological bases of negotiation and provides insights into how inter-brain dynamics shape collaborative decision-making.

Keywords: decision-making; group orientation; EEG hyperscanning; single-brain; inter-brain

1. Introduction

In real-world workgroups, decision-making conversations can either foster agreement and group cohesion or deepen conflicts and amplify discord. A crucial determinant in steering these conversations toward constructive outcomes is the promotion of negotiation and compromise. Encouraging these strategies has been shown to enhance exploration and the exchange of diverse perspectives, facilitating a more integrative and adaptive approach to decision-making. Notably, dyads that adopt a compromise-oriented mindset tend to achieve more harmonious and mutually satisfactory outcomes than those engaging in competition [1].

Previous research on decision-making conversations highlights two primary factors that shape their trajectory: intentions (as starting points) and communication strategies (which are adopted over the course of the conversation) [2,3]. First, the success of a decision-making conversation often hinges on individuals' underlying intentions—their opinions, goals, and needs. These intentions, in turn, shape the communication strategies adopted to bridge differences and reach agreements. Effective negotiation relies on strategies that align thoughts and emotions, fostering shared understanding and consensus. The interplay between intention and communication strategies ultimately determines the effectiveness of the negotiation process and the quality of the final agreement.

A first key factor of conversational intentions is an individual's decision-making orientation, which can be individual- or group-focused. While individual-oriented people may prioritize personal gains or fixed perspectives, group-oriented individuals are more inclined to integrate multiple viewpoints, making them less susceptible to biases, overconfidence, and cognitive errors [4]. This openness to diverse perspectives not only improves decision accuracy but also fosters the creation of innovative, high-quality solutions [5,6]. In negotiation-driven conversations, group-oriented individuals tend to perceive the goal of acting as a cohesive unit as both socially and personally meaningful, reinforcing their commitment to cooperative dialogue.

The second key factor influencing decision-making conversations is the explicit goal of the interaction, which can be competitive or cooperative. These goals significantly shape the communication strategies employed. A cooperative mindset, strongly associated with group orientation, is characterized by prosocial intentions that prioritize joint success rather than individual gain. Dyads with cooperative goals actively engage with each other's ideas, fostering a dynamic exchange that enhances learning, adaptation, and innovation [7,8]. Research has consistently linked cooperation to greater creativity and cognitive flexibility, which are essential for effective problem-solving and decision-making [9–13].

Critically, cooperation is deeply intertwined with negotiation behaviors [14,15]. Negotiation involves the exchange of information about interests and perspectives, followed by strategic trade-offs that maximize collective benefits [16]. Given its reliance on perspective-taking and shared goal-setting, negotiation thrives in environments where group orientation is strong.

To the best of our knowledge, no previous study has investigated the neurophysiological mechanisms underlying negotiation within group-oriented individuals during real-time, naturalistic decision-making interactions.

In neuroscience, negotiation and cooperative behaviors have been investigated using various tasks, including games, problem-solving activities, creativity exercises, mathematical challenges, and behavioral synchronization [17]. These studies have employed both single-brain and brain-to-brain experimental paradigms, utilizing different neuroscientific tools. While communication between participants is a key element in this research, no study has specifically examined negotiation through a communicative ecologically valid scenario that involves synchrony and turn-taking [18].

Although the importance of studying brain dynamics during real-time social interactions is widely acknowledged, most research has measured the electrophysiological (EEG) brain activity in single subjects rather than in interacting pairs or groups [19]. Regarding the functional significance of EEG frequency bands, delta waves are thought to signal novelty in emotional contexts, responding more to a stimulus's attentional salience than its emotional content [20]. Theta activity has been linked to motivation, emotional processing, task relevance [20], and adaptive behavior in social negotiation [21]. Alpha and beta bands indicate cognitive effort, attention, and engagement in decision-making processes [22], with alpha suppression playing a key role in decision-making by reflecting cognitive resource allocation [23]. Beta waves are associated with emotional significance, attentional control, and cognitive regulation [24,25]. Gamma oscillations integrate sensory and emotional processing and are markers of top-down attention and conscious awareness [26,27]. Both beta and gamma bands also play a role in memory and decision-making processes [28].

Also, the specific localization of EEG frequency bands correlate with different neural functions: frontal delta activity is linked to emotional and motivational processing [20,29], while theta waves in frontal regions reflect attentional engagement and strategic control [21,30]. Reduced parietal alpha activity is associated with enhanced cognitive performance [30], and parietal–occipital beta waves indicate focused attention [31]. Moreover, alpha and beta bands in parieto-occipital areas are associated with neural correlates of sense of agency, defined as the subjective experience of initiating and controlling actions to affect external events [32] as well as index of working-memory performance [33]. Gamma modulations in parieto-occipital regions, instead, may be interpreted as top-down activation of mental representations and the maintenance of auditory and visual information in working memory. This could reflect the increasing cognitive resources required for more effective planning, as previously evidenced [34,35].

While single-brain approaches help localize neural activations related to social processes, they fail to grasp the real-time brain-to-brain dynamics. In this field, the hyperscanning paradigm, which enables the simultaneous recording of brain activity from multiple individuals, may reveal functional connections and interbrain synchronization. This approach shifts the focus from isolated brain activity to dynamic interactions, where behaviors and neural patterns continuously adapt to each other.

Hyperscanning research has demonstrated the critical role of interbrain synchrony in joint attention, communication, coordination [36–38], cooperation [37], prosocial exchange [38] and decision-making tasks. In particular, hyperscanning studies have shown that the dorsolateral prefrontal cortex, specifically in the left side, is a key area of cooperation processes [37] as well as the prefrontal cortex, which is implicated in joint action performance [39]. Moreover, during joint actions in cooperative behavior, a higher synchronization was reported in the delta and theta low frequency bands, as an index of social competencies, empathy, and emotional engagement [39]. Additionally, the temporoparietal junction (TPJ) is involved in social cognition, including perspective-taking, Theory of Mind, and communication [40]. TPJ also plays a role in managing social dynamics, including negotiation processes [41,42]. However, for a comprehensive understanding of cooperative processes, it is essential to consider real, dynamic interactions where partners are actively engaged, as social exchanges naturally occur in face-to-face settings [39].

Building on these premises, this study aims to: (i) examine the single-brain neural correlates of group-oriented individuals during decision-making processes that explicitly involve negotiation; (ii) investigate inter-brain neural synchronization within dyads working toward a shared decision. The participants, paired in dyads, discussed a hypothetical situation where a workgroup member did not align with the team's ideals. Before the interaction, their group decision-making orientation was assessed using the Likert scale. During the negotiation interaction, they were asked to negotiate on the best solution for the hypothetical situation previously presented, while EEG data were continuously recorded to analyze their neural activity.

Given the previous evidence, first at the single-brain level, we hypothesized that group-oriented individuals (for which a group decision-making solution would be a relevant option) will show a significantly increase in low-frequency bands (delta and theta) connected to emotional processing in the frontal brain region compared to parieto-occipital regions [20,29]. This is because frontal delta and theta activity has been associated with emotional engagement, motivation, and goal-directed control, all crucial for effective negotiation and group alignment, especially in group-oriented individuals.

Secondly, a greater high frequency band power (beta and gamma bands [34,35]) is expected in parieto-occipital regions, deputed to cognitive processing and awareness during the negotiation process. Beta and gamma oscillations in these areas are linked to cognitive control, memory, and decision-making, supporting the integration of different perspectives and the monitoring of the negotiation process. These bands have been associated with a greater ability to integrate multiple pieces of information, a key element for those who are group-oriented.

Furthermore, since previous hyperscanning studies showed an increase in delta and theta EEG coherence in the frontal area during cooperation and prosocial exchange [38], we expect to observe a lower dissimilarity in the frontal area compared to other brain regions in this sample of group-oriented dyads during the negotiation process. While frontal low-frequency synchrony is typically observed in joint action and imitation paradigms, cooperative negotiation may involve functionally divergent but complementary regulatory processes, whereby each individual evaluates and modulates their own cognitive and affective strategies in the pursuit of a shared decision [43,44]. Frontal EEG coherence has been linked to interpersonal alignment, empathy, and joint decision-making, suggesting that a greater synchronization in this region reflects a shared mental framework essential for negotiation.

2. Materials and Methods

2.1. Sample

A total of 26 university students (6 males and 20 females; mean age = 25.4 years, standard deviation ± 6.4 years) participated in the study. They were recruited through a non-probabilistic convenience sampling method from [BLIND]. An a priori power analysis was conducted using G*Power (version 3.1.9.7 [45]) to determine the required sample size for the planned repeated-measures analysis of variance (ANOVA), which included three Regions of Interest (ROIs), two groups, and two measurement points. Based on a medium effect size (f = 0.40) derived from prior hyperscanning EEG studies reporting regional differences in spectral activity during social interaction (e.g., [46]), the analysis indicated that a minimum of 24 participants would be required to achieve a statistical power of 0.95 at an alpha level of 0.05. This supports the adequacy of the recruited sample for detecting effects of interest in the present study. This analysis indicated that, for a medium effect size (f = 0.40), an alpha level of 0.05, and a statistical power of 0.95, a minimum of 24 participants was necessary to achieve reliable results across two groups and two measurement points, confirming the sufficiency of the recruited sample.

The participants were randomly assigned to one of two experimental roles, labelled Member 1 and Member 2. This attribution was conducted to form 13 dyads, with each dyad comprising two individuals of the same gender. To control for potential biases and confounding variables, participants were explicitly selected to ensure that no prior familiarity or acquaintance existed between dyad members before the experimental session.

The participants were enrolled in the study on a voluntary basis, with no monetary incentives or alternative compensation offered. Inclusion criteria were rigorously defined to ensure homogeneity in relevant variables: all participants were required to be right-handed, possess normal or corrected-to-normal visual acuity, and meet standardized thresholds for general health and cognitive functioning. Exclusion criteria were systematically applied to eliminate potential confounding factors. These included a self-reported history of psychiatric or neurological disorders, clinically significant depressive symptoms, diminished global cognitive functioning, and documented impairments in short- or long-term memory. Additionally, individuals currently using psychoactive medications with the potential to alter cognitive processes were excluded to maintain the integrity of the experimental outcomes. All participants signed a written informed consent before participating in the study.

Ethical approval for the study was granted by [BLIND]. The study was carried out in full accordance with the ethical standards delineated in the Declaration of Helsinki (2013). Additionally, the study adhered to the General Data Protection Regulation (GDPR, EU Reg. 2016/679).

2.2. Experimental Procedure

The experiment was conducted in a quiet darkened room, where participants, after providing informed consent forms, were seated comfortably next to one another to promote direct interaction for all the duration of the experimental procedure, which took approximately 30 min. Before starting the negotiation task, an EEG eyes-open resting-state baseline of 120 s was detected by the participants using the EEG hyperscanning paradigm. EEG activity was also recorded simultaneously by the participants during the negotiation task, which lasted approximately 3 min (Figure 1A). Specifically, the experimental procedure followed three main steps: (i) the decision-making orientation assessment (step 1), (ii) the negotiation task (step 2), and (iii) the perception of negotiation exchange (step 3).

Step 1: Decision-making orientation assessment.

To verify the decision-making orientation (individual versus group orientation), a real-life decision-making script based on a common group problem was administered to the members through the Qualtrics XM platform (Qualtrics LLC, Provo, UT, USA) before the negotiation interaction. Specifically, this script was presented to take advantage of a hypothetical scenario with high ecological validity. After instructing the participants to identify with the situation that will be presented to them, the following realistic decision-making scenario and instructions were shown to participants:

> *"A member of a workgroup does not appear to be in line with the ideals and style of the group. What does the group do? Think about your real work group.*
>
> *Now rate with a number from 1 to 5 (where 1 stands for "not at all" and 5 for "very much") how close the following decisions are to how your real group would behave."*

Subsequently, eight items about different ways in which the real group would face the situation were presented. The eight items were presented in a random order, with four of them representing a behavior characterized by an individual decision-making orientation (1–4 item, e.g., *"The decision follows the perspective of the most innovative member of the group"*) and the other four a group decision-making orientation (5–8 item, e.g., *The group adopts the decision endorsed by the majority*).

The individual orientation score ($\text{Individual}_{\text{orientation}}$) and the group orientation score ($\text{Group}_{\text{orientation}}$) were calculated by averaging the four individual and four group orientation items.

Step 2: Negotiation task.

After the decision-making orientation assessment, the participants were introduced to the negotiation task, in which they had to negotiate and decide together how to deal with the realistic situation previously proposed in the first step. Specifically, the following instructions were provided:

> *"This study aims to investigate how individuals negotiate their opinions. You will therefore be asked to discuss the previously proposed topic and share your perspectives. Specifically, as a first step, we ask you to individually choose from the previous sentences the one that is closest to your decision and the way your real group would behave."*

After receiving these instructions, the participants had the chance to read again the eight previous statements, and the experimenter recorded each participant's chosen statement as *Member 1's initial choice* and *Member 2's initial choice.*

Following this decision, the two participants engaged in a three-minute timed negotiation exchange, during which they were required to collaboratively decide on a single statement that best reflected their shared perception of their real workgroup, following this instruction:

> *"Now, reflecting on the same scenario you have just read, you must work together to agree on a single statement that represents what your real group means to both of you. Therefore, the statement must be mutually accepted and should best reflect how to act within the group and for the group. You will have a maximum of 3 min to reach an agreement."*

Finally, at the end of the negotiation exchange, the experimenter recorded the final agreed-upon statement as the dyad's *final negotiated choice.* All dyads involved in this study ended the negotiation process within the given time by reaching a common agreement.

Step 3: Perception of the negotiation exchange: manipulation check.

At the end of the negotiation task, the perception of the negotiation exchange inside each dyad was investigated by asking the members to fill in an item presented on a 5-point Likert scale, where 1 stands for "completely in disagreement" and 5 for "completely in agreement".

Specifically, the following two items were created ad hoc to investigate how the negotiation exchange influenced each member's idea: "*The other agent influenced my decision*" and "*The real group idea I had before the discussion has changed*". Moreover, 10 items, which were derived, translated, and modified from the validated Group Questionnaire [47], were administrated to check the negotiation exchange in terms of cohesion, alliance, conflict, climate, and empathic failure.

The behavioral data showed how participants reported that the other member of the dyad influenced his/her decision (M = 3.23, SD = 0.92) and that he/she does not change his/her idea due to the negotiation exchange (M = 2.80, SD = 1.08). Additionally, a sense of cohesion, alliance, positive climate, as well as an absence of conflict and emphatic failure were reported as being above the average (cohesion: M = 4.40, SD = 0.23; alliance: M = 4.18, SD = 0.12; climate: M = 4.30, SD = 0.17; conflict: M = 1.50, SD = 0.13; emphatic failure: M = 1.58, SD = 0.05).

2.3. EEG Data: Acquisition and Processing

EEG data were acquired using a 16-channel direct current (DC) amplifier (SYNAMPS system) integrated with the NEUROSCAN 4.2 software. Recordings were performed during both resting-state and negotiation tasks, with electrode placement adhering to the international 10/20 system [48]. A total of 15 Ag/AgCl electrodes (Fp1, Fp2, F3, F4, Fz, Cz, C3, C4, T7, T8, Pz, P3, P4, O1, O2) were positioned on the scalp via an ElectroCap (Compumedics Germany GmbH, Singen, Germany), with the earlobes serving as reference points (Figure 1B). Additionally, two electrooculographic (EOG) electrodes were positioned at the outer canthus of the left eye to monitor ocular artifacts.

Prior to data collection, the impedance levels of all electrodes were carefully measured and maintained below 5 kΩ to ensure optimal signal integrity. The EEG signal was digitized at a 1000 Hz sampling rate, with a 50 Hz notch filter applied to mitigate electrical interference from power lines. For offline processing, a 0.01–50 Hz infinite impulse response (IIR) bandpass filter with a slope of 48 dB/octave was applied to the data. Continuous EEG recordings were then segmented into two-second epochs, with visual inspection to identify and exclude any artifacts originating from ocular, muscular, or movement-related sources. Artifact-free segments were subjected to spectral analysis using fast Fourier transform (FFT) with a Hamming window, achieving a spectral resolution of 0.5 Hz. This procedure facilitated the computation of average power spectral density (PSD) for both resting-state and negotiation conditions. In addition, a rigorous multistep preprocessing pipeline was applied to enhance signal quality. Raw EEG data were re-referenced to linked earlobes and filtered with a 0.01–50 Hz band-pass filter and a 50 Hz notch filter. An Independent Component Analysis (ICA) was employed to identify and remove components linked to ocular and muscular artifacts. Data were then segmented into 2-s epochs, and epochs contaminated by residual artifacts were excluded following combined automated and manual inspection. These procedures ensured high-fidelity signal integrity across participants and dyads, enabling reliable spectral analyses in all frequency bands of interest.

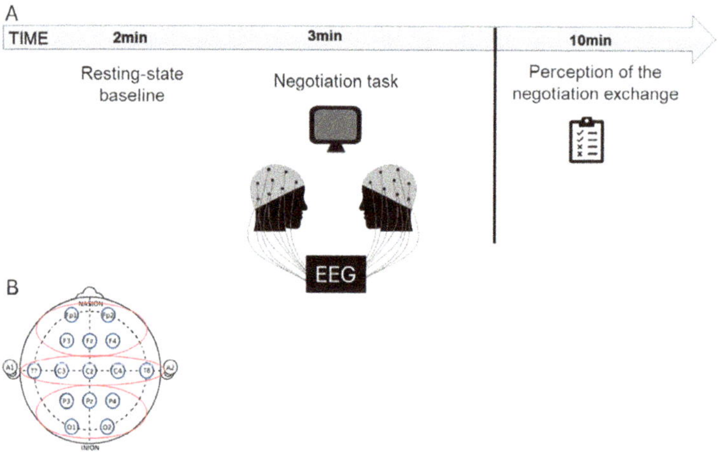

Figure 1. (**A**) The picture displays the timeline of the experimental procedure encompassing the EEG recording during the resting state baseline and the negotiation task. After the negotiation task, participants filled in a scale on the perception of negotiation exchange. (**B**) Following the 10/20 international system, each subject's EEG montage included 15 selected electrodes. For statistical analysis, three anatomically defined regions of interest (ROIs) were considered: the frontal region (Fp1, Fp2, F3, F4), the temporo-central region (T7, T8, C3, C4), and the parieto-occipital region (P3, P4, O1, O2).

Power spectral densities (PSDs) were calculated for the standard EEG frequency bands: delta (0.5–3.5 Hz), theta (4–7.5 Hz), alpha (8–12.5 Hz), beta (13–30 Hz), and gamma (30.5–50 Hz). Signal processing and analysis were conducted using Brain Vision Analyzer 2.0 (Brain Products GmbH, Gilching, Germany). To assess task-induced EEG modulation, the PSD values recorded during the negotiation phase were normalized against the baseline resting-state PSD values. The normalization procedure was performed using the following formula: [Normalized PSD = $(PSD_{task} - PSD_{BL})/PSD_{BL}$], where PSD_{task} represents the power measured during the task and PSD_{BL} corresponds to the baseline power. This normalization was applied separately to each frequency band and electrode location for both the negotiator and receiver within each dyad, allowing for the individualized assessment of task-related changes in EEG activity.

During the negotiation interaction, the participants were instructed to remain relaxed and maintain a steady posture, minimize motor movements, articulate their speech clearly, and avoid whispering and overlapping with the other participant's speech, adhering to turn-taking conventions. Furthermore, the negotiation exchange was video-recorded, confirming the absence of noticeable changes in facial expressions or posture throughout the interaction. Thanks to these guidelines and to the video recordings, it was possible to consider and analyze only negotiation-speaking turns, with conversational pauses and non-negotiation utterances excluded. Accordingly, the data for each individual were defined as the average duration of all speech segments in which they actively participated in negotiation. Any utterances unrelated to the negotiation process, as well as conversational pauses, were omitted from the analysis.

For statistical analysis, three anatomically defined Regions of Interest (ROIs) were identified: frontal (Fp1, Fp2, F3, F4), temporo-central (T7, T8, C3, C4), and parieto-occipital (P3, P4, O1, O2). Within each ROI, power spectral density (PSD) values were calculated as the mean spectral power across the corresponding electrodes. This method allowed for a precise assessment of regional EEG activity across frequency bands, facilitating the comparison of task-related neural dynamics during the negotiation phase relative to the resting-state baseline.

2.4. Statistical Data Analyses

Prior to conducting statistical analyses, it was verified that all 13 dyads had reached an agreement within the three-minute negotiation task.

The statistical data analyses conducted in this study encompassed both behavioral and EEG data evaluations. For the behavioral analysis, a repeated measures ANOVA was performed to examine the group decision-making orientation based on Likert scale scores.

EEG data were analyzed at both the individual and dyadic levels (single-brain and dyads analyses), as outlined below.

2.4.1. Behavioral Data Analyses

As the preliminary analysis, one repeated-measures ANOVA with *Orientation* (2: $Group_{orientation}$, $Individual_{orientation}$) as within-subject factor was computed on the score of the Likert scale to verify the group decision-making orientation of the members.

2.4.2. EEG Data Analysis

For EEG data, two main types of statistical analysis were included: a single-brain brain and an inter-brain analysis.

For the single-brain analysis, five repeated-measures ANOVAs considering *ROIs* (3: frontal, temporo-central, and parieto-occipital) as the within-subject variable were applied to the five different frequency bands (delta, theta, alpha, beta, and gamma) as the distinct dependent variable.

For the inter-brain analysis, EEG activity was analyzed at the dyadic level to quantify the extent to which Member 1 and Member 2 differed in their EEG activation. This was achieved by calculating the Euclidean Distance (PSD) [49–51] between the power spectral density (PSD) values of each frequency band for corresponding ROIs within each dyad. Then, five repeated mixed-measures ANOVAs considering *ROIs* (3: frontal, temporo-central, and parieto-occipital) as within-subject variables were performed considering the Euclidean Distance (PSD) for the five frequency bands (delta, theta, alpha, beta, and gamma) as the distinct dependent variables.

For all repeated-measures ANOVAs, degrees of freedom were adjusted using the Greenhouse-Geisser epsilon to correct for violations of sphericity, as assessed by Mauchly's test of sphericity. When significant effects or interactions were identified, post hoc pairwise comparisons were performed with Bonferroni correction to control Type I error. Effect sizes were reported using eta-squared (η^2), and the significance threshold was set at $\alpha = 0.05$. Prior to conducting the primary analyses, the assumption of normality was checked by assessing skewness, kurtosis, and Q-Q plots. Levene's test was applied to verify the homogeneity of variances.

Descriptive statistics, including means and standard errors, were computed for each condition to facilitate the interpretation of results. All statistical analyses were performed using Jamovi (version 2.6.22; The Jamovi Project, 2022).

3. Results

3.1. Behavioral Results

Descriptive statistics, with mean scores of 4.28 (SD = 0.47) for $Group_{orientation}$ and 2.54 (SD = 0.61) for $Individual_{orientation}$, indicating a predominant preference for collective decision-making strategies in the sample.

Although the post-negotiation items were not subjected to inferential statistical analysis due to their exploratory nature, their descriptive values revealed moderate mutual influence (M = 3.23, SD = 0.92) and low perceived conflict (M = 1.50, SD = 0.13). These data suggest that negotiation occurred in a predominantly cooperative climate, which may help contextualize the neural dynamics observed.

The repeated-measures ANOVA on behavioral data confirmed a significant main effect for *Orientation* ($F[1,25] = 34.57$, $p < 0.001$, $\eta^2 = 0.41$), with higher values for the $Group_{orientation}$ compared to $Individual_{orientation}$.

3.2. EEG Results

3.2.1. Single-Brain Results

The results relating to the EEG frequency bands will be reported in the following sections (Figure 2).

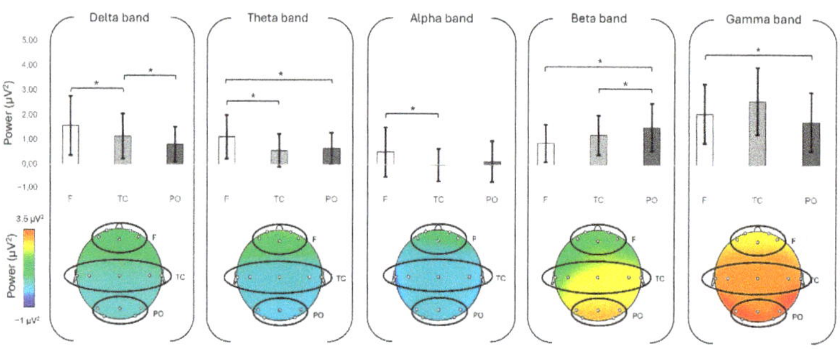

Figure 2. Bar graphs of EEG bands (delta, theta, alpha, beta, and gamma) for each ROI (F = frontal, TC = temporo-central, PO = parieto-occipital). The second row shows topographic maps of the average power in the five EEG frequency bands, highlighting the spatial distribution of EEG activity across the ROIs. The color scale represents the relative intensity of EEG power (blue = low power, red = high power). Bars represent ±1 standard error, and stars (*) mark statistically significant effects.

Delta Band

A significant main effect was found for ROI ($F[2,50] = 10.010$, $p < 0.001$, $\eta^2 = 0.093$), with higher delta activity in the frontal region compared to the parieto-occipital region ($p = 0.002$) and in the temporo-central region compared to the parieto-occipital region ($p = 0.022$).

Theta Band

A significant main effect was found for ROI ($F[2,50] = 9.170$, $p < 0.001$, $\eta^2 = 0.098$), with higher theta activity in the frontal region compared to the temporo-central region ($p = 0.003$) and the parieto-occipital region ($p = 0.020$).

Alpha Band

A significant main effect was found for ROI (F[2,50] = 9.593, $p < 0.001$, $\eta^2 = 0.062$), with higher alpha activity in the frontal region compared to the temporo-central region ($p < 0.001$).

Beta Band

A significant main effect was found for ROI (F[2,50] = 10.916, $p < 0.001$, $\eta^2 = 0.092$), with higher beta activity in the parieto-occipital region compared to the frontal region ($p = 0.002$), as well as in the parieto-occipital compared to the temporo-central region ($p = 0.01$).

Gamma Band

A significant main effect was found for ROI (F[2,50] = 4.439, $p = 0.017$, $\eta^2 = 0.058$), with higher gamma activity in the parieto-occipital region compared to the frontal region ($p = 0.008$).

3.2.2. Inter-Brain Results

Delta Band

A significant main effect was found for ROI (F[2,24] = 7.574, $p = 0.003$, $\eta^2 = 0.193$), with higher dissimilarity in the frontal region compared to the parieto-occipital region ($p = 0.040$) and in the temporo-central region compared to the parieto-occipital region ($p = 0.016$) (Figure 3).

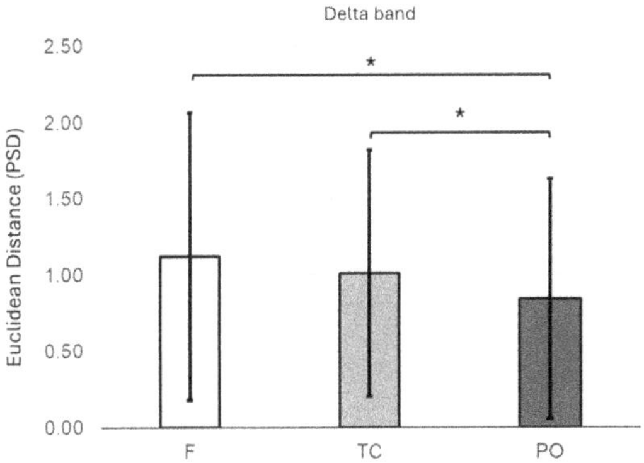

Figure 3. The bar chart shows the Euclidean Distance between the power spectral density (PSD) values across the three ROIs (F = frontal, TC = temporo-central, PO = parieto-occipital). Bars represent ±1 standard error and stars (*) mark statistically significant effects.

No significant main effect was found for ROI or other factors in the theta, alpha, beta, and gamma bands (all $p > 0.05$).

4. Discussion

This study investigated the single-brain EEG correlates of group-oriented individuals during a decision-making process with an explicit negotiation goal, as well as the inter-brain neural activity of dyads converging on the decision. The findings contribute to the growing body of literature on social neuroscience by elucidating how negotiation processes in dyads shape individual and inter-brain neural dynamics. Specifically, negotiation was studied through a communicative ecologically valid scenario that involves synchrony and turn-taking. The modulation of different EEG frequency bands and inter-brain synchronization during negotiation allowed to explore the extent of shared cognitive and emotional states between interacting partners.

The behavioral findings confirmed that participants demonstrated a clear preference for group-oriented decision-making over individualistic approaches. This behavioral predisposition is mirrored by the neurophysiological patterns observed during the negotiation task. The increase in frontal delta and theta power likely reflects the affective–motivational engagement and strategic regulation characteristic of cooperative social cognition, while the enhanced parieto-occipital beta and gamma activity suggests a reliance on integrative and monitoring processes necessary for converging toward a shared solution. Together, these findings support the notion that group-oriented decision-making is instantiated through a specific configuration of distributed cortical activity during negotiation. This aligns with prior research suggesting that individuals with a cooperative orientation tend to prioritize collective interests and shared perspectives in decision-making contexts [1]. Interestingly, participants reported that their partners influenced their decisions to a moderate extent, but their initial perceptions of group dynamics remained relatively stable. This suggests that, while negotiation fosters an exchange of ideas, individuals may integrate new perspectives without completely altering their initial viewpoints. Additionally, high ratings for cohesion and alliance, coupled with low scores for conflict and empathic failure, indicate that the negotiation process was largely cooperative and prosocial in nature. This interpersonal pattern may also highlight the observed inter-brain dissociation in delta power, which we interpret as reflecting individual regulatory engagement within a non-adversarial, co-constructed exchange.

The EEG findings reveal distinct neural patterns associated with negotiation, providing insights into the cognitive and emotional mechanisms underlying group-oriented decision-making.

Firstly, the single-brain analysis demonstrated increased delta and theta activity in the frontal region compared to the parieto-occipital region. This is consistent with previous studies linking frontal delta and theta oscillations to emotional engagement, motivation, and goal-directed control [20,29]. Moreover, former hyperscanning works demonstrated the key role of frontal areas in cooperation process [37] as well as the prefrontal cortex, implicated in joint action performance [39]. These findings suggest that negotiation involves heightened emotional and cognitive effort, particularly in this sample of individuals committed to group-oriented goals.

Moreover, alpha activity was higher in the frontal region than in the temporo-central region during the negotiation task. The suppression of oscillations within the alpha frequency range is an index of cortical excitability [52] and enhances the efficiency of cognitive processing [53]. Moreover, previous work on social bargaining showed how alpha oscillations could serve as a mechanism by which different brain areas flexibly interact in order to adapt ongoing behavior in socially demanding contexts [21].

Temporoparietal regions are also involved in social cognition and communication [40], managing social dynamics, including negotiation processes [41,42]. Given that alpha suppression in posterior regions has been previously associated with enhanced cognitive performance [30], a possible interpretation can be that this reduction in alpha power in the temporo-central regions, compared to the frontal ones, marks the cognitive effort derived from the negotiation process in our sample of participants.

The study also found that beta and gamma activity was more prominent in the parieto-occipital region compared to the frontal and temporo-central regions. Since beta and gamma oscillations are associated with cognitive control, memory integration, and decision-making [34,35], their increased presence in parieto-occipital areas suggests that negotiation requires the integration of multiple perspectives and the monitoring of shared goals. Elevated beta and gamma power in these regions may reflect working memory demands and the multimodal integration of verbal and social cues, alongside the ongoing evaluation of semantic relevance—key processes for managing complex, dialogic interactions in which information must be actively maintained and dynamically updated.

Secondly, at the dyadic level, the analysis revealed significant patterns of dissimilarity within the dyads. Specifically, and differently from what hypothesized, the Euclidean Distance analysis indicated greater dissimilarity in delta band activity in the frontal region compared to other brain regions. The increased dissimilarity in delta band activity likely reflects divergent cognitive and affective states between interactants rather than uniform synchronization. Delta oscillations, particularly in frontal regions, are implicated in large-scale neural coordination, executive control, and predictive coding in social contexts [54]. The observed increase in frontal delta dissimilarity, contrary to classic models of inter-brain synchrony during cooperation, may reflect individualized engagement of top-down regulatory processes such as conflict monitoring, prediction updating, or fairness evaluation—each governed independently in real time. This dissociation aligns with emerging views that cooperation does not always entail neural alignment but may also involve parallel regulatory roles within a shared social frame. While previous studies have reported frontal delta synchrony during joint tasks, the negotiation paradigm used in this paper emphasizes strategic divergence and autonomous cognitive control. Thus, the inter-brain dissimilarity in delta activity may reflect asynchronous but complementary activation of conflict regulation and decision evaluation circuits, independently engaged by each partner under shared but non-identical cognitive loads [55]. Unlike mid- and high-frequency bands, delta activity is critical for integrating top-down regulatory mechanisms essential for decision-making and social evaluation [54,56]. The higher dyadic dissimilarity in frontal delta activity, compared to the temporo-central and parieto-occipital regions, may reflect the distinct functional roles of these areas in social interactions. The temporo-central and parieto-occipital regions facilitate the processing of social cues, such as speech prosody and facial expressions, supporting higher synchronization through perceptual-motor mirroring [57]. In contrast, the frontal cortex is engaged in executive functions, including conflict monitoring, strategic adaptation, and inhibition of prepotent responses, which are inherently more individualized and context dependent [58]. As a result, lower coherence in frontal delta activity may indicate asymmetric cognitive control processes, where interactants independently regulate uncertainty, strategic shifts, and fairness perceptions [59]. This neural dissociation is consistent with the behavioral profile reported by participants, characterized by low conflict and moderate influence. Rather than reflecting interpersonal discord, frontal inter-brain dissimilarity may index self-regulated cognitive engagement within a shared, non-adversarial task structure. Hyperscanning research suggests that inter-brain synchronization does not necessarily reflect functional alignment but may arise from cooperative or adversarial neural dynamics [60,61]. The observed lower delta synchronization in the frontal regions of the dyads may thus signal increased cognitive effort in conflict resolution, suppression of competing responses, or the negotiation of divergent interpretations.

Despite the study's contributions to understanding the neural dynamics of negotiation, several methodological limitations should be acknowledged to guide future research.

Firstly, the interpretation of frontal delta dissimilarity as an index of cognitive conflict regulation is supported by research on low-frequency frontal oscillations and executive control. However, the current study did not include time-locked behavioral markers of conflict dynamics or parametric measures of decisional divergence, and the sample size limited the ability to test direct neuro-behavioral correlations. Future research should integrate continuous measures of negotiation dynamics, synchronized behavioral annotations, and temporally resolved EEG analyses to better understand the neural mechanisms of conflict regulation during negotiation.

Secondly, although the study was adequately powered for detecting medium within-subject effects, the relatively small sample limits broader generalization and the exploration of interaction-level variability.

Thirdly, although gender was not a variable of interest in the present study, the sample's imbalance (20 females, 6 males) may limit generalizability, particularly given documented sex differences in negotiation styles, emotional regulation, and neural synchrony patterns. Future studies should implement stratified dyadic designs or include gender as a planned covariate to test its specific impact.

Finally, while designed to simulate a realistic group decision, the absence of real-world stakes may have limited affective arousal, potentially influencing the magnitude of neural activation during negotiation.

An interesting avenue for future research would be to examine how individual differences, such as personality traits, empathy levels, or prior negotiation experience, influence the observed neural patterns. It is possible that more experienced negotiators exhibit different frontal delta dynamics, reflecting more efficient strategic adaptation. Similarly, individuals with higher trait empathy may show reduced inter-brain dissimilarity, indicating a greater tendency to align with their partner's cognitive and emotional states. Investigating these moderating factors could provide a more nuanced understanding of how neural mechanisms underpin real-world negotiation behavior. In addition, future studies could benefit from incorporating a time-resolved analysis of the negotiation process. Segmenting interactions into distinct phases—such as initial disagreement, deadlock, and resolution—could help uncover more detailed temporal dynamics of neural engagement and clarify how cognitive and affective processes evolve throughout the negotiation.

5. Conclusions

These findings provide first evidence for understanding the EEG neural basis of negotiation in group-oriented decision-making. The observed frontal delta and theta activity highlight the role of emotional engagement and motivation in shaping negotiation strategies, while parieto-occipital beta and gamma oscillations underscore the cognitive demands associated with integrating diverse perspectives. Furthermore, the presence of greater frontal dissimilarity in the delta band suggests that negotiation is not a process of complete alignment at the neural level, but rather it requires increased cognitive effort in conflict resolution, suppression of competing responses, or the negotiation of divergent interpretations. This has potential applications in various applied settings, such as the organizational one, where negotiation appears essential in group decision-making dynamics and knowing the neural bases of this process can promote empowering interventions in professionals. Indeed, beyond its theoretical contributions, this study has potential applications in domains such as professional negotiation training, conflict resolution, and teamwork optimization. The finding that negotiation involves distinct cognitive and affective regulation processes suggests that interventions aimed at enhancing strategic flexibility and emotional regulation could improve negotiation outcomes. Furthermore, the observed inter-brain dissimilarity in frontal delta activity highlights the importance of developing training programs that foster adaptive cognitive control rather than mere synchrony, particularly in high-stakes decision-making contexts.

In conclusion, this study provides first evidence that negotiation in group-oriented decision-making is supported by distinct neural mechanisms at both the individual and dyadic levels. These findings advance our understanding of the neurophysiological foundations of negotiation and offer new perspectives on how neural alignment contributes to effective group interactions.

Future research should explore additional factors that may influence neural synchronization during negotiation, such as personality traits, cultural differences, and the complexity of decision-making scenarios. Moreover, extending the investigation to larger and more diverse groups could provide deeper insights into how neural dynamics scale in multi-person interactions.

Author Contributions: L.A.: conceptualization, methodology, investigation, visualization, writing—original draft, writing—review and editing; K.R.: data curation, investigation, formal analysis, software, visualization, writing—original draft, writing—review and editing; C.A.: data curation, investigation, formal analysis, software, visualization, writing—original draft, writing—review and editing; A.D.: writing—review and editing, visualization; M.B.: conceptualization, project administration, supervision, resources, funding acquisition, validation, writing—original draft; writing—review and editing. All authors have read and agreed to the published version of the manuscript.

Funding: This work was funded by the D1 funding of the Catholic University of the Sacred Heart, Milan, Italy.

Institutional Review Board Statement: The study's experimental design and procedures were reviewed and approved by the Ethics Committee of the Department of Psychology at the Catholic University of the Sacred Heart, Milan, Italy (approval code: 125/24—Valutare il Decision Making: consapevolezza e metacognizione decisionale; approval date: 23 July 2024), and according to the GDPR—Reg. UE 2016/679 and its ethical guidelines. The participants provided their written informed consent to participate in this study.

Informed Consent Statement: Informed consent was obtained from all subjects involved in the study.

Data Availability Statement: The data presented in this study are available on request from the corresponding author. The data are not publicly available due to ethical reasons for sensitive personal data protection (requests will be evaluated according to the GDPR—Reg. UE 2016/679 and its ethical guidelines).

Conflicts of Interest: The authors declare no conflicts of interest.

References

1. Speer, S.P.H.; Sened, H.; Mwilambwe-Tshilobo, L.; Falk, E.B.; Tsoi, L.; Burns, S.M.; Tamir, D.I. Finding Agreement: FMRI-Hyperscanning Reveals That Dyads Explore in Mental State Space to Align Opinions. *bioRxiv* **2024**. [CrossRef]
2. Brett, J.M. Culture and Negotiation. *Int. J. Psychol.* **2000**, *35*, 97–104. [CrossRef]
3. Brett, J.; Thompson, L. Negotiation. *Organ. Behav. Hum. Decis. Process.* **2016**, *136*, 68–79. [CrossRef]
4. Kugler, T.; Kausel, E.E.; Kocher, M.G. Are Groups More Rational than Individuals? A Review of Interactive Decision Making in Groups. *WIREs Cogn. Sci.* **2012**, *3*, 471–482. [CrossRef] [PubMed]
5. Paulus, P.B.; Nijstad, B.A. *Group Creativity*; Oxford University Press: Oxford, UK, 2003; ISBN 9780195147308.
6. Balconi, M.; Vandelli, G.V.; Angioletti, L. Be Creative to Innovate! EEG Correlates of Group Decision-Making in Managers. *Sustainability* **2024**, *16*, 2175. [CrossRef]
7. Hargadon, A.B.; Bechky, B.A. When Collections of Creatives Become Creative Collectives: A Field Study of Problem Solving at Work. *Organ. Sci.* **2006**, *17*, 484–500. [CrossRef]
8. Vera, D.; Crossan, M. Improvisation and Innovative Performance in Teams. *Organ. Sci.* **2005**, *16*, 203–224. [CrossRef]
9. Bittner, J.V.; Bruena, M.; Rietzschel, E.F. Cooperation Goals, Regulatory Focus, and Their Combined Effects on Creativity. *Think. Ski. Creat.* **2016**, *19*, 260–268. [CrossRef]
10. Carmeli, A.; Dutton, J.E.; Hardin, A.E. Respect as an Engine for New Ideas: Linking Respectful Engagement, Relational Information Processing and Creativity among Employees and Teams. *Hum. Relat.* **2015**, *68*, 1021–1047. [CrossRef]
11. Gelfo, F. Does Experience Enhance Cognitive Flexibility? An Overview of the Evidence Provided by the Environmental Enrichment Studies. *Front. Behav. Neurosci.* **2019**, *13*, 150. [CrossRef]
12. Hon, A.H.Y.; Bloom, M.; Crant, J.M. Overcoming Resistance to Change and Enhancing Creative Performance. *J. Manag.* **2014**, *40*, 919–941. [CrossRef]
13. Laureiro-Martínez, D.; Brusoni, S. Cognitive Flexibility and Adaptive Decision-making: Evidence from a Laboratory Study of Expert Decision Makers. *Strateg. Manag. J.* **2018**, *39*, 1031–1058. [CrossRef]

14. De Dreu, C.K.W.; Weingart, L.R.; Kwon, S. Influence of Social Motives on Integrative Negotiation: A Meta-Analytic Review and Test of Two Theories. *J. Pers. Soc. Psychol.* **2000**, *78*, 889–905. [CrossRef]
15. Deutsch, M. The Resolution of Conflict. *Am. Behav. Sci.* **1973**, *17*, 248. [CrossRef]
16. Weingart, L.R.; Thompson, L.L.; Bazerman, M.H.; Carroll, J.S. TACTICAL BEHAVIOR AND NEGOTIATION OUTCOMES. *Int. J. Confl. Manag.* **1990**, *1*, 7–31. [CrossRef]
17. Czeszumski, A.; Liang, S.H.-Y.; Dikker, S.; König, P.; Lee, C.-P.; Koole, S.L.; Kelsen, B. Cooperative Behavior Evokes Interbrain Synchrony in the Prefrontal and Temporoparietal Cortex: A Systematic Review and Meta-Analysis of FNIRS Hyperscanning Studies. *eNeuro* **2022**, *9*, ENEURO.0268-21.2022. [CrossRef]
18. Ikegami, T.; Iizuka, H. Turn-Taking Interaction as a Cooperative and Co-Creative Process. *Infant Behav. Dev.* **2007**, *30*, 278–288. [CrossRef]
19. Montague, P.R.; Berns, G.S.; Cohen, J.D.; McClure, S.M.; Pagnoni, G.; Dhamala, M.; Wiest, M.C.; Karpov, I.; King, R.D.; Apple, N.; et al. Hyperscanning: Simultaneous FMRI during Linked Social Interactions. *Neuroimage* **2002**, *16*, 1159–1164. [CrossRef]
20. Knyazev, G.G. Motivation, Emotion, and Their Inhibitory Control Mirrored in Brain Oscillations. *Neurosci. Biobehav. Rev.* **2007**, *31*, 377–395. [CrossRef]
21. Billeke, P.; Zamorano, F.; López, T.; Rodriguez, C.; Cosmelli, D.; Aboitiz, F. Someone Has to Give in: Theta Oscillations Correlate with Adaptive Behavior in Social Bargaining. *Soc. Cogn. Affect. Neurosci.* **2014**, *9*, 2041–2048. [CrossRef]
22. Rovelli, K.; Angioletti, L.; Acconito, C.; Balconi, M. Neurosciences Tell Us How to Be Adaptable, Creative, and Proactive Agents in Decision Making: A Pilot Study. *J. Neurosci. Psychol. Econ.* **2024**, *17*, 30–45. [CrossRef]
23. Haegens, S.; Cousijn, H.; Wallis, G.; Harrison, P.J.; Nobre, A.C. Inter- and Intra-Individual Variability in Alpha Peak Frequency. *Neuroimage* **2014**, *92*, 46–55. [CrossRef] [PubMed]
24. Güntekin, B.; Emek-Savaş, D.D.; Kurt, P.; Yener, G.G.; Başar, E. Beta Oscillatory Responses in Healthy Subjects and Subjects with Mild Cognitive Impairment. *NeuroImage Clin.* **2013**, *3*, 39–46. [CrossRef]
25. Bastos, A.M.; Vezoli, J.; Bosman, C.A.; Schoffelen, J.M.; Oostenveld, R.; Dowdall, J.R.; DeWeerd, P.; Kennedy, H.; Fries, P. Visual Areas Exert Feedforward and Feedback Influences through Distinct Frequency Channels. *Neuron* **2015**, *85*, 390–401. [CrossRef] [PubMed]
26. Balconi, M.; Rovelli, K. Does Emotional Valence Affect Cognitive Performance and Neurophysiological Response during Decision Making? A Preliminary Study. *Front. Neurosci.* **2024**, *18*, 1408526. [CrossRef] [PubMed]
27. Lutz, A.; Lachaux, J.-P.; Martinerie, J.; Varela, F.J. Guiding the Study of Brain Dynamics by Using First-Person Data: Synchrony Patterns Correlate with Ongoing Conscious States during a Simple Visual Task. *Proc. Natl. Acad. Sci. USA* **2002**, *99*, 1586–1591. [CrossRef]
28. Balconi, M.; Angioletti, L.; Acconito, C. Self-Awareness of Goals Task (SAGT) and Planning Skills: The Neuroscience of Decision Making. *Brain Sci.* **2023**, *13*, 1163. [CrossRef]
29. Balconi, M.; Lucchiari, C. EEG Correlates (Event-Related Desynchronization) of Emotional Face Elaboration: A Temporal Analysis. *Neurosci. Lett.* **2006**, *392*, 118–123. [CrossRef]
30. Astolfi, L.; Toppi, J.; De Vico Fallani, F.; Vecchiato, G.; Cincotti, F.; Wilke, C.; Yuan, H.; Mattia, D.; Salinari, S.; He, B.; et al. Imaging the Social Brain by Simultaneous Hyperscanning during Subject Interaction. *IEEE Intell. Syst.* **2011**, *26*, 38–45. [CrossRef]
31. Gola, M.; Magnuski, M.; Szumska, I.; Wróbel, A. EEG Beta Band Activity Is Related to Attention and Attentional Deficits in the Visual Performance of Elderly Subjects. *Int. J. Psychophysiol.* **2013**, *89*, 334–341. [CrossRef]
32. Bu-Omer, H.M.; Gofuku, A.; Sato, K.; Miyakoshi, M. Parieto-Occipital Alpha and Low-Beta EEG Power Reflect Sense of Agency. *Brain Sci.* **2021**, *11*, 743. [CrossRef] [PubMed]
33. Chang, W.-S.; Liang, W.-K.; Li, D.-H.; Muggleton, N.G.; Balachandran, P.; Huang, N.E.; Juan, C.-H. The Association between Working Memory Precision and the Nonlinear Dynamics of Frontal and Parieto-Occipital EEG Activity. *Sci. Rep.* **2023**, *13*, 14252. [CrossRef]
34. Crivelli, D.; Rovelli, K.; Balconi, M. Different Types of Social Feedback Can Enhance or Reduce Performance, and Induce or Alleviate Psychosocial Stress: An Exploratory Study of the Underlying Neurophysiological Mechanisms. *Adapt. Hum. Behav. Physiol.* **2025**, *11*, 3. [CrossRef]
35. Kaiser, J.; Lutzenberger, W. Induced Gamma-Band Activity and Human Brain Function. *Neuroscientist* **2003**, *9*, 475–484. [CrossRef] [PubMed]
36. Czeszumski, A.; Eustergerling, S.; Lang, A.; Menrath, D.; Gerstenberger, M.; Schuberth, S.; Schreiber, F.; Rendon, Z.Z.; König, P. Hyperscanning: A Valid Method to Study Neural Inter-Brain Underpinnings of Social Interaction. *Front. Hum. Neurosci.* **2020**, *14*, 39. [CrossRef]
37. Balconi, M.; Vanutelli, M.E. Cooperation and Competition with Hyperscanning Methods: Review and Future Application to Emotion Domain. *Front. Comput. Neurosci.* **2017**, *11*, 86. [CrossRef]
38. Balconi, M.; Fronda, G.; Vanutelli, M.E. When Gratitude and Cooperation between Friends Affect Inter-Brain Connectivity for EEG. *BMC Neurosci.* **2020**, *21*, 14. [CrossRef]

39. Balconi, M.; Vanutelli, M.E. EEG Hyperscanning and Behavioral Synchronization during a Joint Actions. *Neuropsychol. Trends* **2018**, *2018*, 23–47. [CrossRef]
40. Decety, J.; Lamm, C. The Role of the Right Temporoparietal Junction in Social Interaction: How Low-Level Computational Processes Contribute to Meta-Cognition. *Neuroscientist* **2007**, *13*, 580–593. [CrossRef]
41. Landsiedel, J.; Daughters, K.; Downing, P.E.; Koldewyn, K. The Role of Motion in the Neural Representation of Social Interactions in the Posterior Temporal Cortex. *Neuroimage* **2022**, *262*, 119533. [CrossRef]
42. Rovelli, K.; Balconi, M. Mind in Others' Shoes: Neuroscientific Protocol for External Referent Decision Awareness (ERDA) in Organizations. *Neuroscience* **2025**, *567*, 249–260. [CrossRef]
43. Froese, T.; Loh, C.L.; Putri, F. Inter-Brain Desynchronization in Social Interaction: A Consequence of Subjective Involvement? *Front. Hum. Neurosci.* **2024**, *18*, 1359841. [CrossRef]
44. Pérez, A.; Carreiras, M.; Duñabeitia, J.A. Brain-to-Brain Entrainment: EEG Interbrain Synchronization While Speaking and Listening. *Sci. Rep.* **2017**, *7*, 4190. [CrossRef] [PubMed]
45. Faul, F.; Erdfelder, E.; Lang, A.-G.; Buchner, A. G*Power 3: A Flexible Statistical Power Analysis Program for the Social, Behavioral, and Biomedical Sciences. *Behav. Res. Methods* **2007**, *39*, 175–191. [CrossRef] [PubMed]
46. Balconi, M.; Acconito, C.; Angioletti, L. A Preliminary EEG Study on Persuasive Communication towards Groupness. *Sci. Rep.* **2025**, *15*, 6242. [CrossRef] [PubMed]
47. Krogel, J.; Burlingame, G.; Chapman, C.; Renshaw, T.; Gleave, R.; Beecher, M.; MacNair-Semands, R. The Group Questionnaire: A Clinical and Empirically Derived Measure of Group Relationship. *Psychother. Res.* **2013**, *23*, 344–354. [CrossRef]
48. Jasper, H.H. The Ten-Twenty Electrode System of International Federation EEG. *Electroencephalogr. Clin. Neurophysiol.* **1958**, *10*, 371–375.
49. Balconi, M.; Allegretta, R.A.; Acconito, C.; Saquella, F.; Angioletti, L. The Functional Signature of Decision Making Across Dyads During a Persuasive Scenario: Hemodynamic fNIRS Coherence Measure. *Sensors* **2025**, *25*, 1880. [CrossRef]
50. Balconi, M.; Bortolotti, A. Emotional Face Recognition, Empathic Trait (BEES), and Cortical Contribution in Response to Positive and Negative Cues. The Effect of RTMS on Dorsal Medial Prefrontal Cortex. *Cogn. Neurodyn.* **2013**, *7*, 13–21. [CrossRef]
51. Vanutelli, M.E.; Gatti, L.; Angioletti, L.; Balconi, M. May the Best Joint-Actions Win: Physiological Linkage During Competition. *Appl. Psychophysiol. Biofeedback* **2018**, *43*, 227–237. [CrossRef]
52. Sauseng, P.; Klimesch, W.; Gerloff, C.; Hummel, F.C. Spontaneous Locally Restricted EEG Alpha Activity Determines Cortical Excitability in the Motor Cortex. *Neuropsychologia* **2009**, *47*, 284–288. [CrossRef] [PubMed]
53. Klimesch, W. Alpha-Band Oscillations, Attention, and Controlled Access to Stored Information. *Trends Cogn. Sci.* **2012**, *16*, 606–617. [CrossRef]
54. Hirvonen, J.; Monto, S.; Wang, S.H.; Palva, J.M.; Palva, S. Dynamic Large-Scale Network Synchronization from Perception to Action. *Netw. Neurosci.* **2018**, *2*, 442–463. [CrossRef] [PubMed]
55. Angioletti, L.; Acconito, C.; Balconi, M. An EEG Hyperscanning Study during Persuasion toward Groupness. The Frontal Brain Area Activation as a Function of Role. *Soc. Neurosci.* **2024**, *19*, 340–353. [CrossRef]
56. Nácher, V.; Ledberg, A.; Deco, G.; Romo, R. Coherent Delta-Band Oscillations between Cortical Areas Correlate with Decision Making. *Proc. Natl. Acad. Sci. USA* **2013**, *110*, 15085–15090. [CrossRef] [PubMed]
57. Straube, B.; Wroblewski, A.; Jansen, A.; He, Y. The Connectivity Signature of Co-Speech Gesture Integration: The Superior Temporal Sulcus Modulates Connectivity between Areas Related to Visual Gesture and Auditory Speech Processing. *Neuroimage* **2018**, *181*, 539–549. [CrossRef]
58. Friedman, N.P.; Robbins, T.W. The Role of Prefrontal Cortex in Cognitive Control and Executive Function. *Neuropsychopharmacology* **2022**, *47*, 72–89. [CrossRef]
59. Basharpoor, S.; Heidari, F.; Molavi, P. EEG Coherence in Theta, Alpha, and Beta Bands in Frontal Regions and Executive Functions. *Appl. Neuropsychol. Adult* **2021**, *28*, 310–317. [CrossRef]
60. Zhang, M.; Jia, H.; Wang, G. Interbrain Synchrony of Team Collaborative Decision-Making: An FNIRS Hyperscanning Study. *Front. Hum. Neurosci.* **2021**, *15*, 702959. [CrossRef]
61. Hao, S.; Lina, L.; Xiaoqin, W.; Cenlin, Z. Group Identity Modulates Interbrain Synchronization during Repeated Lottery Contest. *J. Neurosci. Psychol. Econ.* **2024**, *17*, 1–18. [CrossRef]

applied sciences

MDPI

Article

Functional Connectome Fingerprinting Through Tucker Tensor Decomposition

Vitor Carvalho [1,2], Mintao Liu [1,2], Jaroslaw Harezlak [3,4], Ana María Estrada Gómez [1,*,†] and Joaquín Goñi [1,2,5,*,†]

1 Edwardson School of Industrial Engineering, Purdue University, West Lafayette, IN 47907, USA; vcostade@purdue.edu (V.C.); liu2549@purdue.edu (M.L.)
2 Purdue Institute for Integrative Neuroscience, Purdue University, West Lafayette, IN 47907, USA
3 Institute of Mathematics, University of Wroclaw, 50-384 Wroclaw, Poland; harezlak@iu.edu
4 Department of Epidemiology and Biostatistics, Indiana University Bloomington, Bloomington, IN 47405, USA
5 Weldon School of Biomedical Engineering, Purdue University, West Lafayette, IN 47907, USA
* Correspondence: amestrad@purdue.edu (A.M.E.G.); jgonicor@purdue.edu; (J.G.)
† These authors contributed equally to this work.

Abstract: The human functional connectome (FC) is a representation of the functional couplings between brain regions derived from blood oxygen level-dependent (BOLD) signals. Over the past decade, studies related to FC fingerprinting have sought to uncover functional patterns that enable uniquely identifying individuals across repeated scanning sessions, hence demonstrating the stability and distinctiveness of functional brain organization. In this study, it is hypothesized that tensor decomposition techniques, given their ability to project high-dimensional data into lower-dimensional spaces, enable detecting the brain fingerprint with high accuracy. A mathematical framework based on Tucker decomposition is presented to uncover the FC fingerprint of 426 unrelated participants from the Young-Adult Human Connectome Project (HCP) Dataset. An analysis of how brain parcellation granularity, decomposition rank, and scan length relate to within- and between-condition (resting state-task) fingerprinting was conducted. Relative to FC matrices as well as to Principal Components Analysis (PCA), tensor decomposition significantly increases the functional connectome's fingerprint. For parcellation granularity of 214 in the within-condition setting, an improvement of 11–36% was seen across all fMRI conditions. Similarly, a substantial improvement, ranging from 43 to 72%, was observed in the between-condition setting relative to FC matrices. Compared to matching rates obtained directly on FCs and when applying other data-driven decomposition methods, Tucker decomposition led to higher or the same level of matching rates for all analyses. Furthermore, in the context of between-condition fingerprinting, results from the proposed framework suggest that partially sampling time points from resting-state time series is sufficient to uncover FC fingerprints with high accuracy.

Keywords: functional connectome; tensor decomposition; fingerprinting; dimensionality reduction

1. Introduction

Human brain connectomics, driven by the increasing availability of large-scale neuroimaging datasets [1], has emerged in recent years as a prominent field of research. This field has the potential to address many of the open questions about the structure and function of the human brain. Notably, connectomics-based analyses have revealed meaningful differences between healthy and disease conditions [2,3]. However, to further assess the reliability of such findings and to capture individual-specific characteristics that may be

 https://doi.org/10.3390/app15094821

overlooked in group-level studies, the concept of a "brain connectivity fingerprint" has gained growing interest [4–6].

Functional connectivity fingerprinting of the brain refers to the ability to identify an individual's functional connectome (FC) from a set of FCs in repeated fMRI imaging sessions. The existence of a brain fingerprint has been established in the last decade with work done with data from functional magnetic resonance imaging (fMRI) and electroencephalography (EEG) [7,8]. Such studies have shown that the functional connectome of the brain varies between individuals, therefore, serving, to some extent, as a *fingerprint*. In the literature, different studies of FC fingerprinting have been conducted with varying approaches, such as principal component analysis (PCA) [9,10], sparse dictionary learning (SDL) [11], geodesic distance in regularized FC [12] and correlation distance in FC tangent space projections [13]. In [9], the authors show that individual connectivity profiles can be reconstructed through an optimal linear combination of PCA-derived orthogonal components. In [10], the authors perform PCA in a subset of "learning" FCs to obtain an eigenspace in which the "validation" FCs are projected, thus enabling the identification of the fMRI condition or the participant to which the "validation" FC belongs. In [11], SDL is used to refine FC profiles, leading to a higher distinctiveness in FCs relative to raw connectivity. PCA and SDL work on 2-dimensional data, while Tucker decomposition is designed for higher-dimensional data structures called tensors, essentially allowing it to analyze complex relationships across multiple variables in a dataset by decomposing it into a core tensor and factor matrices along each dimension. In essence, Tucker decomposition may be thought of as a "higher-order PCA".

Using the fact that FCs estimated as correlation matrices lie on or inside a Symmetric Positive Definite (SPD) manifold, Venkatesh and colleagues proposed using geodesic distance to compare FCs [14]. In a follow-up study [12], the authors explored how optimally regularized FCs maximize the individual fingerprint of participants, measured by the geodesic distance between their FCs. A limitation of this approach is that the geodesic distance between FCs of size $M \times M$ is a map $\mathbb{R}^{M\times M} \times \mathbb{R}^{M\times M} \to \mathbb{R}$. Hence, even though the geodesic distance provides a global measure of similarity between FCs, it does not directly highlight the specific features that make the individual's functional connectivity unique. As an alternative to using FCs, tangent FCs have demonstrated a high capacity to predict cognition and behavior [15–17]. Most recently, ref. [13] analyzed the effects of tangent FCs with respect to fingerprinting. In that study, a high degree of fingerprinting was achieved not only across sessions of a unique participant, but also for matching the sessions of twins.

To simultaneously overcome the drawbacks of the studies mentioned above, we propose utilizing tensor decomposition for FC fingerprinting. Tensor decomposition enables projecting high-dimensional data into a lower-dimensional space, preserving its structure while independently extracting meaningful information from each dimension.

Tensors are multidimensional arrays with applications in various fields, including signal processing, computer vision, and neuroscience [18]. In brain connectomics, tensors enable modeling and analyzing the functional and structural connections within the brain by reducing the dimensionality of complex, interrelated, and high-dimensional data through tensor decomposition. In [19], the authors studied the dynamics of FCs to understand the process of formation and dissolution of brain functional networks through tensor decomposition techniques. In another study [20], it was demonstrated how the analysis of the tensor components enables the extraction of unbiased and interpretable descriptions of single-trial dynamics across many trials through low-dimensional representations of neural data. In [21], the authors discuss challenges associated with interpreting the brain connectivity patterns derived from tensor decomposition. To our knowledge, only [22] has considered using tensors in brain connectivity fingerprinting analysis in the literature. However, their study was based on structural connectivity. The task of identifying subjects through their functional connectomes has the additional challenge of dynamic changes and functional reconfigurations happening at a fast rate in response to cognitive stimuli.

Through this study, the effectiveness of tensor based methods in uncovering FC fingerprints is assessed. The adopted fingerprinting framework can be broadly described as follows:

1. Each participant underwent a total of sixteen data acquisition sessions, two for each of the considered fMRI conditions. Using the BOLD time series from each acquisition, an FC matrix of dimension Number of Brain Regions $\times$ Number of Brain Regions is estimated, thus yielding a total of sixteen FC matrices for each participant.
2. For each condition, a tensor of dimensions Number of Brain Regions $\times$ Number of Brain Regions $\times$ Number of Participants, is constructed by concatenating all participants' FCs derived in the first acquisition session. Similarly, a second tensor is obtained for all conditions by concatenating all participants' FCs derived in the second acquisition session.
3. The obtained tensors are then decomposed via Tucker decomposition [23], yielding a core tensor and three factor matrices. The first two factor matrices contain cohort-level functional connectivity information and will later on be referred to as "brain parcellation factor matrices". The third factor matrix contains participant-specific information and will be later addressed as the "participants factor matrix". The participants factor matrix, acting as a "fingerprint" of each participant, is used to match participants FCs corresponding to different data acquisition sessions. The accuracy with which different sessions are able to be correctly matched is quantified by a metric denominated matching rate [24].

The aims of this study are: (i) assess the impact of Tucker decomposition on functional connectome fingerprinting in within- and between-fMRI condition settings for different parcellation granularities, (ii) estimate optimal levels of compression of brain parcellation-specific and participant-specific information that maximize fingerprinting, and (iii) analyze how sampling in resting-state prior to Tucker decomposition affects fingerprinting.

The remainder of the article is organized as follows. In Materials and Methods, we: (i) describe the data set used in this study as well as the preprocessing procedures; (ii) introduce the adopted tensor notation; (iii) introduce Tucker decomposition; and (iv) describe matching rate, the fingerprinting measurement used in this study. In Results, we: (i) provide a comparative analysis of fingerprints across parcellation granularities, (ii) compare the fingerprinting performance of different dimensionality reduction methods across parcellation granularities, (iii) disclose optimal levels of compression of parcellation-specific and participant-specific information that maximize within- and between-condition fingerprints, and (iv) present the findings for the different strategies of sampling time points in resting-state time series. In the Discussion, we (i) discuss our findings and (ii) highlight some limitations of our study and make suggestions for future work. Lastly, in Conclusions, we summarize the presented work.

2. Materials and Methods

In this section, a description of the dataset, tensor notation, and fingerprinting framework is presented. First, there is an overview of the HCP dataset, then brain parcellation granularities are considered, and preprocessing procedures are presented. Then, the adopted tensor notation is introduced, along with commonly used tensor decomposition techniques. Lastly, we introduce the metric used to quantify FC fingerprinting, and a detailed description of the fingerprinting framework.

2.1. Dataset Description

The HCP Dataset [25] has been widely used as a standard in the neuroimaging literature for a broad range of research domains [7,26–29] due to its large-scale, high-quality, and open-access data gathered from a large and diverse cohort of participants. This study utilizes data from eight fMRI conditions included in the Young-Adult Human Connectome Project (HCP) dataset. The fMRI dataset used in this study is from the publicly available Human Connectome Project (HCP). The original study was approved by the Washington University Institutional Review Board. Per HCP protocol, written informed consent statement was obtained from all subjects by the HCP Consortium. To minimize the impact of hereditary influences on fingerprinting, a subset of 426 unrelated individuals was selected. The demographic information of the participants can be found in Table 1. The fMRI conditions analyzed in this study include resting-state (RS), Emotion processing , Gambling (GAM), Language (LAN), Motor (MOT), Relational processing (REL), Social cognition (SOC), and Working Memory (WM). Each participant completed two sessions per condition, corresponding to separate left-to-right (LR) and right-to-left (RL) acquisitions, which are designated as test and retest sessions. Resting-state scans were conducted over four sessions ("REST1" and "REST2") on two separate days, though only the two REST1 sessions were considered in this study.

Table 1. Demographic information of the 426 unrelated participants from HCP's Young Adult Dataset.

Demographic Information (n = 426, 203 Males)	Mean (SD)	Range
Age	28.67 (3.78)	22–36
Years of Education	14.99 (1.77)	11–17

2.2. Brain Parcellations

In this study, we utilized the Schaefer parcellation functional brain atlas to analyze the human cortex [30]. This parcellation was derived from resting-state fMRI data of 1489 participants, and registered with surface alignment. It was generated using a gradient-weighted Markov random field approach that integrates local gradient and global similarity methods. The Schaefer parcellation is available at nine levels of granularity, ranging from 100 to 900 regions in increments of 100, in both volumetric and grayordinate space. Since the grayordinate versions of these parcellations are already in the same surface space as the HCP fMRI data, mapping them onto the fMRI scans is straightforward. Moreover, surface-based mapping ensures better alignment between the parcellations and fMRI data compared to volumetric mapping. Therefore, we applied surface-based mapping to project the Schaefer parcellations onto the fMRI data. To ensure comprehensive brain coverage, we incorporated 14 subcortical regions into each parcellation, as provided in the HCP dataset (file: Atlas_ROI2.nii.gz). This file was converted from a NIFTI to a CIFTI format using the HCP Workbench 1.5 software (Saint Louis, MI, USA.) (wb_command -cifti-createlabel). As a result, the Schaefer-200 parcellation, for instance, ultimately included 214 brain regions.

2.3. Preprocessing

The preprocessing of fMRI data followed the "minimal" preprocessing pipeline provided by the HCP, which includes artifact removal, motion correction, and alignment to a standardized template [31]. Further details on this pipeline can be found in previous studies [26,32]. We enhanced this minimal pipeline by incorporating additional preprocessing steps, as described in [13]. Specifically, for resting-state fMRI data, we applied the following procedures: (i) regressed out the global gray matter signal from voxel-wise time series [31], (ii) implemented a first-order Butterworth bandpass filter in both forward and reverse directions within the frequency range of 0.001–0.08 Hz [31], and (iii) z-scored and averaged voxel time series within each brain region, while excluding time points that deviated more than three standard deviations from the mean (processed using the Workbench software, wb_command -cifti-parcellate). The same preprocessing steps were applied to all task-based fMRI conditions. However, the bandpass filter was adjusted to a broader frequency range (0.001–0.250 Hz) [33], as the optimal filtering range remains uncertain [34].

2.4. Estimation of Whole-Brain FCs

The functional connectivity between pairs of brain regions was estimated by computing Pearson's correlation (*corr* MATLAB 2023a function), which results in a symmetric $M \times M$ correlation matrix, with M being the number of brain regions for a given parcellation. Throughout this article, this correlation matrix is referred to as FC. For each participant, we computed a whole-brain FC for each of the two sessions (test and retest), each fMRI condition (all seven tasks and resting state), and all parcellation granularities.

2.5. Tensor Notation

We refer to multidimensional data structures as tensors. Mathematically, tensors can be described as objects that lie in $\mathbb{R}^{I_1 \times I_2 \times \cdots \times I_N}$ and originate from the tensor product of N vector spaces. The number N of vector spaces from which a tensor originates defines the order (or mode) of the tensor. Throughout this work, we adopt the following notation: scalars are denoted by lower or upper case letters, e.g., x or X, vectors are denoted by boldface lowercase letters, e.g., $\mathbf{x}$, matrices are denoted by boldface capital letters, e.g., $\mathbf{X}$, and high-order tensors are denoted by boldface, calligraphic letters, e.g., $\boldsymbol{\mathcal{X}}$. Entries of a tensor $\boldsymbol{\mathcal{X}} \in \mathbb{R}^{I_1 \times I_2 \times \cdots \times I_N}$ are denoted by lowercase letters with subscripts, e.g., $x_{i_1,i_2,\ldots,i_N}$, with $1 \leq i_n \leq I_n$ for all $1 \leq n \leq N$. Tensor fibers are constructed by fixing every index of the tensor but one. Therefore, tensors have as many fibers as dimensions. For example, third-order tensors have column (mode-1), row (mode-2), and tube (mode-3) fibers, which are denoted by $\mathbf{x}_{:jk}$, $\mathbf{x}_{i:k}$, and $\mathbf{x}_{ij:}$, respectively, with colons denoting all the entries of a mode. A tensor slice is defined by fixing all entries from the tensor except two. We can define the slices $\mathbf{X}_{i,:,:}$, $\mathbf{X}_{:,j,:}$, and $\mathbf{X}_{:,:,k}$ for a third-order tensor.

The process of reshaping a tensor into a matrix is known as matricization. The mode-n matricization of a tensor $\boldsymbol{\mathcal{X}} \in \mathbb{R}^{I_1 \times I_2 \times \cdots \times I_N}$ is given by the matrix $\mathbf{X}_{(n)}$, and its columns correspond to the mode-n fibers of $\boldsymbol{\mathcal{X}}$. The *n-mode* matrix product between a tensor $\boldsymbol{\mathcal{X}} \in \mathbb{R}^{I_1 \times I_2 \times \cdots \times I_N}$ and a matrix $\mathbf{Y} \in \mathbb{R}^{J \times I_n}$ is denoted by $\boldsymbol{\mathcal{X}} \times_n \mathbf{Y}$ and is of size $I_1 \times \cdots \times I_{n-1} \times J \times I_{n+1} \times \cdots \times I_N$. The outer product between two vectors $\mathbf{x} \in \mathbb{R}^m$ and $\mathbf{y} \in \mathbb{R}^n$ is represented by

$$x \circ y = \begin{bmatrix} x_1 y_1 & x_1 y_2 & \cdots & x_1\, y_n \\ x_2 y_1 & x_2 y_2 & \cdots & x_2\, y_n \\ \vdots & \vdots & \ddots & \vdots \\ x_m y_1 & x_m y_2 & \cdots & x_m y_n \end{bmatrix}$$

and is of dimension $\mathbb{R}^{m \times n}$, while the inner product of $\boldsymbol{\mathcal{X}} \in \mathbb{R}^{I_1 \times I_2 \times \cdots \times I_N}$ and $\boldsymbol{\mathcal{Y}} \in \mathbb{R}^{I_1 \times I_2 \times \cdots \times I_N}$ is defined as

$$\langle \boldsymbol{\mathcal{X}}, \boldsymbol{\mathcal{Y}} \rangle = \sum_{i_1=1}^{I_1} \sum_{i_2=1}^{I_2} \cdots \sum_{i_N=1}^{I_N} x_{i_1,i_2,\cdots,i_N} y_{i_1,i_2,\cdots,i_N}$$

and the norm of $\boldsymbol{\mathcal{X}}$ is defined as $||\boldsymbol{\mathcal{X}}|| = \sqrt{\langle \boldsymbol{\mathcal{X}}, \boldsymbol{\mathcal{X}} \rangle}$.

2.6. Tensor Decomposition

In recent years, tensors have become increasingly popular in the fields of signal processing, machine learning, and neuroscience for their capacity to model complex high-order relationships among objects [35–37]. Tensor decomposition enables projecting high-dimensional data into a lower-dimensional space while preserving the original structure of the data. For the purpose of brain fingerprinting, tensor decomposition has the potential of extracting unique features from each participant's fMRI data acquisition session, thus facilitating subject distinctiveness. Several tensor decomposition algorithms can be found in the literature, each with their own characteristics and applications. The most commonly used ones are the CANDECOMP/PARAFAC (CP) [38,39] decomposition and the Tucker decomposition [23].

The CP decomposition of a tensor $\boldsymbol{\mathcal{X}} \in \mathbb{R}^{I_1 \times I_2 \times \cdots \times I_N}$ factorizes it into a sum of R rank-one tensors, where a rank-one tensor denotes the outer product between N vectors. Equation (1) shows the CP decomposition of $\boldsymbol{\mathcal{X}}$:

$$\boldsymbol{\mathcal{X}} \approx \sum_{r=1}^{R} \lambda_r \boldsymbol{a_r}^{(1)} \circ \boldsymbol{a_r}^{(2)} \cdots \circ \boldsymbol{a_r}^{(N)} \tag{1}$$

where R denotes the rank of the decomposition. The vectors $\boldsymbol{a_r}^{(1)} \in \mathbb{R}^{I_1}$, $\boldsymbol{a_r}^{(2)} \in \mathbb{R}^{I_2}, \ldots, \boldsymbol{a_r}^{(N)} \in \mathbb{R}^{I_N}$ are typically assumed to be normalized, with a corresponding scaling factor of λ_r. Equation (1) can also be expressed in the simplified form $[\![\boldsymbol{\lambda}; A^{(1)}, A^{(2)}, \cdots, A^{(N)}]\!]$, in which $A^{(i)} = [\boldsymbol{a}_1^{(i)} \quad \boldsymbol{a}_2^{(i)} \quad \cdots \quad \boldsymbol{a}_R^{(i)}]$, for $i = 1, 2, \cdots, N$, are referred to as factor matrices, and $\boldsymbol{\lambda} \in \mathbb{R}^R$ is a vector containing all scaling factors λ_r, for $r = 1, 2, \ldots, R$.

The Tucker decomposition [23] decomposes a tensor into a core tensor multiplied by a matrix along each of the tensor modes. For $\boldsymbol{\mathcal{X}} \in \mathbb{R}^{I_1 \times I_2 \times \cdots \times I_N}$, its *n-rank*, denoted by $rank_n(\boldsymbol{\mathcal{X}})$, is defined as the column rank of its mode-n matricization $\boldsymbol{X}_{(n)}$. In other words, the *n-rank* is the number of linearly independent vectors that span the basis of the *mode-n* fibers of $\boldsymbol{\mathcal{X}}$. Equation (2) shows the Tucker decomposition of $\boldsymbol{\mathcal{X}}$.

$$\boldsymbol{\mathcal{X}} \approx \sum_{r_1=1}^{R_1} \sum_{r_2=1}^{R_2} \cdots \sum_{r_n=1}^{R_N} g_{r_1 r_2 \cdots r_n} A_{r_1}^{(1)} \circ A_{r_2}^{(2)} \circ \cdots \circ A_{r_n}^{(N)} \equiv [\![\boldsymbol{\mathcal{G}}; A^{(1)}, A^{(2)}, \cdots, A^{(N)}]\!] \tag{2}$$

where $A^{(1)} \in \mathbb{R}^{I_1 \times R_1}, A^{(2)} \in \mathbb{R}^{I_2 \times R_2}, \cdots, A^{(N)} \in \mathbb{R}^{I_N \times R_N}$ are column-wise orthonormal factor matrices, and $R_1, R_2, \ldots, R_N$ are the ranks of the decomposition, where $R_n \leq rank_n(\boldsymbol{\mathcal{X}}) \leq I_n$ for $n = 1, \ldots, N$. If $R_n < rank_n(\boldsymbol{\mathcal{X}})$, we refer to the decomposition as a truncated Tucker decomposition and refer to it as a rank-$(R_1, R_2, \ldots, R_N)$ decomposition. The tensor $\boldsymbol{\mathcal{G}} \in \mathbb{R}^{R_1 \times R_2 \times \cdots \times R_N}$ is referred to as the core tensor and its entries represent the level of interaction between the different factors. Note that the CP decomposition can be understood as a special case of the Tucker decomposition when the Tucker's core is reduced to a hyper-diagonal tensor (all non-diagonal entries are equal to zero) and $R = R_1 = R_2 = \cdots = R_N$. For simplicity, consider the third-order tensor $\boldsymbol{\mathcal{X}} \in \mathbb{R}^{I_1 \times I_2 \times I_3}$. The Tucker decomposition of $\boldsymbol{\mathcal{X}}$ is obtained by solving the optimization problem 3, which seeks to minimize the norm of the difference between the true and estimated tensors. The decision variables of this problem are the core tensor $\boldsymbol{\mathcal{G}}$, and the factor matrices $A^{(1)}, A^{(2)}$, and $A^{(3)}$.

$$\min_{\mathcal{G},A^{(1)},A^{(2)},A^{(3)}} \quad ||\mathcal{X} - \mathcal{G} \times_1 A^{(1)} \times_2 A^{(2)} \times_3 A^{(3)}|| \tag{3}$$
$$\text{s.t.} \quad \mathcal{G} \in \mathbb{R}^{R_1 \times R_2 \times R_3}$$
$$A^{(1)} \in \mathbb{R}^{I_1 \times R_1} \text{ and column-wise orthonormal}$$
$$A^{(2)} \in \mathbb{R}^{I_2 \times R_2} \text{ and column-wise orthonormal}$$
$$A^{(3)} \in \mathbb{R}^{I_3 \times R_3} \text{ and column-wise orthonormal}$$

2.7. Tucker Decomposition of Functional Connectomes

For each of the eight fMRI conditions analyzed in this study, we represent the data as a third-order tensor $\mathcal{X} \in \mathbb{R}^{M \times M \times N}$. Here, M is equal to the granularity of the brain parcellation and $N = 426$ corresponds to the total number of participants. Given the symmetry of FC matrices, we refer to $\mathcal{X}$ as a semi-symmetric tensor, meaning it remains invariant under permutation of two (or more) indices. In our case, $x_{i,j,k} = x_{j,i,k}$ for $k = 1, \ldots, 426$. All analyses performed in this work have an input of a semi-symmetric tensor $\mathcal{X}$ constructed by concatenating participants' FCs obtained in one fMRI scanning session (either test or retest).

Once the data have been structured as a semi-symmetric tensor, we can produce a low-rank estimation of the FCs through either of the previously mentioned tensor decomposition methods. Due to the lack of interactions between components, the results of CP decomposition are generally easier to interpret [21] compared to Tucker decomposition. However, this lack of interaction often leads CP to produce less accurate approximations of the original tensor, as measured by the L_2 norm. In contrast, Tucker decomposition leverages its core tensor to capture interactions between components, enabling it to approximate the original tensor with greater precision [40]. Considering the interpretability/accuracy trade-off in the context of brain fingerprinting, we focus on Tucker decomposition.

Several methods have been developed to estimate the Tucker decomposition. Notably, we highlight Sequentially Truncated Higher-Order Singular Value Decomposition (ST-HOSVD) [41,42] and Higher-Order Orthogonal Iteration (HOOI) [43]. For a given tensor $\mathcal{X} \in \mathbb{R}^{I_1 \times I_2 \times \cdots \times I_N}$, ST-HOSVD sequentially computes the factor matrices via truncated eigenvalue decompositions of the mode-n matricizations (for $n = 1, 2, \ldots, N$) of $\mathcal{X}$, while iteratively updating $\mathcal{X}$ at each step. The output of the algorithm is a core tensor and a set of column-wise orthogonal factor matrices. Similarly to PCA, the components from the factor matrices capture most of the variance across each of the tensor modes. The Tucker estimation via ST-HOSVD of $\mathcal{X}$ is shown in Algorithm 1.

Algorithm 1 Sequentially Truncated Higher-Order Singular Value Decomposition (ST-HOSVD)

Input: $\mathcal{X} \in \mathbb{R}^{I_1 \times I_2 \times \cdots \times I_N}, \{R_1, R_2, \ldots, R_N\}$

```
for i ← 1 to N do
    Matricize the tensor along mode i to obtain X_(i)
    Compute the Gram matrix Z_(i) = X_(i) × X_(i)^T
    Compute the eigenvalue decomposition of Z_(i) and sort the eigenvalues in descending order
    A^(i) ← R_i eigenvectors corresponding to the sorted R_i eigenvalues of Z_(i)
    X ← X ×_i A^(i)T
end for
G ← X
```

Output: $\mathcal{G} \in \mathbb{R}^{R_1 \times R_2 \times \cdots \times R_N}, A^{(1)} \in \mathbb{R}^{I_1 \times R_1}, A^{(2)} \in \mathbb{R}^{I_2 \times R_2}, \cdots, A^{(N)} \in \mathbb{R}^{I_N \times R_N}$

On the other hand, HOOI utilizes an Alternating Least Squares (ALS) approach to estimate each of the factor matrices by sequentially solving sub-problems of the form in (4):

$$\begin{aligned} \max_{\boldsymbol{A}^{(n)}} \quad & ||\boldsymbol{\mathcal{X}} \times_1 \boldsymbol{A}^{(1)} \times_2 \boldsymbol{A}^{(2)} \cdots \times_N \boldsymbol{A}^{(N)}|| \\ \text{s.t.} \quad & \boldsymbol{A}^{(n)} \in \mathbb{R}^{I_n \times R_N} \text{ and column-wise orthonormal} \end{aligned} \tag{4}$$

where the factor matrices $\boldsymbol{A}^{(1)} \times_2 \boldsymbol{A}^{(2)} \cdots \times_N \boldsymbol{A}^{(N)}$ are commonly initialized via the ST-HOSVD of $\boldsymbol{\mathcal{X}}$. By iteratively optimizing each factor matrix while keeping the others fixed, HOOI provides a better fit as measured by the norm of the difference between the true and estimated tensors compared to ST-HOSVD, but for a higher computational cost. However, since HOOI is not guaranteed to converge to a global optimum nor to a stationary point [43,44] and does not provide substantial fingerprinting improvements in comparison to ST-HOSVD (see Table 2), ST-HOSVD is the chosen algorithm to compute the Tucker decomposition of FCs.

Table 2. Matching rate comparison between (baseline) FCs and distinct dimensionality reduction methods for parcellation granularity of 414. Bold values highlight the highest matching rates among all techniques for each fMRI condition, which are presented in ascending order of scanning length ("TPs" represent the number of time points in the time series of each condition).

fMRI Conditions	FCs	PCA	CP	ST-HOSVD	HOOI
Emotion (166 TPs)	0.57	0.63	0.64	**0.90**	**0.90**
Relational (222 TPs)	0.79	0.84	0.90	**0.99**	**0.99**
Gambling (243 TPs)	0.84	0.86	0.93	**0.98**	**0.98**
Social (264 TPs)	0.88	0.88	0.92	**0.97**	**0.97**
Motor (274 TPs)	0.82	0.82	0.75	**0.96**	**0.96**
Language (306 TPs)	0.91	0.90	0.92	**0.98**	**0.98**
Working Memory (395 TPs)	0.86	0.84	0.92	**0.97**	**0.97**
Resting-state (1190 TPs)	0.92	0.97	0.99	**1.00**	**1.00**

When applied to tensors that exhibit partial (or full) symmetries, HOSVD preserves the symmetric structure of the tensor [41]. Hence, for a tensor consisting of one session (e.g., test sessions) of participants' FC matrices, the Tucker decomposition of $\boldsymbol{\mathcal{X}} \in \mathbb{R}^{M \times M \times N}$ can be reformulated as:

$$\begin{aligned} \min_{\boldsymbol{\mathcal{G}},\boldsymbol{B},\boldsymbol{P}} \quad & ||\boldsymbol{\mathcal{X}} - \boldsymbol{\mathcal{G}} \times_1 \boldsymbol{B} \times_2 \boldsymbol{B} \times_3 \boldsymbol{P}|| \\ \text{s.t.} \quad & \boldsymbol{\mathcal{G}} \in \mathbb{R}^{R_1 \times R_1 \times R_2} \\ & \boldsymbol{B} \in \mathbb{R}^{M \times R_1} \text{ and column-wise orthonormal} \\ & \boldsymbol{P} \in \mathbb{R}^{N \times R_2} \text{ and column-wise orthonormal} \end{aligned} \tag{5}$$

where the factor matrices $\boldsymbol{B}$ and $\boldsymbol{P}$ obtained via HOSVD contain, respectively, brain parcellation and participant-specific information and the ranks R_1 and R_2 express the compression levels of brain parcellation and participant-specific information. While solving the optimization problem presented in (5), the brain parcellation ranks were chosen with a step size of $\frac{(Parcellation - 14)}{4}$. In addition to the previous parcellation ranks, we also performed a full-rank decomposition. Thus, for a parcellation granularity of 414, for example, the brain parcellation ranks were set to $R_1 = \{100, 200, 300, 400, 414\}$. In contrast, for all parcellation granularities explored, the participant ranks were set to $R_2 = \{50, 100, 150, 200, 250, 300, 350, 400, 426\}$.

Under the hypothesis that the functional connectivity patterns of a participant are, to some extent, reproducible across scanning sessions, we fix the core tensor $\boldsymbol{\mathcal{G}}$ and brain parcellation factor matrix B derived from the Tucker decomposition of tensor $\mathcal{X}$ and estimate the participant factor matrix Q of the tensor $\mathcal{Y}$ comprising FCs from another data acquisition session (e.g., retest session). By doing so, we aim to detect a consistent presence of underlying cohort-level functional connectivity patterns across different data acquisition sessions for each participant. The optimization problem shown in (6)

$$\min_{Q} \quad ||\mathcal{Y} - \mathcal{G} \times_1 B \times_2 B \times_3 Q|| \tag{6}$$

and admits a closed-form solution given by

$$Q = \mathbf{Y}_{(3)} \times [(\mathcal{G} \times_1 B \times_2 B)_{(3)}]^{\dagger}$$

where † denotes the Moore–Penrose inverse [45] of a matrix, and $\mathbf{Y}_{(3)}$ and $(\mathcal{G} \times_1 B \times_2 B)_{(3)}$ denote the *mode*-3 matricization of $\mathcal{Y}$ and $(\mathcal{G} \times_1 B \times_2 B)$, respectively.

2.8. Fingerprinting Quantification

To quantify fingerprinting, we used a measure denominated matching rate [24] for an identifiability matrix $\mathbf{I} \in \mathbb{R}^{N \times N}$, where $i_{j,k}$ denotes the Pearson's correlation between the j-th row of the participant factor matrix $\mathbf{P}_{j,:}$, and the k-th row of the participant factor matrix $Q_{k,:}$. The main diagonal entries of $\mathbf{I}$ represent similarity levels between different imaging sessions of the same participant. By hypothesis, we expect those entries to be higher than the off-diagonal entries, which represent the similarity level between different imaging sessions of different participants. Matching rate is a variation of ID_{rate} [7] that accounts for the fact that each participant is present only once in the test and retest sets. ID_{rate} (7) is the average frequency at which a participant's test session is most highly correlated to their retest session, and their retest session is most highly correlated to their test session (note that one does not necessarily imply the other). For matching rates, we impose that once a test session is paired with a retest session, it can no longer be chosen for a new pairing. The relative frequency of successful participants matching in both directions is then averaged, yielding a value in the range $[0, 1]$, where 0 indicates a failure to correctly match any of the participant's FCs, and 1 indicates success in matching all participant's FCs correctly. An algorithmic description of the computation of the matching rate is presented in Algorithm 2.

$$ID_{rate} = \frac{\frac{\text{Number of correctly paired test-retest samples}}{\text{Total number of participants}} + \frac{\text{Number of correctly paired retest-test samples}}{\text{Total number of participants}}}{2} \tag{7}$$

Algorithm 2 Matching Rate Computation

```
Input: I ∈ ℝ^{N×N}
I_mask ← I
MR_1 ← 0
for i ← 1 to N do
    [maxValPerCol, rowIndices] ← max(I_mask, [], 1)
    [~, colIndex] ← max(maxValPerCol)
    if (colIndex = rowIndices(colIndex)) then
        MR_1 ← MR_1 + 1
    end if
    I_mask(:, colIndex) ← -inf
    I_mask(rowIndices(colIndex), :) ← -inf
end for
I_mask ← I^T
MR_2 ← 0
for i ← 1 to N do
    [maxValPerCol, rowIndices] ← max(I_mask, [], 1)
    [~, colIndex] ← max(maxValPerCol)
    if (colIndex = rowIndices(colIndex)) then
        MR_2 ← MR_2 + 1
    end if
    I_mask(:, colIndex) ← -inf
    I_mask(rowIndices(colIndex), :) ← -inf
end for
Avg_MR ← (MR_1 + MR_2) / (2×N)
Output: Avg_MR
```

2.9. Fingerprinting Framework Adapted to Tucker Decomposition

The proposed fingerprinting framework consists of five key steps: (i) given a data acquisition session (either test or retest) of an fMRI condition, construct a tensor that contains all participants FCs; (ii) decompose the tensor via Tucker decomposition to obtain a core tensor, a brain parcellation factor matrix, and a participant factor matrix; (iii) estimate the other session's participant factor matrix based on the decomposition of the given session; (iv) obtain an identifiability matrix by computing pairwise Pearson's correlation between the rows of both participant factor matrices; and (v) calculate the matching rate for the obtained identifiability matrix. A schematic representation of our framework is presented in Figure 1.

In the following section, we discuss how matching rate is affected by parcellation granularity, decomposition rank, scanning length of fMRI conditions, and under within- and between-condition scenarios.

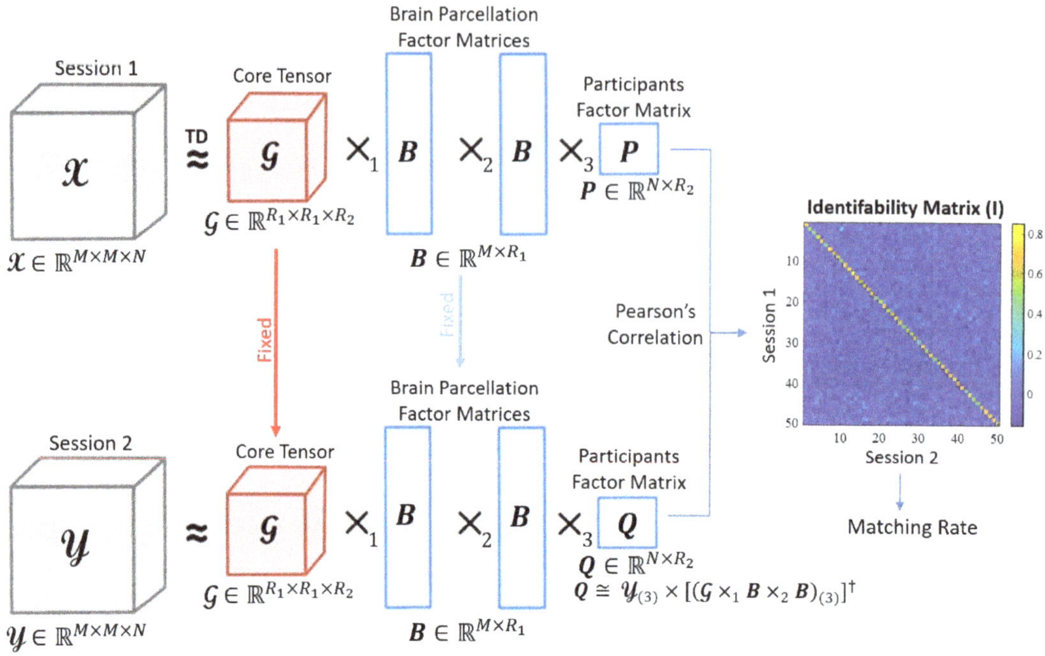

Figure 1. Graphic representation of the participant identifiability framework. For an input tensor $\mathcal{X}$ consisting of FCs from one imaging session, we compute its Tucker decomposition to obtain a core tensor $\mathcal{G}$, two identical brain parcellation factor matrices B, and a participants' factor matrix $\boldsymbol{P}$. Fixing $\mathcal{G}$ and B factor matrices, we estimate the participant's factor matrix $\boldsymbol{Q}$ for the FCs from another imaging session $\mathcal{Y}$. We then obtain an identifiability matrix $\boldsymbol{I}$ by computing row-wise Pearson's correlation between $\boldsymbol{P}$ and $\boldsymbol{Q}$, and obtain a matching rate score through $\boldsymbol{I}$.

3. Results

For all eight fMRI conditions, the proposed fingerprinting framework was applied to two main settings. First, when test and retest FCs correspond to the same fMRI condition (within-condition fingerprinting). Second, when combining resting-state FCs with task FCs (between-condition fingerprinting). For the aforementioned settings, we further investigated the impact of brain parcellation rank and participant rank on Tucker decomposition and subsequent fingerprinting, and whether scanning length duration has a significant effect on matching rates.

3.1. Parcellation Granularity Effect on Fingerprinting

Finer parcellation granularities offer a more detailed representation of the functional connectome by dividing the brain into more regions. However, this comes with a trade-off, as the time series for each region may be less reliable due to being derived from a smaller number of voxels. Hence, FCs fingerprints are expected to be dependent on the parcellation granularity of the data. Figure 2 displays the within-condition fingerprinting results of maximum-rank ST-HOSVD decompositions (i.e, for parcellation 614, the brain parcellation rank R_1 and participant rank R_2 were chosen to be 614 and 426, respectively) for parcellation granularities of $114, 214, 314, 414, 514, 614, 714, 814$, and 914. As the parcellation granularity increases, matching rates plateau for all fMRI conditions starting at parcellation granularity of 414. Therefore, we give a stronger emphasis to parcellation granularity of 414 in the remaining analyses of this study. It is noteworthy that the longest (resting-state) and shortest (Emotion) fMRI conditions display, respectively, the highest and lowest matching rate values across all parcellation granularities. However, tasks with intermediate scanning lengths do not follow such trend.

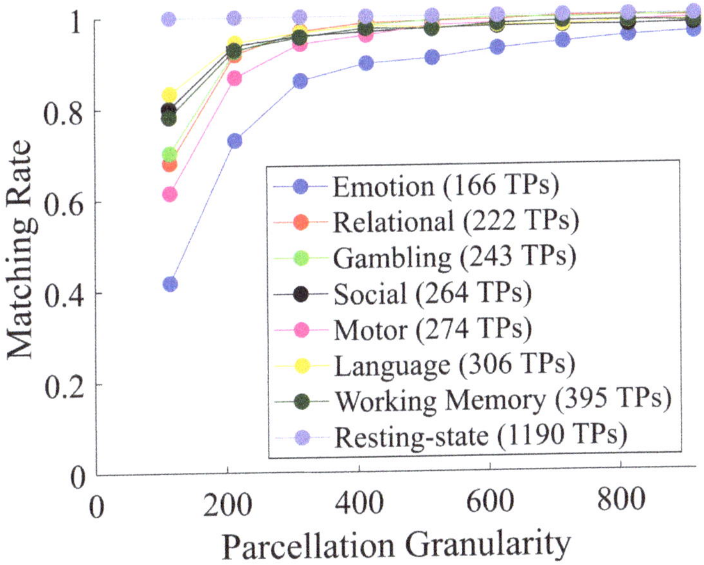

Figure 2. Fingerprinting gradient as measured by matching rate for all fMRI conditions. For each parcellation granularity, the brain parcellation rank and participant rank were set to their maximum value.

3.2. Dimensionality Reduction Methods in Fingerprinting

In Table 2, a comparison between different dimensionality reductions methods versus a baseline (consisting of vectorized FCs) for parcellation granularity of 414 is shown. When computing PCA, the number of principal components used to obtain the identifiability matrix was chosen to be the maximizer of subject identifiability as introduced in the work of Amico and Goñi (2018) [9]. The reported matching rates for CP decomposition are the highest among rank-R decompositions, with $R \in \{100, 200, 300, 400, 414\}$. Equivalently, the reported matching rates for Tucker-based methods are the highest among all considered combinations of brain parcellation rank R_1 and participant rank R_2. Among the presented dimensionality reduction methods, Tucker decomposition (ST-HOSVD and HOOI) demonstrates, by a considerable margin, the highest fingerprint.

Considering the high dimensionality of neuroimaging data, it is crucial to consider the scalability of Tucker-based methods for the purpose of fingerprinting. In Table 3, the runtime (seconds) of the most computationally demanding decomposition of a tensor of dimensions $PG \times PG \times 426$, where PG denotes the parcellation granularity, is presented for each dimensionality reduction method. For CP and Tucker-based methods, the highest runtime corresponds to a decomposition of maximum possible rank. For example, for a tensor of dimensions $414 \times 414 \times 426$, the CP rank is set to $R = 414$, and the ranks for ST-HOSVD and HOOI are set to $R_1 = R_2 = 414, R_3 = 426$. Equivalently, a matrix of dimensions $\binom{414}{2} \times 852$ (columns represent the upper triangular entries of vectorized FCs) all 852 principal component are computed for PCA. As the dimensionality of the data increases with the parcellation granularity, PCA tends to become less computationally expensive than ST-HOSVD. However, the increase in computational cost is negligible compared to the effectiveness of ST-HOSVD in uncovering fingerprints. For reference, the reported runtimes were measured on a Lenovo Legion 5i laptop equipped with an Intel Core i7 processor, 32GB of RAM, and an NVIDIA GeForce RTX 3070 Ti GPU.

Table 3. Comparison between the worst runtimes of different dimensionality reduction methods. For each parcellation granularity PG, a tensor of dimension $PG \times PG \times 426$ was decomposed by CP Decomposition, ST-HOSVD, and HOOI. Equivalently, PCA was applied to a matrix of dimension $\binom{PG}{2} \times 852$, where the rows consist of vectorized upper triangular FCs, and the columns correspond to the participants' test and retest sessions. Bold values indicate the shortest runtime across methods.

Method	Runtime (Seconds) per Parcellation Granularity								
	114	214	314	414	514	614	714	814	914
PCA	**0.38**	0.85	1.93	3.74	6.26	**8.75**	**12.45**	21.23	**24.71**
CP Decomposition	2.15	12.88	45.82	108.23	158.75	222.00	288.06	378.45	465.32
ST-HOSVD	**0.38**	**0.73**	**1.58**	**3.20**	**5.15**	8.78	13.24	**19.55**	27.62
HOOI	0.87	1.76	3.81	6.41	11.70	18.03	26.51	36.20	72.07

3.3. Evaluating the Impact of Brain Parcellation Rank and Participant Rank on Fingerprinting

To obtain a holistic view of how brain parcellation rank and participant rank affect fingerprinting, we computed matching rates under different combinations of both ranks for all fMRI conditions. Results shown in Figure 3 indicate that higher brain parcellation ranks led to higher matching rates compared to lower brain parcellation ranks. However, the impact of the participant rank on matching rates depends on the fMRI condition. Specifically, with Emotion and Motor tasks we can achieve near-optimal matching rates with a participant rank of 300 or higher, while for all the other fMRI tasks, we achieve near-optimal matching rates earlier, starting at a participant rank of 150. Resting-state matching rates were the highest among all fMRI conditions, reaching optimal scores starting at a participant rank 100, with 100% matching accuracy. In contrast, Emotion had the lowest matching rate.

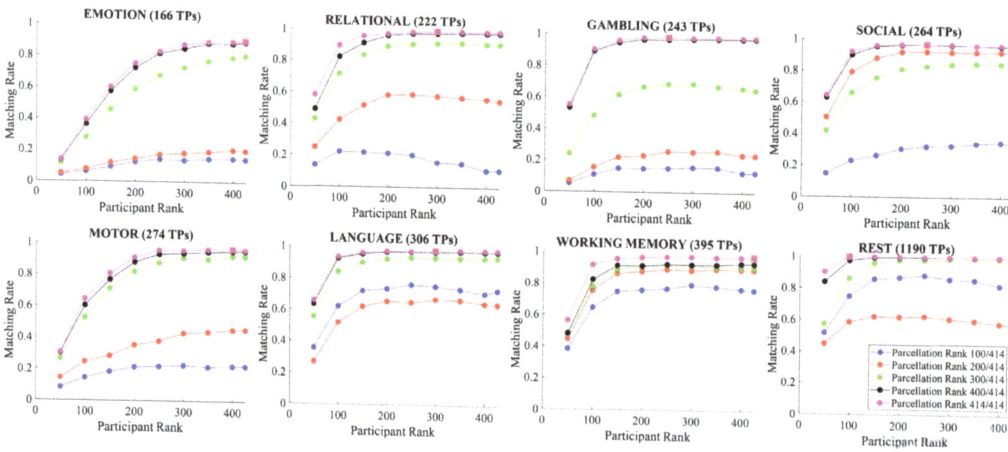

Figure 3. Brain parcellation rank and participant rank effects on matching rate for all eight fMRI conditions, ordered by scan length (represented by the number of time points or "TPs" in their time series) for granularity of Schaefer parcellation 414. For each condition, the fingerprinting framework is applied in both test→retest and retest→test directions. Each curve illustrates the average matching rate computed across both settings at each combination of brain parcellation ranks and participant ranks.

3.4. Within-Condition Fingerprinting

The concept of within-condition fingerprinting reflects the ability to correctly match two scans of the same participant when all evaluated scans belong to the same fMRI condition. The proposed fingerprinting framework yields a substantial matching rate increase across all considered parcellation granularities relative to vectorized original FCs, which were adopted as baseline. For original FCs, identifiability matrices were obtained by computing pairwise Pearson's correlation between test and retest vectorized upper triangular original FCs.

In Figure 4A, it is shown, for each condition, the highest matching rate obtained for all possible combinations of brain parcellation rank and participant rank. The highest matching rate increase with respect to the original FCs was observed for parcellation granularity of 214, for which the fingerprinting framework generated matching rate improvements ranging from 11% (Language) to 36% (Emotion). Resting-state matching rates were the highest in all parcellations, achieving 100% for all parcellation granularities.

Given the mismatch in scan duration between resting-state and tasks, a matching rate comparison between fMRI tasks and resting-state when resting-state time series are cropped to match the duration of tasks is shown in Figure 4B. To do so, and in order to avoid possible biases when using the first time-points, resting-state FCs are estimated by using the L central time points, where L is the duration of the corresponding task. For conditions with short scanning length (Emotion, Relational, and Gambling), resting-state displays a lower fingerprint than the fMRI tasks in all parcellation granularities. The previous is also true for Social and Language tasks and parcellation granularity 214.

The fingerprinting framework was carried out in two ways. First, by inputting a tensor consisting of test FCs while aiming to estimate the participants' factor matrix of the retest session. Second, by inputting a tensor consisting of retest FCs while aiming to estimate the participants factor matrix of the test session. The results of both procedures were averaged for all combinations of ranks considered. The reported values correspond to the highest of such averages.

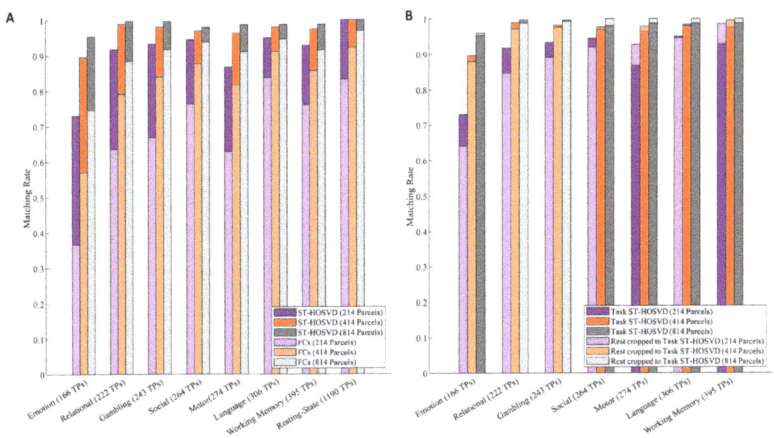

Figure 4. (**A**). Within-condition fingerprinting comparisons between highest matching rates obtained with ST-HOSVD and FCs directly for all fMRI conditions and parcellation granularities 214, 414, and 814. (**B**). Within-condition comparisons between highest matching rates obtained when resting-state FCs are estimated with time series of length equal to those of fMRI tasks. Specifically, the middle time points from resting-state time series are chosen to compute FCs. For example, when resting-state is shortened to match the scanning length of Emotion, resting-state FCs were computed with time points $\left(\frac{1190}{2} - \frac{166}{2}\right) : \left(\frac{1190}{2} + \frac{166}{2}\right)$.

3.5. *Between-Condition Fingerprinting*

Between-condition fingerprinting measures the degree to which we can match, for each participant, a scan from one fMRI condition to their scan from another fMRI condition. Using our fingerprinting framework, we estimate the participant factor matrix from a tensor consisting of task (test) fMRI by inputting a tensor consisting of resting-state (retest) FCs. Given that the scanning length of resting-state is substantially longer than the length of all fMRI tasks, we consider two scenarios when estimating resting-state FCs: (i) using the full resting-state time series and (ii) matching (hence reducing) the number of timepoints in the resting-state time series to the duration of each task (e.g., when pairing with Emotion task, resting-state FCs are computed using 166 out of the total 1190 time points). In the latter case, we further explore two strategies for sampling time points of the resting-state scan: (i) randomly sample time points, and (ii) randomly sample a starting time point in the range of $[1, 1190 - \text{length(task)}]$ and take the randomly sampled starting time point and its length(task) $- 1$ consecutive time points.

3.5.1. Between-Condition Fingerprinting with Resting-State Full Scanning Length

The between-condition matching rates when resting-state FCs are estimated using the time series of the full scanning length are shown in Figure 5. Matching rates ranged from 76% (Relational) to 91% (Language). In Figure 5, we observe a high sensitivity to brain parcellation rank, with 414 being the optimal rank. Comparatively, matching rate also benefits from a higher participant rank. However, the degree to which this occurs is task-dependent (i.e., a participant rank of 350 is nearly optimal for Emotion, but for Relational there is a clear benefit in going up to rank 426).

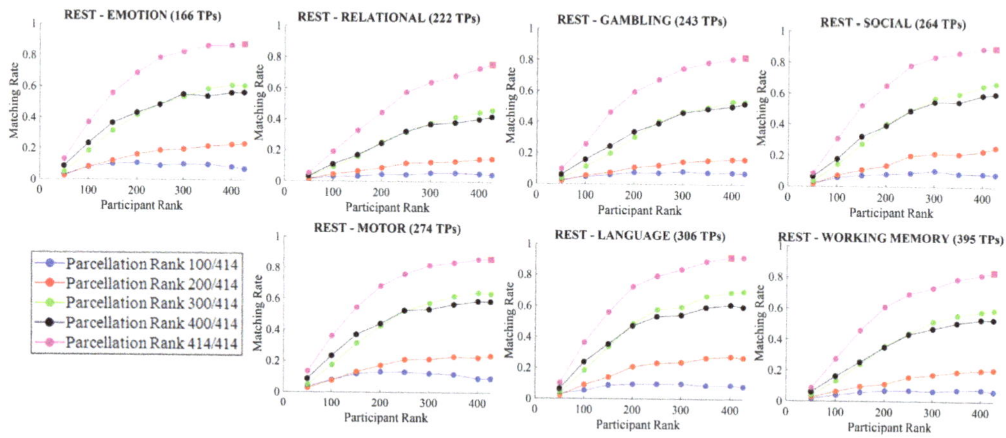

Figure 5. Parcellation rank and participant rank effects on between-condition matching rate for granularity of Schaefer parcellation 414. The fingerprinting framework was carried out by computing the ST-HOSVD of a tensor consisting of concatenated resting-state retest FCs (estimated using the entire scanning length of resting-state BOLD time series) from all participants. The obtained core tensor and brain parcellation factor matrix were then used to estimate the participants' factor matrix of a tensor consisting of concatenated test FCs of fMRI tasks. A similar trend is observed in all plots, with parcellation ranks of 100 and 200 being insufficient to uncover fingerprints, while the small increase in parcellation rank from 400 to 414 results in a substantial improvement in matching rates.

In Figure 6, we observe a major increase in between-condition matching rates using Tucker decomposition when compared to original FCs. The procedure used for computing the identifiability matrix for original FCs in the between-condition setting was analogous to the one in the within-condition setting. Matching rate improvements obtained with Tucker ranged from 43% (Relational) to 72% (Language).

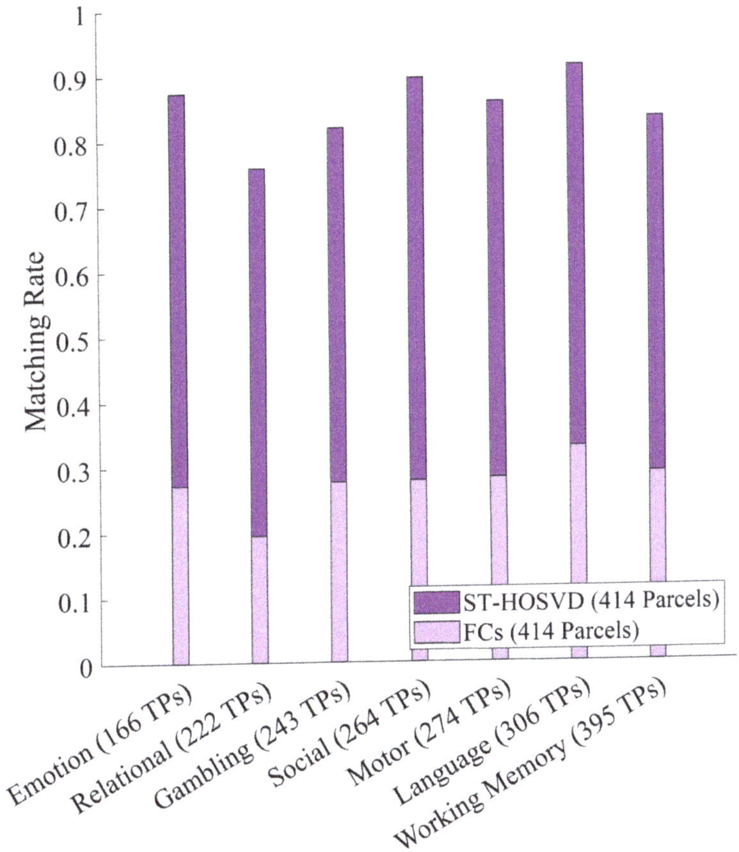

Figure 6. Between-condition comparison between highest matching rate obtained with the ST-HOSVD based fingerprinting framework and original FC matrices for parcellation granularity of 414. Despite being the shortest task and the one with worse within-condition fingerprints, Emotion displays one of the highest between-condition fingerprints. Contrarily, Working Memory possesses one of the highest fingerprints in the within-condition setting, but one of the lowest in the between-condition setting.

3.5.2. Between-Condition Fingerprinting with Resting-State Matched-to-Task Scanning-Length

In order to account for scanning length effects, the between-condition matching rate results were obtained when resting-state FCs are estimated by matching the time-series length to each fMRI task. Results are shown in Figure 7. For this analysis, we fixed the brain parcellation rank to 414, which was the optimal choice in terms of fingerprinting, as shown in Figure 5. For each participant rank across all between-condition settings, we computed matching rates with 100 different samplings of time points for the resting-state time series. In Figure 7A, we display the results of randomly sampling an initial time point and choosing the remaining time points to be consecutive to it. In Figure 7B, we display the results of randomly sampling all time points. When comparing both sampling strategies, randomly sampling all time points is a more robust and effective approach in terms of fingerprinting, as measured by the standard deviation and mean of each box plot.

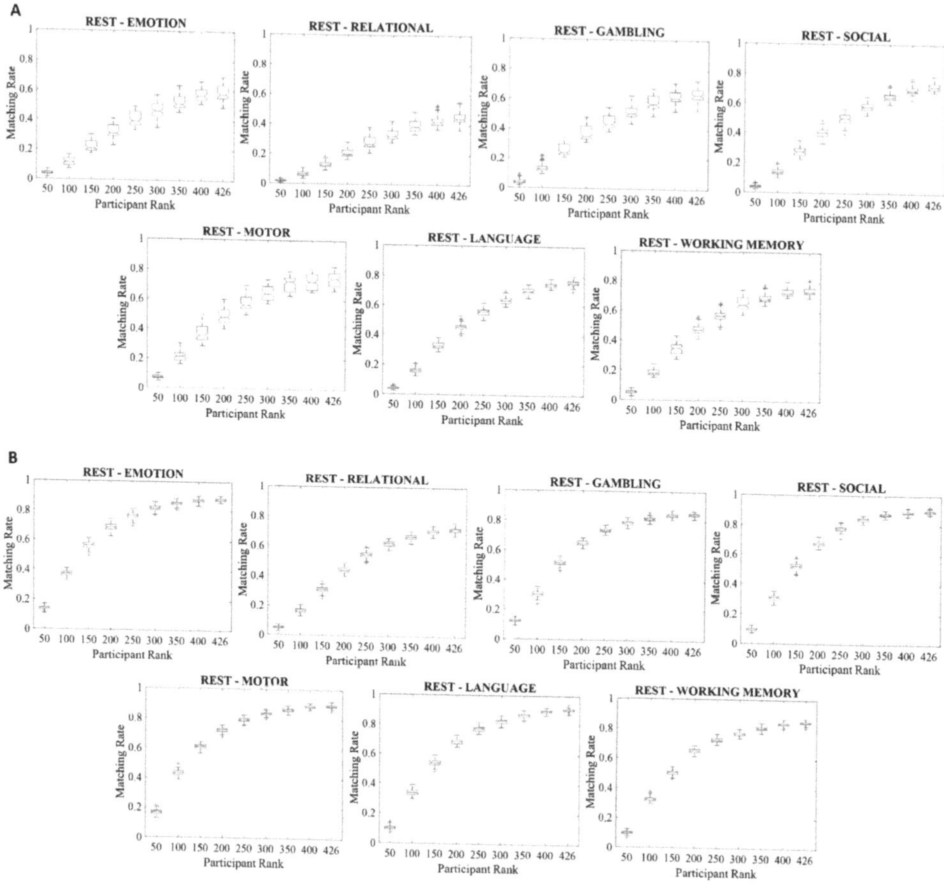

Figure 7. Between-condition matching rates obtained via ST-HOSVD when resting-state FCs are estimated using time series of equal length to those of each fMRI task for parcellation granularity of 414. The brain parcellation rank was set to 414, and 100 random sampling trials were conducted to select the time points used to estimate resting-state FCs for each participant rank. (**A**). Rest (retest)–task (test) matching rate results when, starting with a randomly sampled time point, consecutive time points are used to construct resting-state FCs. In this scenario, matching rates grow at an approximately linear rate with participant rank. (**B**). Matching rates when randomly sampled time points are used to construct resting-state FCs. Here, matching rates exhibit a bounded exponential pattern, reaching a plateau at a participant rank of 350.

3.6. Analysis of Misclassified Participants

Tucker decomposition led to a substantial improvement of matching rates for all fMRI conditions when compared to using FCs directly. Figure 8 provides additional evidence of how this is achieved. Figure 8A,B show, for Emotion and resting-state conditions and parcellation granularity of 414, the identifiability matrices obtained from FCs and ST-HOSVD along with the within- and between-participant distributions. When compared to results on FCs, similarities derived from ST-HOSVD tend to be diminished between participants (centered around 0), while remaining nearly the same within participants. In Figure 8C–E, the following cases are shown: in Figure 8C, a participant whose Emotion test and retest FCs are incorrectly matched using both vectorized FCs and ST-HOSVD; in Figure 8D, a participant whose Emotion test and retest FCs are incorrectly matched using vectorized FCs but correctly matched using ST-HOSVD; and in Figure 8E, a participant whose resting-state test and retest FCs are incorrectly matched using vectorized FCs but correctly matched using ST-HOSVD.

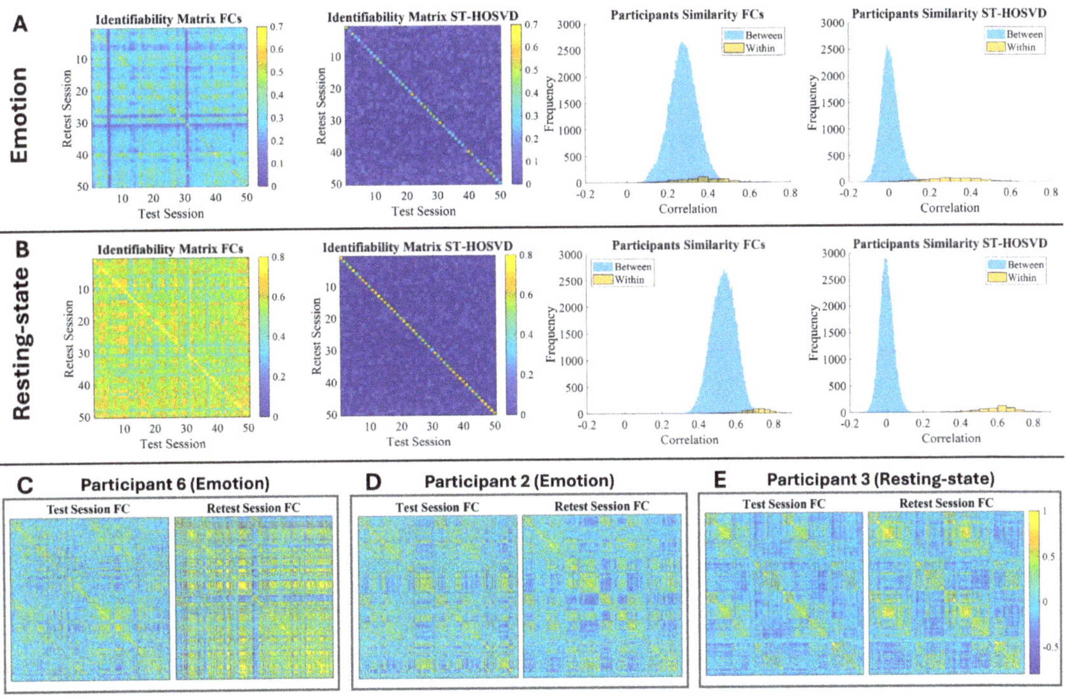

Figure 8. Comparison between tasks that possess the lowest (Emotion) and highest (Resting-state) within-condition fingerprints for parcellation granularity of 414. (**A**). From left-to-right: first 50 × 50 block of the 426 × 426 identifiability matrix derived from vectorized FCs, first 50 × 50 block of the 426 × 426 identifiability matrix derived from the proposed fingerprinting framework based on ST-HOSVD, distribution of within-participant (main diagonal entries) and between-participant (off-diagonal entries) sessions of the FCs identifiability matrix, distribution of within-participant and between-participant sessions of the ST-HOSVD identifiability matrix. (**B**). Analogous resting-state identifiability matrices and distributions. (**C**). Participant whose Emotion test and retest FCs are incorrectly matched when computing the identifiability matrix via vectorized FCs and ST-HOSVD. (**D**,**E**). Participants whose Emotion and resting-state FCs are incorrectly matched when computing the identifiability matrix using vectorized FCs but correctly matched when computing the identifiability matrix via ST-HOSVD.

4. Discussion

Table 2 shows that under all conditions, Tucker decomposition-based methods (ST-HOSVD and HOOI) consistently achieve the highest matching rates compared to FCs and other data-driven techniques. Both methods significantly outperform PCA and CP, particularly in conditions with fewer time points, such as Emotion and Relational, where FCs, whether used directly or in combination with PCA, under-perform. The results highlight the effectiveness of Tucker decomposition techniques in enhancing fingerprinting accuracy even for fMRI conditions with a short scanning length.

The results presented in Figure 2 provide meaningful insights into the influence of parcellation granularity on fingerprinting accuracy. The observed matching rate plateau at higher granularities indicates that beyond a certain threshold, increasing the parcellation resolution offers diminishing matching rate returns, which could be due to an increase in noise in the overall data of each participant. This is consistent with prior findings by Finn et al. (2015) [7], who showed that measures from FCs estimated with long scan sessions provide meaningful information about individuals even with moderate parcellation granularity. The significantly higher matching rates for resting-state compared to tasks aligns with previous research showing that resting-state data captures more stable and individualized connectivity patterns due to its longer scanning length and more consistent brain-wide activity [9]. However, it has also been shown that when accounting for scanning length, resting-state has a lower fingerprint than tasks [13]. Interestingly, the lower performance observed for Emotion reflects the challenges of identifying individuals from shorter or more transient brain states, where brief tasks led to less reliable fingerprinting. The lack of a strict linear relationship between task length and matching rates across intermediate conditions suggests that certain cognitive tasks elicit more distinct and stable connectivity patterns, regardless of their duration. This nuanced relationship highlights the complexity of functional connectome dynamics, emphasizing the need for both sufficient data length and appropriate parcellation resolution to maximize fingerprinting reliability.

As shown in Figure 3, compressing the brain parcellation information is detrimental to fingerprinting, as the highest matching rates were obtained with a brain parcellation rank of 414. In contrast, compressing the participants specific dimension overall preserves matching rate, as shown by the high matching rates obtained with participant rank as low as 150 for Relational, Gambling, Social, Language, Working Memory, and Resting-State, and have a rank of 300 for Emotion and Motor. The results from this analysis imply that the dimension of the participants information, represented by the participants rank, can be considerably compressed while preserving fingerprints. This indicates that there is some redundancy in inter-participant variability, with not every participant adding new information to the data. However, it is important to emphasize that the compressibility of the participants dimension is likely cohort-size dependent, and must, therefore, be reexamined when dealing with different cohort sizes.

The within-condition results shown in Figure 4A confirm the presence of functional connectome fingerprints and demonstrate that the proposed framework is particularly effective in uncovering them, compared to non-decomposed functional connectomes. Notably, the obtained fingerprinting results were substantially higher than those obtained with FCs, especially for parcellation granularity 214, where fingerprinting performance was notably low (ranging from 37% to 83%). Noting that resting-state achieves matching rates of 100% in Figure 4A, in Figure 4B, a comparison between tasks and resting-state is carried out when the FCs of both are computed with an equal number of time-points. The higher matching rates obtained from Emotion, Relational, and Gambling in comparison to resting-state suggest that, to some degree, the fingerprints derived from resting-state stem from the considerably longer duration of resting-state relative to fMRI tasks.

Unlike during fMRI tasks, participants in the resting-state condition are not engaged with any specific stimulus and, therefore, resting-state FCs encode "baseline" functional couplings among brain regions. Therefore, between-condition analyses allow us to assess the extent to which these couplings—captured by the resting-state brain parcellation factor matrix—can be recovered when participants engage in tasks. Even though extracting fingerprints in this setting is inherently more challenging compared to the within-condition setting, we were able to substantially increase matching rates relative to using original FCs. In Figure 5, it is shown that results are more sensitive to the parcellation and participant ranks, with a clear benefit in setting both to be the maximum value possible in many of the tasks. Furthermore, in Figure 6 it is shown that ST-HOSVD provided an even greater improvement over matching rates relative to FCs compared to the within-condition fingerprints. This supports the rationale that baseline functional couplings exist and can also be effectively uncovered while participants are engaged in a task.

We also explored the effect of reducing the resting-state BOLD time series duration to match the duration of each task. Doing so enables determining whether the entire scanning length is necessary to extract "key features" that facilitate obtaining between-condition fingerprints. From both sampling procedures carried out in this study, it is clear that randomly sampling time points is the superior strategy for fingerprinting. Comparing the results of Figure 7A,B, we see that randomly sampling time points of resting-state scans is not only more effective than sampling them consecutively, but also as effective as constructing resting-state FCs using the full scan length (as shown in Figure 5). This result aligns with the fingerprinting improvements seen when sub-sampling frames of edge-based time-series [46].

The results presented in Figure 8 highlight the effectiveness of the proposed ST-HOSVD-based framework in enhancing FC fingerprinting relative to using FCs directly. Panels A and B illustrate the identifiability matrices and participant similarity distributions for Emotion and resting-state, respectively. The ST-HOSVD framework considerably improves the separation between within-participant and between-participant similarity, as evident by the clearer diagonal structure in the identifiability matrices and the sharper peak of the within-participant similarity distribution. This indicates that the participants factor matrix derived from ST-HOSVD better captures the individual-specific features of FC patterns compared to vectorized FCs. Panels C, D, and E, respectively, display qualitative examples in which the proposed framework fails to correctly match the Emotion FCs of one participant, but succeeds in matching the FCs of participants that cannot be matched using FCs directly. Visually speaking, it is easy to see similarities between the FCs from participants 2 and 3, however the same cannot be said about the FCs from participant 6. While these results demonstrate the improved fingerprinting capability of the ST-HOSVD-based framework, potential limitations exist. Factors such as variations in FC stability across tasks or inter-individual differences in connectivity patterns could influence the framework's performance. Additionally, differences in cohort size, scanner parameters, or preprocessing pipelines may affect the generalizability of the results, and should be further analyzed in future studies.

Both CP and Tucker decomposition are commonly used tensor decomposition techniques for dimensionality reduction and feature extraction purposes. In the context of fingerprinting, CP falls short due to two key reasons. First, the assumption that the original high-dimensional data can be reconstructed using non-interacting components is too restrictive, as we know that there are innate interactions between brain regions under a functional connectivity standpoint [47]. Second, due to CP decomposition being a single-rank decomposition, we cannot freely explore the dynamics of compressing the different dimensions of the data. Conversely, the Tucker's core plays a pivotal role in capturing the interactions between components while giving us the flexibility to explore how different levels of compression of the brain parcellation and participants' information affect fingerprinting, thus overcoming both drawbacks of CP decomposition. However, drawing neuroscientific insights directly from the factor matrices derived from Tucker decomposition of FCs is non-trivial due to the existence of the core tensor [21], which captures several interactions between the components from each factor matrix.

FC fingerprinting, while having the potential to provide valuable insights in clinical and forensic settings, raises significant ethical concerns regarding privacy, bias, and potential misuse. The sensitive nature of neuroimaging data, which can reveal information about an individual's cognitive state, mental health, or even predispositions to certain conditions, makes it highly vulnerable to privacy breaches. Without proper anonymization, such data could be exploited for unauthorized profiling or discrimination. To safeguard privacy, robust anonymization techniques, such as de-identification and differential privacy, should be implemented. Additionally, data security measures, including encryption during storage and transmission, as well as strict access controls, are essential to prevent unauthorized usage. Bias is another concern, as models trained on non-representative datasets may lead to inaccurate or unfair identifications, particularly in forensic applications. Ensuring diverse, unbiased datasets and regularly auditing algorithms for fairness can help mitigate this risk. Addressing these ethical implications is crucial to prevent the misuse of neuroimaging fingerprinting and protect individual rights.

Our study has limitations. As discussed above in detail, interpreting the factor matrices derived from Tucker decomposition is not straightforward due to the presence of a core tensor that captures complex interactions between all factor matrices [21]. Additionally, the proposed fingerprinting framework does not allow for incremental updates to the core and factor matrices when FCs from new participants are introduced. Rather, the entire Tucker decomposition and fingerprinting framework would need to be recomputed. Our study leads to several avenues for further research. When preprocessing fMRI BOLD data, there is a large number of pipelines, steps, and parameters that can be used, with each specific configuration possibly leading to different FC estimations and ultimately differences in fingerprinting. Further work could assess the specific impact of such decisions (e.g., global signal regression, bandpass filter) on the association between Tucker decomposition and matching rates. Also, while Pearson's correlation is the most widely used coupling method for fMRI time-series to estimate functional connectivity, other alternatives such as mutual information should be considered in order to assess the impact of different coupling methodologies when using decomposition methods to assess fingerprinting. To improve the interpretability of Tucker decomposition, future work could explore strategies to extract meaningful neuroscientific insights from the core tensor and factor matrices. One promising approach is to impose sparsity constraints on the core tensor using L_0 or L_1 regularization, which could help isolate dominant functional connectivity patterns shared across individuals. Simultaneously, the participant factor matrix would reveal the contribution of each underlying pattern to an individual's FC, thereby enhancing interpretability. Another possible path for deriving neuroscientific insights is to perform post-hoc statistical analyses by correlating the components from the factor matrices with cognitive or behavioral measures. Doing so would help bridge the gap between Tucker-based decompositions methods and neuroscientific interpretation, allowing for a better understanding of how extracted patterns relate to individual differences in cognition, behavior, or clinical state.

5. Conclusions

In this work, a mathematical framework based on Tucker decomposition is presented for functional connectome fingerprinting. The ability to project high-dimensional functional connectivity data to lower dimensional spaces enables separately extracting cohort-level functional patterns, encoded by the brain parcellation factor matrix, and participant-specific patterns, contained in the participants factor matrix. Through within- and between-condition analyses, it is shown that the participant factor matrix serves, to a great extent, as a fingerprint of the participants. This is supported by the obtained matching rates ranging from 90 to 100% in the within-condition setting, which are consistently higher than those obtained from FCs directly or PCA, and 76 to 92% in the between-condition setting for the intermediate parcellation granularity of 414. Furthermore, the results from this study suggest that fingerprinting of rest vs task is highly dependent on scanning length (see Figure 4B). For up to 275 TPs (3 min 18 s), resting-state FCs had a smaller matching rate than their corresponding tasks FCs (Emotion, Relational, Gambling, Social). Beyond 275 TPs, resting-state FCs showed a higher matching rate than their corresponding task FCs

(Motor, Language, Working Memory). Overall, tensor-based methods demonstrate a high potential to uncover functional connectivity fingerprints. Our work has implications in understanding how data-driven decomposition methods enable us to assess fingerprinting with respect to: (i) brain parcellation granularity from coarse grain (114) to fine grain parcellations (914); (ii) dimensionality reduction for brain parcellation rank and for participants rank; (iii) scanning length of the fMRI condition; (iv) between and within fMRI condition.

Author Contributions: Conceptualization, V.C., A.M.E.G. and J.G.; methodology, V.C., M.L., J.H., A.M.E.G. and J.G.; formal analysis, V.C., A.M.E.G. and J.G.; investigation, V.C., M.L., J.H., A.M.E.G. and J.G.; data curation, V.C. and M.L.; writing—original draft preparation, V.C., A.M.E.G. and J.G.; writing—review and editing, V.C., M.L., J.H., A.M.E.G. and J.G.; supervision, A.M.E.G. and J.G.; funding acquisition J.H. and J.G. All authors have read and agreed to the published version of the manuscript.

Funding: This research was funded by National Institutes of Health (US), Award ID: CTSI CTR EPAR2169 (J.G). National Institutes of Health (US) R01 AA029607 (J.G), National Institutes of Health (US), Award ID: P60AA07611 (J.H and J.G), National Institutes of Health (US), Award ID: R01NS112303 (J.H and J.G).

Institutional Review Board Statement: Not applicable.

Informed Consent Statement: Not applicable.

Data Availability Statement: Upon acceptance of this manuscript, Matlab code to perform Tucker Decomposition and subsequent fingerprinting analyses will be available at the CONNplexity Lab website at the Publications section: https://engineering.purdue.edu/ConnplexityLab/publications (accessed on 12 February 2025).

Acknowledgments: Data were provided by the Human Connectome Project, WU-Minn Consortium (Principal Investigators: David Van Essen and Kamil Ugurbil; 1U54MH091657) funded by the 16 NIH Institutes and Centers that support the NIH Blueprint for Neuroscience Research; and by the McDonnell Center for Systems Neuroscience at Washington University.

Conflicts of Interest: The authors declare no conflicts of interest.

Abbreviations

The following abbreviations are used in this manuscript:

fMRI	Functional Magnetic Resonance Imaging
BOLD	Blood-Oxygen-Level-Dependent
FC	Functional Connectome
PCA	Principal Component Analysis
CP decomposition	CANDECOMP/PARAFAC decomposition
ST-HOSVD	Sequentially Truncated Higher-Order Singular Value Decomposition
HOSVD	Higher-Order Singular Value Decomposition
HOOI	Higher-Order Orthogonal Iteration
TPs	Time points
PG	Parcellation granularity

References

1. Behrens, T.E.; Sporns, O. Human connectomics. *Curr. Opin. Neurobiol.* **2012**, *22*, 144–153. [CrossRef] [PubMed]
2. Yao, G.; Wei, L.; Jiang, T.; Dong, H.; Baeken, C.; Wu, G.R. Neural mechanisms underlying empathy during alcohol abstinence: Evidence from connectome-based predictive modeling. *Brain Imaging Behav.* **2022**, *16*, 2477–2486. [CrossRef] [PubMed]
3. Fornito, A.; Zalesky, A.; Breakspear, M. The connectomics of brain disorders. *Nat. Rev. Neurosci.* **2015**, *16*, 159–172. [CrossRef] [PubMed]
4. Satterthwaite, T.D.; Xia, C.H.; Bassett, D.S. Personalized neuroscience: Common and individual-specific features in functional brain networks. *Neuron* **2018**, *98*, 243–245. [CrossRef]

5. Gratton, C.; Laumann, T.O.; Nielsen, A.N.; Greene, D.J.; Gordon, E.M.; Gilmore, A.W.; Nelson, S.M.; Coalson, R.S.; Snyder, A.Z.; Schlaggar, B.L.; et al. Functional brain networks are dominated by stable group and individual factors, not cognitive or daily variation. *Neuron* **2018**, *98*, 439–452. [CrossRef]
6. Seitzman, B.A.; Gratton, C.; Laumann, T.O.; Gordon, E.M.; Adeyemo, B.; Dworetsky, A.; Kraus, B.T.; Gilmore, A.W.; Berg, J.J.; Ortega, M.; et al. Trait-like variants in human functional brain networks. *Proc. Natl. Acad. Sci. USA* **2019**, *116*, 22851–22861. [CrossRef]
7. Finn, E.S.; Shen, X.; Scheinost, D.; Rosenberg, M.D.; Huang, J.; Chun, M.M.; Papademetris, X.; Constable, R.T. Functional connectome fingerprinting: Identifying individuals using patterns of brain connectivity. *Nat. Neurosci.* **2015**, *18*, 1664–1671. [CrossRef]
8. Fraschini, M.; Hillebrand, A.; Demuru, M.; Didaci, L.; Marcialis, G.L. An EEG-based biometric system using eigenvector centrality in resting state brain networks. *IEEE Signal Process. Lett.* **2014**, *22*, 666–670. [CrossRef]
9. Amico, E.; Goñi, J. The quest for identifiability in human functional connectomes. *Sci. Rep.* **2018**, *8*, 8254. [CrossRef]
10. Abbas, K.; Amico, E.; Svaldi, D.O.; Tipnis, U.; Duong-Tran, D.A.; Liu, M.; Rajapandian, M.; Harezlak, J.; Ances, B.M.; Goñi, J. GEFF: Graph embedding for functional fingerprinting. *NeuroImage* **2020**, *221*, 117181. [CrossRef]
11. Cai, B.; Zhang, G.; Hu, W.; Zhang, A.; Zille, P.; Zhang, Y.; Stephen, J.M.; Wilson, T.W.; Calhoun, V.D.; Wang, Y.P. Refined measure of functional connectomes for improved identifiability and prediction. *Hum. Brain Mapp.* **2019**, *40*, 4843–4858. [CrossRef] [PubMed]
12. Abbas, K.; Liu, M.; Venkatesh, M.; Amico, E.; Kaplan, A.D.; Ventresca, M.; Pessoa, L.; Harezlak, J.; Goñi, J. Geodesic distance on optimally regularized functional connectomes uncovers individual fingerprints. *Brain Connect.* **2021**, *11*, 333–348. [CrossRef] [PubMed]
13. Abbas, K.; Liu, M.; Wang, M.; Duong-Tran, D.; Tipnis, U.; Amico, E.; Kaplan, A.D.; Dzemidzic, M.; Kareken, D.; Ances, B.M.; et al. Tangent functional connectomes uncover more unique phenotypic traits. *iScience* **2023**, *26*, 107624. [CrossRef]
14. Venkatesh, M.; Jaja, J.; Pessoa, L. Comparing functional connectivity matrices: A geometry-aware approach applied to participant identification. *NeuroImage* **2020**, *207*, 116398. [CrossRef]
15. Ng, B.; Dressler, M.; Varoquaux, G.; Poline, J.B.; Greicius, M.; Thirion, B. Transport on Riemannian manifold for functional connectivity-based classification. In Proceedings of the Medical Image Computing and Computer-Assisted Intervention–MICCAI 2014: 17th International Conference, Boston, MA, USA, 14–18 September 2014; Proceedings; Part II 17; Springer: Berlin/Heidelberg, Germany, 2014; pp. 405–412.
16. Pervaiz, U.; Vidaurre, D.; Woolrich, M.W.; Smith, S.M. Optimising network modelling methods for fMRI. *Neuroimage* **2020**, *211*, 116604. [CrossRef] [PubMed]
17. Dadi, K.; Rahim, M.; Abraham, A.; Chyzhyk, D.; Milham, M.; Thirion, B.; Varoquaux, G.; Initiative, A.D.N.; Alzheimer's Disease Neuroimaging Initiative. Benchmarking functional connectome-based predictive models for resting-state fMRI. *NeuroImage* **2019**, *192*, 115–134. [CrossRef]
18. Kolda, T.G.; Bader, B.W. Tensor Decompositions and Applications. *SIAM Rev.* **2009**, *51*, 455–500. [CrossRef]
19. Mahyari, A.G.; Zoltowski, D.M.; Bernat, E.M.; Aviyente, S. A tensor decomposition-based approach for detecting dynamic network states from EEG. *IEEE Trans. Biomed. Eng.* **2016**, *64*, 225–237. [CrossRef]
20. Williams, A.H.; Kim, T.H.; Wang, F.; Vyas, S.; Ryu, S.I.; Shenoy, K.V.; Schnitzer, M.; Kolda, T.G.; Ganguli, S. Unsupervised discovery of demixed, low-dimensional neural dynamics across multiple timescales through tensor component analysis. *Neuron* **2018**, *98*, 1099–1115. [CrossRef]
21. Mokhtari, F.; Laurienti, P.J.; Rejeski, W.J.; Ballard, G. Dynamic functional magnetic resonance imaging connectivity tensor decomposition: A new approach to analyze and interpret dynamic brain connectivity. *Brain Connect.* **2019**, *9*, 95–112. [CrossRef]
22. Zhang, Z.; Allen, G.I.; Zhu, H.; Dunson, D. Tensor network factorizations: Relationships between brain structural connectomes and traits. *Neuroimage* **2019**, *197*, 330–343. [CrossRef] [PubMed]
23. Tucker, L.R. Some mathematical notes on three-mode factor analysis. *Psychometrika* **1966**, *31*, 279–311. [CrossRef] [PubMed]
24. Chiêm, B.; Abbas, K.; Amico, E.; Duong-Tran, D.A.; Crevecoeur, F.; Goñi, J. Improving Functional Connectome Fingerprinting with Degree-Normalization. *Brain Connect.* **2022**, *12*, 180–192. [CrossRef]
25. Van Essen, D.C.; Smith, S.M.; Barch, D.M.; Behrens, T.E.; Yacoub, E.; Ugurbil, K.; Wu-Minn HCP Consortium. The WU-Minn human connectome project: An overview. *Neuroimage* **2013**, *80*, 62–79. [CrossRef]
26. Glasser, M.F.; Sotiropoulos, S.N.; Wilson, J.A.; Coalson, T.S.; Fischl, B.; Andersson, J.L.; Xu, J.; Jbabdi, S.; Webster, M.; Polimeni, J.R.; et al. The minimal preprocessing pipelines for the Human Connectome Project. *Neuroimage* **2013**, *80*, 105–124. [CrossRef] [PubMed]
27. He, T.; Kong, R.; Holmes, A.J.; Nguyen, M.; Sabuncu, M.R.; Eickhoff, S.B.; Bzdok, D.; Feng, J.; Yeo, B.T. Deep neural networks and kernel regression achieve comparable accuracies for functional connectivity prediction of behavior and demographics. *NeuroImage* **2020**, *206*, 116276. [CrossRef]
28. Bijsterbosch, J.D.; Woolrich, M.W.; Glasser, M.F.; Robinson, E.C.; Beckmann, C.F.; Van Essen, D.C.; Harrison, S.J.; Smith, S.M. The relationship between spatial configuration and functional connectivity of brain regions. *eLife* **2018**, *7*, e32992. [CrossRef]

29. Tavor, I.; Jones, O.P.; Mars, R.B.; Smith, S.; Behrens, T.; Jbabdi, S. Task-free MRI predicts individual differences in brain activity during task performance. *Science* **2016**, *352*, 216–220. [CrossRef]
30. Schaefer, A.; Kong, R.; Gordon, E.M.; Laumann, T.O.; Zuo, X.N.; Holmes, A.J.; Eickhoff, S.B.; Yeo, B.T. Local-global parcellation of the human cerebral cortex from intrinsic functional connectivity MRI. *Cereb. Cortex* **2018**, *28*, 3095–3114. [CrossRef]
31. Power, J.D.; Mitra, A.; Laumann, T.O.; Snyder, A.Z.; Schlaggar, B.L.; Petersen, S.E. Methods to detect, characterize, and remove motion artifact in resting state fMRI. *Neuroimage* **2014**, *84*, 320–341. [CrossRef]
32. Smith, S.M.; Beckmann, C.F.; Andersson, J.; Auerbach, E.J.; Bijsterbosch, J.; Douaud, G.; Duff, E.; Feinberg, D.A.; Griffanti, L.; Harms, M.P.; et al. Resting-state fMRI in the human connectome project. *Neuroimage* **2013**, *80*, 144–168. [CrossRef] [PubMed]
33. Amico, E.; Arenas, A.; Goñi, J. Centralized and distributed cognitive task processing in the human connectome. *Netw. Neurosci.* **2019**, *3*, 455–474. [CrossRef] [PubMed]
34. Cole, M.W.; Bassett, D.S.; Power, J.D.; Braver, T.S.; Petersen, S.E. Intrinsic and task-evoked network architectures of the human brain. *Neuron* **2014**, *83*, 238–251. [CrossRef] [PubMed]
35. Sidiropoulos, N.D.; De Lathauwer, L.; Fu, X.; Huang, K.; Papalexakis, E.E.; Faloutsos, C. Tensor decomposition for signal processing and machine learning. *IEEE Trans. Signal Process.* **2017**, *65*, 3551–3582. [CrossRef]
36. Cong, F.; Lin, Q.H.; Kuang, L.D.; Gong, X.F.; Astikainen, P.; Ristaniemi, T. Tensor decomposition of EEG signals: A brief review. *J. Neurosci. Methods* **2015**, *248*, 59–69. [CrossRef]
37. Martınez-Montes, E.; Valdés-Sosa, P.A.; Miwakeichi, F.; Goldman, R.I.; Cohen, M.S. Concurrent EEG/fMRI analysis by multiway partial least squares. *NeuroImage* **2004**, *22*, 1023–1034. [CrossRef]
38. Carroll, J.D.; Chang, J.J. Analysis of individual differences in multidimensional scaling via an N-way generalization of "Eckart-Young" decomposition. *Psychometrika* **1970**, *35*, 283–319. [CrossRef]
39. Harshman, R.A. *Foundations of the PARAFAC Procedure: Models and Conditions for an "Explanatory" Multimodal Factor Analysis*; University Microfilms: Ann Arbor, MI, USA, 1970.
40. Li, X.; Xu, D.; Zhou, H.; Li, L. Tucker tensor regression and neuroimaging analysis. *Stat. Biosci.* **2018**, *10*, 520–545. [CrossRef]
41. De Lathauwer, L.; De Moor, B.; Vandewalle, J. A multilinear singular value decomposition. *SIAM J. Matrix Anal. Appl.* **2000**, *21*, 1253–1278. [CrossRef]
42. Vannieuwenhoven, N.; Vandebril, R.; Meerbergen, K. A new truncation strategy for the higher-order singular value decomposition. *SIAM J. Sci. Comput.* **2012**, *34*, A1027–A1052. [CrossRef]
43. De Lathauwer, L.; De Moor, B.; Vandewalle, J. On the best rank-1 and rank-(r 1, r 2,. . . , rn) approximation of higher-order tensors. *SIAM J. Matrix Anal. Appl.* **2000**, *21*, 1324–1342. [CrossRef]
44. Kroonenberg, P.M.; De Leeuw, J. Principal component analysis of three-mode data by means of alternating least squares algorithms. *Psychometrika* **1980**, *45*, 69–97. [CrossRef]
45. Penrose, R. A generalized inverse for matrices. In *Mathematical Proceedings of the Cambridge Philosophical Society*; Cambridge University Press: Cambridge, UK, 1955; Volume 51, pp. 406–413.
46. Cutts, S.A.; Faskowitz, J.; Betzel, R.F.; Sporns, O. Uncovering individual differences in fine-scale dynamics of functional connectivity. *Cereb. Cortex* **2023**, *33*, 2375–2394. [CrossRef] [PubMed]
47. Fox, M.D.; Snyder, A.Z.; Vincent, J.L.; Corbetta, M.; Van Essen, D.C.; Raichle, M.E. The human brain is intrinsically organized into dynamic, anticorrelated functional networks. *Proc. Natl. Acad. Sci. USA* **2005**, *102*, 9673–9678. [CrossRef]

applied sciences

MDPI

Case Report

Hypergraph Analysis of Functional Brain Connectivity During Figurative Attention

Alexander N. Pisarchik [1,*], Natalia Peña Serrano [2], Walter Escalante Puente de la Vega [1] and Rider Jaimes-Reátegui [3]

1 Centro de Tecnología Biomédica, Universidad Politécnica de Madrid, Campus de Montegancedo, Pozuelo de Alarcón, 28223 Madrid, Spain; walter.escalante@ctb.upm.es
2 Instituto de Fisiología Celular, Universidad Nacional Autónoma de México, Coyoacán, Mexico City 04510, Mexico; nattps13@gmail.com
3 Departamento de Ciencias Exactas y Tecnología, Centro Universitario de los Lagos, Universidad de Guadalajara, Enrique Díaz de León 1144, Colonia Paseos de la Montaña, Lagos de Moreno 47463, Mexico; rider.jaimes@academicos.udg.mx
* Correspondence: alexander.pisarchik@upm.es

Abstract: Hypergraph analysis extends traditional graph theory by enabling the study of complex, many-to-many relationships in networks, offering powerful tools for understanding brain connectivity. This case study introduces a novel methodology for constructing both graphs and hypergraphs of functional brain connectivity during figurative attention tasks, where subjects interpret the ambiguous Necker cube illusion. Using a frequency-tagging approach, we simultaneously modulated two cube faces at distinct frequencies while recording electroencephalography (EEG) responses. Brain connectivity networks were constructed using multiple measures—coherence, cross-correlation, and mutual information—providing complementary insights into functional relationships between regions. Our hypergraph analysis revealed distinct connectivity patterns associated with attending to different cube orientations, including previously unobserved higher-order relationships between brain regions. The results demonstrate bilateral cortico–cortical interactions and suggest integrated processing hubs that may coordinate visual attention networks. This methodological framework not only advances our understanding of the neural basis of visual attention but also offers potential applications in attention monitoring and clinical assessment of attention disorders. While based on a single subject, this proof-of-concept study establishes a foundation for larger-scale investigations of brain network dynamics during ambiguous visual processing.

Keywords: hypergraph; brain connectivity; cognitive neuroscience; visual perception; electroencephalography (EEG); coherence; cross-correlation; mutual information; neuronal networks

1. Introduction

Understanding brain connectivity in response to external stimuli is essential for uncovering the mechanisms behind information processing, perception, attention, and decision-making. Functional connectivity, which identifies active brain regions exhibiting correlated frequency, phase, and amplitude, can be analyzed in both frequency and time domains [1]. When a sufficiently large population of neurons synchronizes, their collective electrical activity becomes detectable through brain imaging techniques such as electroencephalography (EEG). To examine the emergence and dynamic evolution of functional brain networks, researchers often apply complex network theory, grounded in mathematical graph

https://doi.org/10.3390/app15073833

theory [2,3]. By representing brain regions as nodes and their interactions as links, this approach has facilitated the study of biorhythmic connections across different brain areas [4,5]. It has also proven valuable in identifying individual cognitive differences [6], diagnosing neurological disorders [7–9], and assessing aging-related neural changes [10–13].

EEG-based functional connectivity networks have become essential tools for understanding individual variability in brain activation patterns during cognitive tasks [14,15]. These networks provide valuable insights into how different brain regions interact, shedding light on the neural mechanisms underlying attention and other cognitive processes. Traditional graph-theoretic approaches model functional connectivity by representing brain regions as vertices and their functional interactions as edges [4]. However, these methods primarily capture second-order relationships—interactions between pairs of brain regions—while neglecting higher-order relationships that are fundamental to complex neural processing.

Since cognitive processes such as attention involve dynamic and multifaceted interactions across multiple brain regions, conventional functional connectivity networks fail to provide a complete picture by only capturing pairwise relationships. To overcome this limitation, hypergraph-based connectivity models have been introduced. Unlike traditional graphs, hypergraphs allow for the representation of multi-node interactions through hyperedges, enabling a richer and more comprehensive characterization of brain connectivity.

Hypergraphs have recently gained prominence as an advanced framework for investigating brain connectivity [16–30]. Unlike conventional graphs that are limited to pairwise relationships, hypergraphs extend graph theory to model complex, multi-region interactions simultaneously [31,32]. This approach offers a more nuanced and holistic view of functional brain networks by capturing the intricate, dynamic interactions among multiple neural regions. Through hypergraph analysis, researchers can gain deeper insights into how brain regions integrate, synchronize, and adapt to support cognitive functions, providing novel perspectives on neural organization and functionality.

Hypergraphs representing brain connectivity have been constructed using neurophysiological data from various imaging modalities. Most studies have focused on functional magnetic resonance imaging (fMRI) data, particularly in the resting state [16–24]. Structural MRI has also been employed for hypergraph-based connectivity modeling [25]. Additionally, researchers have developed hypergraphs using multimodal data that integrate MRI with fluorodeoxyglucose positron emission tomography (PET) and cerebrospinal fluid analysis [29]. Beyond MRI-based studies, both invasive intracranial EEG [26] and noninvasive EEG [27,28] have been used to construct hypergraphs. Magnetoencephalography (MEG) data have also been explored in hypergraph-based connectivity studies [30].

The versatility of hypergraph-based connectivity modeling has led to applications in diverse areas, including emotion recognition [22,27], motor imagery [28], and the diagnosis of neurological disorders such as mild cognitive impairment [18,21,29], schizophrenia [20], autism spectrum disorder [24], Alzheimer's disease [25,29], and epileptic seizure detection [26]. Many of these studies employ correlation-based metrics to measure functional connectivity, with Pearson correlation being a widely used approach for quantifying neural interactions [33]. Given the importance of attentional processes in cognitive health, hypergraph-based connectivity models offer a promising avenue for understanding the neural mechanisms underlying attention and related cognitive functions.

Attention is a crucial cognitive function, and its impairment is associated with various neurological and psychiatric disorders, including cognitive decline, dementia, attention deficit hyperactivity disorder, and autism spectrum disorder. These conditions affect a broad range of cognitive and behavioral abilities, such as learning, problem-solving, communication, and social interaction. Individuals with attention deficits may struggle to express their experiences or understand others' perspectives, which can hinder their development of essential social and learning skills. Electroencephalography (EEG) is a widely used modality for assessing attention. While many studies have focused on the relationship between P-300 event-related potentials and attention in EEG signals, others have explored spectral analysis as an alternative approach [34–38]. The use of hypergraphs to model brain connectivity could provide deeper insights into the neural mechanisms underlying attention deficits by capturing the complex, multi-dimensional relationships within brain networks. This approach may complement traditional EEG-based methods, offering a more comprehensive understanding of attention-related impairments and their neural correlates.

This study is motivated by the growing interest in machine learning techniques for the early diagnosis and prediction of neurodegenerative diseases based on hypergraphs of brain connectivity. The primary objective of this study is to develop a method for constructing hypergraphs of brain connectivity to enhance our understanding of brain functionality during figurative attention. Unlike traditional pairwise graphs, hypergraphs provide a more comprehensive representation by capturing the most significant connections and disconnections across multiple brain lobes, identified through different connectivity measures such as coherence, correlation, and mutual information. To investigate this, we use the Necker cube—an ambiguous figure requiring attentional focus for orientation interpretation (left- or right-facing). EEG data is recorded while the participant performs three cognitive tasks: (i) passive viewing without interpreting the cube's orientation (MI), (ii) imagining the cube as left-oriented (MLV), and (iii) imagining it as right-oriented (MRV). Task (i) serves as a baseline for involuntary attention, whereas tasks (ii) and (iii) assess voluntary attention. By comparing brain connectivity patterns across these conditions, we aim to identify changes in connectivity associated with voluntary and involuntary attention. These insights contribute to understanding how figurative attention influences brain network dynamics, particularly in sustaining focus on a specific perceptual interpretation.

A graphical framework of our study is presented in Figure 1.

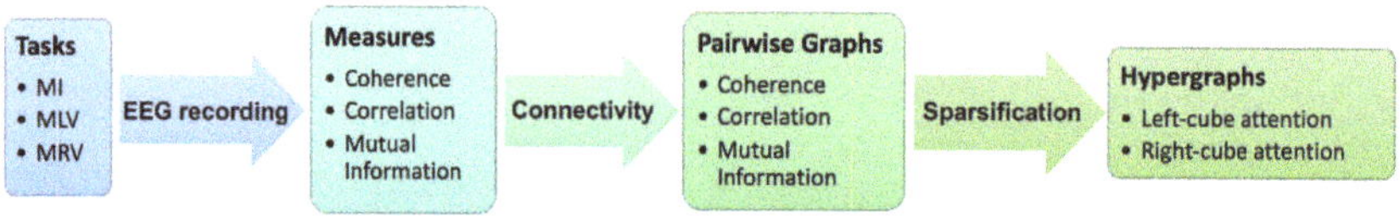

Figure 1. Main steps for constructing hypergraphs of brain connectivity during figurative attention. EEG data are recorded while the subject performs three cognitive tasks: MI, MLV, and MRV. Based on three connectivity measures (coherence, correlation, and mutual information), corresponding pairwise graphs are constructed. A sparsification procedure is then applied to generate hypergraphs that capture higher-order interactions associated with attention to the left-oriented and right-oriented cube.

To the best of our knowledge, this work presents the first hypergraph-based analysis of functional brain connectivity associated with attention. We demonstrate how hypergraphs can be constructed to represent brain connectivity patterns associated with attention, which may, in the future, contribute to the diagnosis and prediction of attention-related disorders. Using EEG data from a subject exhibiting strong attentional engagement, we construct both pairwise graphs and hypergraphs based on three important connectivity measures—coherence [39], cross-correlation [40], and mutual information [41]. Our analysis investigates how functional connectivity differs when the subject actively focuses on one interpretation of the Necker cube's orientation compared to a neutral attentional state.

The remainder of this paper is organized as follows: Section 2 introduces the mathematical foundations underlying our methods. Section 3 details the experimental paradigm and methodology. Section 4 describes the procedures for analyzing the experimental EEG data. Sections 5 and 6 present results on the construction of attention-based pairwise graphs and hypergraphs, respectively. Section 7 provides an in-depth discussion of our findings, followed by key conclusions in Section 8. Additional details on graph construction are included in the Appendices.

2. Mathematical Basis

In this section, we provide the most important definitions of pairwise graphs and hypergraphs, as well as the measures of brain connectivity that we utilize in this study: coherence, cross-correlation, and mutual information.

2.1. Main Definitions

A *plain graph* G is an abstract mathematical structure composed of a finite set of n vertices (or nodes) V and a set of m edges (or links) E, where each edge represents an interaction between a pair of vertices. Mathematically, the plain graph is defined as

$$G = (V, E), \ \ V = \{v_1, \dots, v_i, \dots, v_n\}, \ \ E = \{v_1, \dots, v_j, \dots, v_m\}. \tag{1}$$

The plain graph is commonly represented by its adjacency matrix A, which can be either binary (indicating the presence or absence of a connection between two vertices) or numerical (reflecting the strengths or weights of the connections between edges). A graph with labels assigned to its edges and vertices is referred to as a *labeled graph* . A label of edge j is its weight $w(e_j)$ (or connectivity measure), i.e., $\mathrm{Lab}(e_j) = w(e_j)$, while a label of vertex i is the sum of weights of all edges which connect this vertex with other nodes, i.e.,

$$\mathrm{Lab}(v_i) = \sum_j^m w(e_j)_{\{e_j \sim v_i\}}, \tag{2}$$

where $w(e_j)_{\{e_j \sim v_i\}}$ is the indicator function, equal to weight $w(e_j)$ of edge j if this edge is incident to vertex i, and 0 otherwise.

In contrast to a plain graph, where an edge connects only two vertices, a *hypergraph* generalizes this concept by allowing a single edge, called a hyperedge, to connect more than two vertices [31,32]. Formally, a hyperedge is defined as a collection of links that connect two or more edges with significantly correlated temporal or spectral profiles. Thus, a hypergraph is defined by a set of such hyperedges and nodes, offering a powerful framework for modeling and analyzing complex relationships, particularly in systems such as brain networks.

To illustrate this, consider a brain neural network that is conventionally partitioned into n distinct regions, referred to as nodes or vertices. Signals from these regions can be recorded using various neuroimaging modalities, such as EEG, MEG, or fMRI. When the brain is exposed to a stimulus, neuronal activity in specific regions associated with the stimulus is either activated or deactivated, and information about this process is encoded in the corresponding vertices. By capturing the relationships between these regions, a hypergraph $\mathcal{H}$ is defined by a set of n vertices $\mathcal{V}$ and a set of m hyperedges $\mathcal{E}$ as

$$\mathcal{H} = (\mathcal{V}, \mathcal{E}), \ \mathcal{V} = \{v_1, v_2, \dots, v_i, \dots, v_n\}, \ \mathcal{E} = \{e_1, e_2, \dots, e_j, \dots, e_m\}, \tag{3}$$

where each hyperedge $e_j \in \mathcal{E}$ represents a specific measure of connectivity. In this work, we consider widely used measures of brain connectivity, such as coherence, cross-correlation, and mutual information. These measures can be computed for specific spectral bands and time intervals of interest, allowing for a detailed characterization of functional relationships within the brain.

A hypergraph is characterized by its *order* and *size*. The order refers to the number of vertices $|\mathcal{V}| = n$, while the size corresponds to the number of hyperedges $|\mathcal{E}| = m$. Furthermore, both vertices and hyperedges are associated with a degree (Deg), which quantifies their connectivity within the hypergraph. Specifically, the degree of a vertex $v \in \mathcal{V}$ is defined as the number of hyperedges incident to it, whereas the degree of a hyperedge $v \in \mathcal{E}$ is defined as the number of vertices incident to it. Mathematically, these degrees can be expressed as follows:

$$\mathrm{Deg}(v) = \sum_{e \in \mathcal{E}} 1_{\{v \in e\}} \text{ and } \mathrm{Deg}(e) = \sum_{v \in \mathcal{V}} 1_{\{v \in e\}}, \tag{4}$$

where $1_{\{\cdot\}}$ is the indicator function, equal to 1 if the condition is satisfied and 0 otherwise. These degree measures provide insights into the topological structure of the hypergraph, highlighting the centrality of vertices and the complexity of hyperedges in capturing higher-order interactions.

If a vertex v is covered by two or more hyperedges, we say that these hyperedges are adjacent through v ($e_i \sim v_j$). Similarly, the vertices covered by a hyperedge e are considered adjacent through e ($v_j \sim e_i$).

A hypergraph $\mathcal{H}$ can be represented in several equivalent forms, including an incidence graph, an incidence matrix $\mathcal{H} = \{0,1\}^{|\mathcal{V}| \times |\mathcal{E}|}$, and a hyperedge weight matrix $\mathcal{W}$.

An *incidence graph* (also known as a bipartite graph representation) is a way to represent a hypergraph by transforming it into a standard graph. In this representation, the hypergraph's vertices and hyperedges are treated as two distinct sets of nodes in a bipartite graph.

The *incidence matrix* is defined as

$$\mathcal{H}[i][j] = \begin{cases} 1, & \text{if } v_i \in e_j, \\ 0, & \text{otherwise,} \end{cases} \tag{5}$$

where $v_i \in \mathcal{V}$ represents the i-th vertex and $e_j \in \mathcal{E}$ represents the j-th hyperedge. This matrix captures the membership of vertices in hyperedges, providing a binary representation of the hypergraph structure.

A *hyperedge weight matrix* $\mathcal{W}$ encodes the significance of each hyperedge in the hypergraph. Specifically, the weight of each hyperedge e_j is located at the corresponding diagonal entry of $\mathcal{W}$, such that

$$\mathrm{diag}(\mathcal{W}) = [w(e_1), w(e_2), \ldots, w(e_{|\mathbb{E}|})], \tag{6}$$

where $w(e_j)$ represents the weight of the j-th hyperedge. In this work, the weight $w(e_j)$ is defined as the degree of the hyperedge e_j, i.e., $\mathrm{Deg}(e_j)$. This weighting scheme reflects the topological importance of each hyperedge within the hypergraph.

A hypergraph can be modified by transforming vertices into edges ($\mathcal{V} \to \mathcal{E}$) and vice versa. This transformation is called "duality of hypergraphs" and is denoted as $\mathcal{H} = (\mathcal{V}, \mathcal{E}) \to \mathcal{H}^* = (\mathcal{E}, \mathcal{X})$, where $\mathcal{X} = \{x_1, x_2, \ldots \ldots, x_n\}$, with x_i representing the set of all edges $\mathcal{E}$ incident to vertex v_i. The degree of vertices in a *dual hypergraph* $\mathcal{H}^*$ is equal to the degree of hyperedges in $\mathcal{H}$, while the degree of the hyperedges remains unchanged from $\mathcal{H}$. This relationship can be expressed as follows:

$$\mathrm{Deg}(v_i) = \mathrm{Deg}(x_i), \ \mathrm{Deg}(e_i) = \mathrm{Deg}(e_i). \tag{7}$$

Several hypergraphs can be represented as *multilayer hypergraphs*, which generalize both hypergraphs and multilayer networks to model complex relationships in multi-dimensional or multi-modal data. This framework extends traditional hypergraphs by incorporating multiple layers, where each layer represents a distinct context, domain, or interaction type among entities.

A multilayer hypergraph is formally defined as follows:

$$\mathcal{H} = (\mathcal{V}, \mathcal{E}, \mathcal{L}), \quad (8)$$

where $\mathcal{L}$ is the set of layers. Each layer represents a different context, interaction type, or domain. Each layer represents a different context or interaction type, and hyperedges can be associated with one or more layers, enabling the representation of complex, heterogeneous relationships beyond simple pairwise connections.

Key properties of multilayer hypergraphs include the following: (i) cross-layer connectivity—nodes can belong to multiple layers, allowing hyperedges to span across different layers, (ii) layer-specific interactions—each layer captures a distinct type of interaction, such as coherence, correlation, or mutual information, and (iii) heterogeneous hyperedges—hyperedges can vary in size and exist across multiple layers, enabling flexible modeling of higher-order relationships. By leveraging these properties, multilayer hypergraphs provide a powerful framework for analyzing brain connectivity across different cognitive states or regions.

Another common approach to representing higher-order relationships is the *clique expansion* of a hypergraph. In this transformation, each hyperedge is replaced by a clique—a complete subgraph—connecting all vertices within that hyperedge. This method is widely used to approximate hypergraphs with graphs. The clique expansion algorithm constructs a graph $\mathcal{G}$ from the original hypergraph $\mathcal{H}$ by replacing each hyperedge with edges between every pair of its vertices [42]. Thus, the vertices in a hyperedge e form a clique in $\mathcal{G}$. In the weighted clique expansion, each edge within the clique corresponding to hyperedge e is assigned a weight $w(e)$, reflecting the discriminative model used in the transformation.

2.2. Summarization

Graphs of real datasets are often very massive. For example, the graph constructed on the basis of the MEG data has 306 nodes, each representing time series recorded by magnetic sensors [30,43]. The manipulation of such large networks requires high computational capacity and extensive time. To decrease the communicational cost, a graph summarization technique is used [44]. Summarization methods allow one to reduce the graph size by extracting the most important information from the original graph. Then, the resultant summary graph can be queried, analyzed, and understood more efficiently using existing tools and algorithms. The resultant graph can more easily visualize the dataset that is originally too large to load into memory. In addition, real graph data are frequently large scale and considerably noisy, with many hidden, unobserved, or erroneous links and labels. Such noise hinders analysis by increasing the workload of data processing and hiding the more important information. Summarization serves to filter out noise and reveal specific features in the data.

Summarization is application-dependent and can be made using various methods. The most popular summarization techniques are node grouping and edge grouping. While the node-grouping methods recursively aggregate nodes into "supernodes" based on an application-dependent optimization function, which can be based on structure and/or attributes, e.g., clustering techniques and map each densely connected cluster to a supernode, edge-grouping methods aggregate edges into hyperedges. In this paper, we apply the latter approach that compresses neighborhoods around high-degree nodes, accelerating query processing and enabling direct operations on the compressed graph. The edge-grouping method was used by Maccioni and Abadi [45], who introduced "compressor nodes", which represent common connections of high-degree nodes. They assumed that high-degree nodes are surrounded by redundant information that can be synthesized and eliminated. To provide global guarantees and reduce the scope of compressor handling during query processing, dedensification only occurs when every node has at most one outgoing edge to a compressor node, and every high-degree node has incoming edges coming only from a compressor node. These guarantees are then used to create query processing algorithms that enable direct pattern matching queries on the compressed graph.

Other summarization techniques imply simplification or sparsification. These methods streamline an input graph by removing less "important" nodes or edges, resulting in a sparsified graph. In the brain connectivity network, some edges are more indicative of predicting cognitive performance. Therefore, the node grouping layer is designed to "hide" the non-indicative ('noisy') edges by grouping them into a cluster (supernode), thus highlighting the indicative edges. Finally, there are influence-based approaches that aim to discover a high-level description of the influence propagation in large-scale graphs. Techniques in this category formulate the summarization problem as an optimization process in which some quantity related to information influence is maintained.

The output of a summarization procedure can take one of two forms: (i) a sparsified graph, which retains only a subset of nodes and/or edges from the original graph, or (ii) a hypergraph composed of selected nodes united by hyperedges. Unlike traditional graphs, hypergraphs allow nodes to belong to multiple hyperedges, enabling the representation of complex, higher-order relationships within the data. In this paper, we integrate both approaches, leveraging sparsified graphs as a foundation to construct hypergraphs that model brain connectivity associated with attention. This combined methodology enables the capture of intricate neural interactions while maintaining computational efficiency, providing a robust framework for analyzing attention-related brain networks.

2.3. Coherence

One of the key measures to quantify neuronal synchrony is event-related coherence [39,46,47]. It examines the frequency-domain relationship between two signals, reflecting the extent to which their spectral components are synchronized. Specifically, it assesses the consistency of the relative amplitude and phase between two signals within a given frequency range. Mathematically, it is a linear method that generates a symmetrical matrix, which lacks directional information. When two signals are identical, the coherence value is 1, whereas it approaches 0 as the signals become increasingly dissimilar. Since its introduction, coherence has been widely employed in brain connectivity studies involving both patients and healthy individuals. These studies span a diverse range of applications, including working memory [48], brain lesions [49], hemiparesis [50], resting-state networks [51], schizophrenia [52,53], responses to panic medications [54], and motor imagery [55]. Due to the inherent variability in human brains, distinct patterns of coherent neuronal activity have been observed across individuals. For instance, when exposed to flickering visual stimuli, subjects exhibit coherent responses in the visual cortex at the flicker frequency and its harmonics, with varying sizes of coherent neural networks [43,56].

Coherence between two signals ($X(t)$ and $Y(t)$) is defined as

$$Coh_{XY}(f) = \frac{P_{XY}(f)^2}{P_{XX}(f)P_{YY}(f)}, \tag{9}$$

where $P_{XX}(f)$ and $P_{YY}(f)$ are the auto spectral densities of $X(t)$ and $Y(t)$ signals, respectively, and $P_{XY}(f)$ is the cross-spectral density. The coherence function estimates the extent to which $Y(t)$ may be predicted from $X(t)$ by an optimum linear least squares function.

Coherence values always satisfy $Coh_{XY}(f) \in [0, 1]$. In our study, since we focus on the changes in coherence induced by attention (ΔCoh), these differences can be either positive or negative, thus satisfying $\Delta Coh \in [-1, 1]$. In the subsequent analysis, we will refer to an increase in coherence ($\Delta Coh > 0$) as *coherence* and a decrease in coherence ($\Delta Coh < 0$) as *anticoherence*. This distinction allows us to better characterize the dynamic shifts in neural connectivity associated with attentional processes.

2.4. Cross-Correlation

Unlike previous studies that rely on Pearson correlation for constructing functional connectivity networks [57,58], our approach employs cross-correlation analysis. The Pearson correlation has several limitations, including sensitivity to delays in neural responses and potential confounding effects from other brain regions. To address these issues, we used cross-correlation analysis, selecting the optimal time delay at which correlation is maximized.

Cross-correlation measures the similarity between two time series as a function of the time lag between them. This method provides a robust and sensitive approach for analyzing EEG signals recorded simultaneously from different channels, independent of their amplitudes [40]. By accounting for time lags, cross-correlation allows for the evaluation of relationships between signals not only at the same moment but also at different time points, offering deeper insights into temporal dependencies and synchronization patterns in neural activity.

The normalized cross-correlation between two EEG channels $x(t)$ and $y(t)$, at a given time lag τ, is calculated as follows:

$$Corr_{xy}(\tau) = \frac{\sum_t (x(t) - \overline{x})(y(t+\tau) - \overline{y})}{\sqrt{\sum_t (x(t) - \overline{x})^2 \sum_t (y(t) - \overline{y})^2}}, \quad (10)$$

where $\overline{x}$ and $\overline{y}$ demote the mean values of $x(t)$ and $y(t)$, respectively. For each pair of channels, we identify the time lag τ at which $Corr_{xy}$ attains its maximum value and use this peak correlation value in our subsequent analysis. This approach ensures that we capture the strongest temporal relationship between the signals, providing a robust basis for reconstructing a functional connectivity network.

Cross-correlation values range from -1 and 1. To analyze changes in cross-correlation induced by attention, we define the difference in cross-correlation ($\Delta Corr$), which can take values between -2 and 2. For clarity, we refer to cases where $\Delta Corr > 0$ as *correlation* and cases where $\Delta Corr < 0$ as *anticorrelation*. This distinction helps characterize neural connections that emerge (increased synchrony) and disappear (decreased synchrony) due to attention.

2.5. Mutual Information

Since its introduction by Shannon [59], mutual information has been used across various fields to quantify coupling or information transmission between systems [60]. In neuroscience, several studies have employed mutual information analysis to investigate information transfer in the brain. For example, Jeong et al. [41] applied this method to multi-channel EEG data to assess information flow between cortical regions in Alzheimer's disease (AD) patients.

Cross mutual information (CMI) measures the information gained about one system from observing another, in contrast to auto mutual information, which quantifies mutual information between two parts of he same time series $x(t)$ separated by a lag τ. Unlike traditional correlation functions, which capture only linear dependencies, CMI detects both linear and nonlinear statistical relationships between time series. CMI between measurement x_i from system X and y_j from system Y represents the amount of information x_i provides about y_j. Thus, CMI serves as a measure of dynamical coupling or information transmission between X and Y. When applied to EEG data, it can be interpreted as an indicator of functional connectivity between brain regions. If two systems are entirely independent, their CMI is zero, meaning no information is transmitted between them. Therefore, in our case, CMI quantifies the information flow between different brain areas.

CMI is calculated as

$$Inf(X,Y) = \sum_{x\in X}\sum_{y\in Y} p(x,y)\log_2\left(\frac{p(x,y)}{p(x)p(y)}\right), \tag{11}$$

where $p(x)$ and $p(y)$ are the normalized histograms of the distributions of measurements x and y, respectively, and $p(x,y)$ denotes their joint probability density.

In this study, we compute the time-delayed CMI, $Inf(X(t), Y(t+\tau))$, which quantifies mutual information between EEG signals from each pair of channels as a function of a time delay. To assess information transmission between different cortical areas, we select the peak CMI value within a time delay range of 0 ± 500 ms for each electrode pair.

CMI measures the degree of dependence between X and Y, ranging from 0 to entropy H of X. A value of $Inf(X,Y) = 0$ indicates mutual independence, whereas $Inf(X,Y) = H(X)$ signifies that one signal completely determines the other. Similar to coherence and correlation measures, changes in CMI due to attention (ΔI) can be either positive or negative. Specifically, positive values ($\Delta I > 0$) indicate *information gain*, while negative values ($\Delta I < 0$) correspond to *information loss*.

3. Materials and Experimental Methods

3.1. Participant

In this case study, we analyze the EEG data of a healthy 22-year-old female subject, recorded during experiments using a flickering image paradigm. This study was conducted at the Center for Biomedical Technology, Universidad Politécnica de Madrid, Spain. This subject was selected from a pool of 28 subject participants as the one demonstrating the highest level of attention (subject #2 from [38]). Before the experiment began, the subject provided written informed consent, ensuring anonymized data processing and compliance with data protection regulations. The subject was also briefed on the experiment's objectives and duration. The EEG study followed the ethical guidelines of the Declaration of Helsinki and was approved by the Ethics Committee of the Universidad Politécnica de Madrid (Ethics Approval Code: 2020-096) prior to participant recruitment.

3.2. Stimulus

The subject was presented with a visual stimulus—the Necker cube—displayed as a white line drawing on a black screen. The image, approximately 10×10 cm in size, was shown on a 22-inch liquid crystal monitor with a refresh rate of 60 Hz and luminance of 100 cd/m^2. The Necker cube was chosen for its ambiguous nature, allowing for perceptual alternation between left and right orientations.

The cube's pixel brightness alternated between black (0 in an 8-bit format) and gray (200) following a square-wave modulation. This modulation was applied at frequencies $f_L = 6.67$ Hz and $f_R = 8.57$ Hz to the left and right faces of the cube, respectively, consistent with previous studies [43,56]. These modulation frequencies generated frequency tags in the fast Fourier transform (FFT) spectra of the EEG signal, particularly in spectral regions around $6.15 < f_1 < 6.75$ and $8.00 < f_2 < 8.60$ within the visual cortex. To present the visual stimuli, we developed custom software using the PsychoPy 2024.2.5 and Python 3.12.4 platforms.

3.3. Experimental Setup

The EEG recording setup consisted of a 16-channel LIVEAMP amplifier, a standard wireless 16-channel cap, a reference channel, and a ground channel. The cap was fitted with slim active electrodes from EASYCAP GmbH (Wörthsee, Germany). To enhance conductivity, a conductive gel was applied to each electrode.

The experiments took place in a dimly lit room, with the subject seated comfortably in front of a computer monitor positioned 80 cm away. The visual stimulus subtended an approximate viewing angle of 8 degrees.

The EEG signal was recorded using BrainVision Recorder (version 1.27) from Brain Products GmbH, a highly reliable software for EEG signal acquisition. One of its key features is real-time impedance monitoring, which ensures optimal electrode-to-scalp contact by detecting potential faulty connections and maintaining impedance below the predefined threshold of <10 kΩ. EEG data were sampled at a rate of 500 Hz.

The data acquisition details can be found in Appendix A.

3.4. Experimental Paradigm

The experiment comprised a series of Necker cube presentations, each lasting approximately 30 s, with 20 s intervals between presentations. During these intervals, the subject was allowed to blink. EEG recordings were continuously conducted throughout cube observation to monitor brain activity in real time.

Before the main experiment, the participant underwent a training session using a non-flickering Necker cube to familiarize herself with distinguishing between left- and right-oriented perceptions. The subject was instructed to complete three tasks: (1) observing the first cube image without interpreting its orientation, (2) perceiving the second cube image as left-oriented, and (3) perceiving the third cube image as right-oriented.

In the main experiment, the subject performed similar tasks but with a flickering (modulated) Necker cube. The first task involved passive observation of the cube without interpretation of its orientation (MI: modulated involuntary). In the second task (MLV: modulated left voluntary) and third task (MRV: modulated right voluntary), the subject was instructed to voluntarily perceive the cube as left- or right-oriented, respectively.

Figure 2 illustrates how the subject perceives the left and right cube orientations. The way the cubes are perceived depends on the task (MLV or MRV). Figurative attention to a specific cube orientation leads to the activation and deactivation of different neural pathways. The passive viewing condition (MI) is not included because the brain's perception is unstable, alternating between left- and right-cube orientations.

To minimize artifacts caused by eye movements and blinking, the subject was instructed to maintain focus on a red dot at the center of the cube and to avoid blinking during stimulus presentation. EEG preprocessing further reduced artifacts from involuntary eye movements and blinking.

Each task was repeated twice per participant, with a 2 min break between trials to allow for rest. Before each experiment, electrode conductivity was assessed to ensure optimal signal quality, maintaining electrode impedance below 10 kΩ.

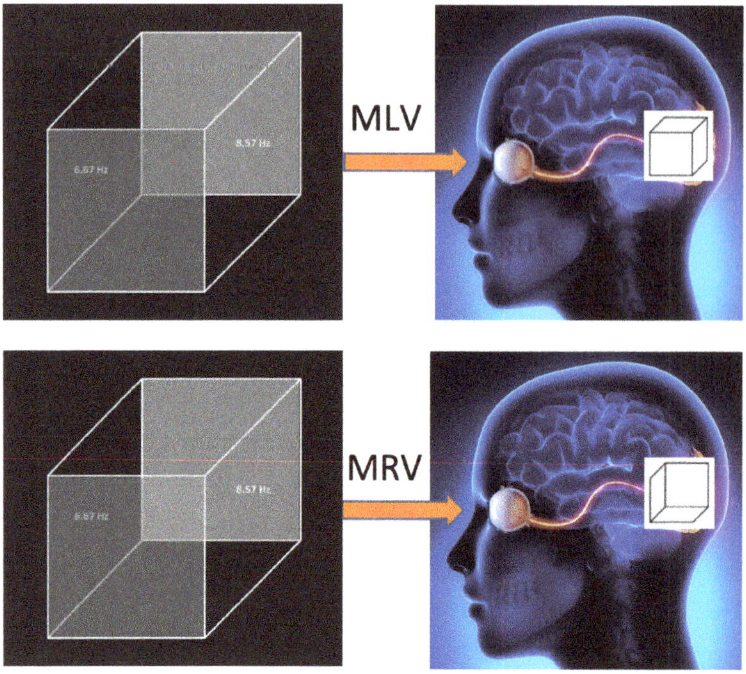

Figure 2. Visual stimulus. In the MLV and MRV tasks, the Necker cube, with its left and right faces modulated at frequencies of 6.67 Hz and 8.57 Hz, is voluntarily perceived as left- or right-oriented, respectively. The right side illustrates the mental images generated by the brain when the subject views the same stimulus.

4. Experimental Data Analysis

The preliminary analysis of the recorded EEG signals was performed using the following: (i) Artifact removal to clean the raw data; (ii) Frequency spectrum analysis to identify relevant frequency regions; (iii) Wavelet analysis to determine time intervals associated with voluntary attention.

Data analysis was conducted using MATLAB (R2024b), Brainstorm [61], and BrainVision Analyzer 2.2.2 from Brain Products GmbH (Gilching, Germany). The EEG dataset utilized in this study is publicly available [62].

4.1. Artifact Removal

The analysis of the raw EEG data started with artifact removal, one of the most critical aspects of data processing. To ensure optimal processing of EEG signals, we identified and eliminated endogenous artifacts (arising from within the subject) and exogenous artifacts (external to the subject), such as interference from nearby electronic equipment, flashing lights, environmental noise, facial muscle movements, and eye blinks.

First, we executed the following steps in data preprocessing: (1) After examining the raw signals, we identified time intervals that delineated the beginning and end of the response to applied visual stimulation. (2) We minimized a continuous global current trend using the method described in [63]. This procedure improved the signal quality and facilitated the detection and correction of ocular artifacts during post-processing. (3) We removed endogenous and exogenous noise and focused on the frequencies of our stimuli. For these aims, we applied infinite impulse response (IIR) filters, specifically, 4th-order zero-phase shift Butterworth filters. Two types of filters were used: (i) band rejection (Notch

Filter) to attenuate or eliminate the electrical network frequency (50 Hz in Spain) and the refresh rate of the monitor used for visual stimulation (60 Hz), with a bandwidth of 0.1 Hz, and (ii) 4th-order bandpass filter (BIF) of 4–20 Hz to concentrate on the frequencies of interest. These filters allowed us to filter out or attenuate unwanted frequency components present in the EEG signals.

Subsequently, we removed ocular artifacts, which can interfere with the detection and analysis of cortical responses as electrical activity resulting from eye movements; blinking can contaminate brain signals. To ensure that these artifacts did not distort the results or complicate the analysis, we carefully examined the brain signals associated with visual stimuli. Since the eyes behave like a dipole, with the cornea serving as the positive part and the retina as the negative part of the dipole, the identification of ocular artifacts is relatively straightforward. Vertical electro-oculogram artifacts (VEOG) and horizontal electro-oculogram artifacts (HEOG) exhibited electrical potentials ranging from 0.4 to 1.0 mV.

To mitigate the impact of ocular artifacts arising from vertical eye movements (VEOG) and lateral eye movements (HEOG) identified by inspection of raw time series, we used the Ocular Correction ICA method in BrainVision Analyzer 2.2.2, using the following algorithms:

(i) Reference to frontal channels: Since our EEG equipment did not include electro-oculography electrodes, we used frontal channels (F7 and F8) as references for the ocular correction ICA. Cumulative quadratic correlation allows the precise removal of relevant components without losing neuronal information. The FP1 channel served as a common VEOG reference to detect vertical artifacts.

(ii) Infomax Extended ICA algorithm: This algorithm improves the efficiency of artifact correction compared to standard ICA while preserving the neuronal signal of interest. Additionally, a Quality Control process is incorporated to ensure the validity of the obtained results. This approach enables the effective identification and reduction of ocular artifacts in EEG signals, thus facilitating the subsequent analysis of cortical responses associated with the visual stimulus by eliminating or mitigating interference from eye movements and blinks.

(iii) Value trigger algorithm: This algorithm facilitated the detection of characteristic patterns, such as blinks. Blinks were identified on the basis of their absolute magnitude, with definitive blink movements determined using the correlation method. The blink detection threshold (blink value trigger) was set at 97%, which means that any value above this threshold was recognized as a blink. The selection of the 97% activation threshold for blink detection was based on experimental analysis using data from 28 participants. During threshold optimization, lower values significantly increased the rate of false positives, compromising the integrity of the neuronal signal. The 97% threshold was identified as the optimal balance between sensitivity and specificity, allowing robust identification of ocular artifacts without affecting the underlying signal. The signal correlation was established at 70% relative to a predefined blink template. This algorithm enhances detection accuracy by ensuring that triggers are only activated for signals with a high morphological similarity to a characteristic blink pattern, thereby reducing the erroneous detection of other artifacts. Additionally, a visual analysis using back-projection (Inverse ICA) was implemented as a validation method. This technique allowed the eliminated components to be projected back into the original domain, facilitating a visual inspection of the processed data. This procedure confirmed that ocular artifact suppression was performed selectively, preserving relevant neuronal activity while minimizing EEG signal distortion.

These values were experimentally determined by considering their influence on attenuating the signal of interest in the spectral regions of f_1 and f_2 in the occipital lobes (O1, Oz, O2 channels).

Lastly, we conducted ICA to find the optimal unmixing matrix to separate independent sources using statistical methods. To assess component independence, methods such as Maximum Negentropy, Kurtosis, Maximum Likelihood, and Mutual Information were employed. The optimization process for the optimal unmixing matrix ended when the maximum time independence of the components was achieved within a maximum number of steps or iterations. For this process, the extended nondeterministic INFOMAX algorithm with a maximum of 512 steps was selected as the optimal configuration. This ICA training method was adjusted after each convergence step to enhance signal separation. Convergence was deemed achieved when the algorithm reached a stable solution, favoring Independent Components (ICs) with Kurtosis (K) values greater than 0 (indicating components with a non-Gaussian distribution) and less than 0 for noise, such as electrical network noise.

Following decomposition into ICs, quality control was performed by examining the number of steps used in decomposition of the ICs and comparing them with components initially flagged as possible ocular artifacts during visual inspection of the raw time series. As a final step, an inverse ICA of ICs and topographic retroprojection were performed to determine the brain lobes to which these possible signals corresponded. The analysis involved a comprehensive evaluation of the properties of IC, including their waveforms over time, energy, and kurtosis ($K < 0$, $K = 0$, $K > 0$).

It should be noted that ICA was performed on full-rank data without any preprocessing steps that could affect its rank [64].

4.2. Spectral Analysis

The FFT method enables the examination of a signal's frequency spectrum, which is essential for identifying frequency tags associated with flickering visual stimuli. In this study, we focus on the tag frequencies corresponding to the two perceived orientations of the Necker cube: f_L for the left-oriented cube and f_R for the right-oriented cube.

To ensure optimal analysis resolution, we implemented an FFT algorithm designed to operate on power-of-2 segmentations. The EEG signals were segmented into 30 s trials based on the duration of each task (MI, MLV, or MRV). Following preprocessing and artifact removal, the FFT was computed for 20 s frames with a 1 s overlap to reduce noise. This configuration maximized spectral resolution, allowing for efficient and precise signal decomposition in the frequency domain. As a result, we could efficiently identify the spectral characteristics associated with each cube orientation when analyzing EEG signals.

The FFT spectra of the MI, MLV, and MRV tests are shown in Figure 3. Distinct spectral peaks can be observed near the modulation frequencies within the ranges $6.15 < f_1 < 6.75$ and $8.00 < f_2 < 8.60$. The brain does not respond on discrete frequencies due to two factors: (i) the modulation signal is not harmonic, and (ii) slight variation in the modulation signal occurs throughout the experiment, influenced by computer load and memory fluctuations.

By comparing the spectrum obtained for involuntary attention (MI) shown in Figure 3a with the spectra for voluntary attention to the left-oriented cube (MLV), shown in Figure 3b, and to the right-oriented cube (MRV), shown in Figure 3c, it can be observed that the spectral amplitude at f_1 dominates the amplitude at f_2 when the subject perceives the cube as left-oriented (Figure 3b), and the spectral amplitude at f_2 dominates the amplitude at f_1 when the subject perceives the cube as right-oriented.

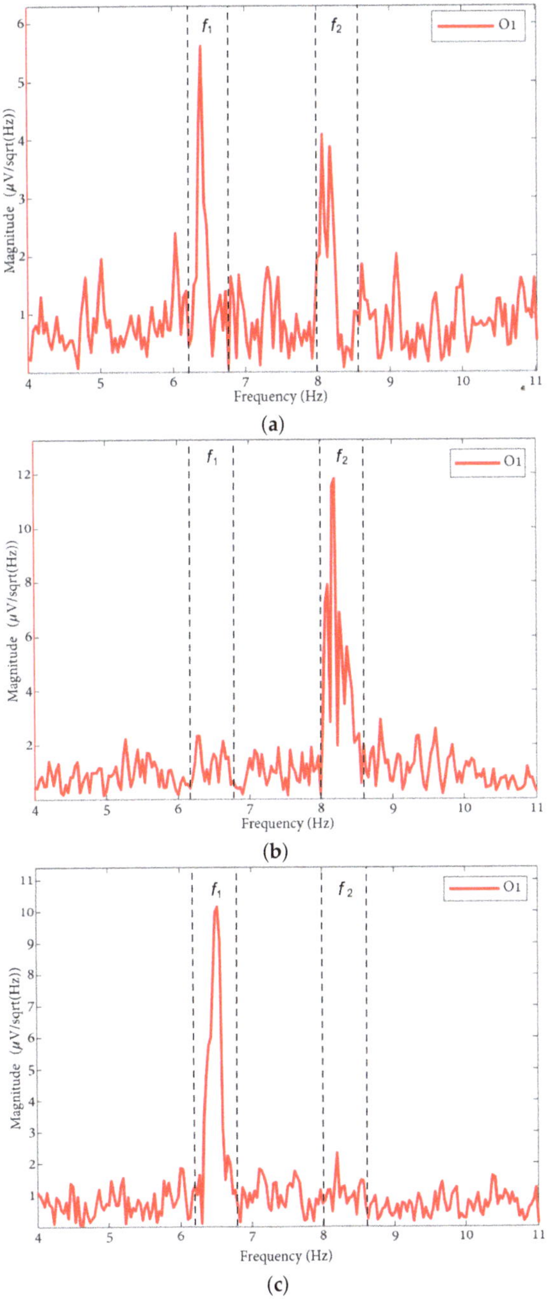

Figure 3. FFT spectra of preprocessed EEG from the O1 channel during (**a**) MI, (**b**) MRV, and (**c**) MLV tasks. The vertical dashed lines bound frequency regions of interest (f_1 and f_2), representing the brain response to stimulus modulation.

4.3. Wavelet Analysis

The wavelet analysis provides a detailed view of how spectral energy fluctuates over time, offering insights into the dynamic nature of decision-making processes as captured through neurophysiological data. The wavelet power spectrum serves as a quantifiable metric, enabling the precise measurement of these temporal changes [43,65–70].

While the spectra depicted in Figure 3 were computed over the entire 30 s EEG time series, it is important to note that sustained attention is not maintained uniformly throughout this period. To identify specific time windows during which attention was focused, we employed wavelet analysis. Using the Morlet wavelet, we were able to pinpoint intervals where attention was predominantly directed toward one interpretation of the cube over another.

The temporal dynamics of the cube's orientation interpretation were explored through continuous wavelet transform (CWT) analysis, implemented using the Brainstorm software (version 3.4.0.0). This method allowed us to investigate how attention shifted over time, revealing the moments when the subject focused attention on specific interpretations of the cube. By leveraging wavelet analysis, we gained a more granular understanding of the temporal patterns underlying attentional processes during the task.

Figures 4a,b depict the variations in wavelet energy corresponding to figurative attention directed towards the left (MLV) and right (MRV) cube orientations, respectively. The wavelet energy was calculated using the following equation:

$$W(f,t) = \sqrt{f}\int_{t-4/f}^{t+4/f} X(t)\psi^*(f,t)dt, \tag{12}$$

where the asterisk (*) denotes the complex conjugate, $X(t)$ represents the analyzed EEG signal, and $\psi(f,t)$ is the complex-valued Morlet wavelet, chosen as the mother wavelet. The Morlet wavelet is defined as follows:

$$\psi(f,t) = \sqrt{f}\pi^{1/4}e^{i\omega_0 f(t-t_0)}e^{f(t-t_0)^2/2} \tag{13}$$

with $\omega_0 = 2\pi f_0$ being the central frequency of the Morlet wavelets, $i = \sqrt{-1}$ the imaginary unit, and t_0 representing the time shift parameter. This formulation allows for a precise analysis of the temporal and spectral characteristics of the EEG signal, capturing the dynamic shifts in attention during the task.

It is evident that the prominence of specific components corresponds to the subject's directed attention towards either the left or right orientation of the cube. The spectral component at f_1 is dominant when the subject perceives the cube as left-oriented (Figure 4a), while f_2 prevails when the cube is interpreted as right-oriented (Figure 4b). As shown in Figure 4, spectral energy at f_1 dominates during the time interval $t_L \in [6.88, 28.12]$ seconds in the MLV test, whereas spectral energy at f_2 dominates during the interval $t_R \in [3.94, 30]$ seconds in the MRV test. In the subsequent analysis, connectivity is calculated within these identified time intervals.

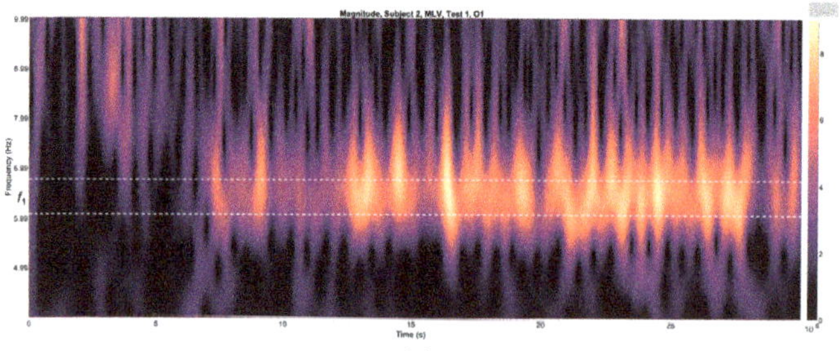

(a)

Figure 4. *Cont.*

right-cube attention ($\Delta Corr_R > 0$). Bold lines highlight the links with the largest correlation differences (Max $\Delta Corr$), which are displayed next to the corresponding links. Orange dots denote the nodes with the highest sums of their link weights, indicated next to these nodes.

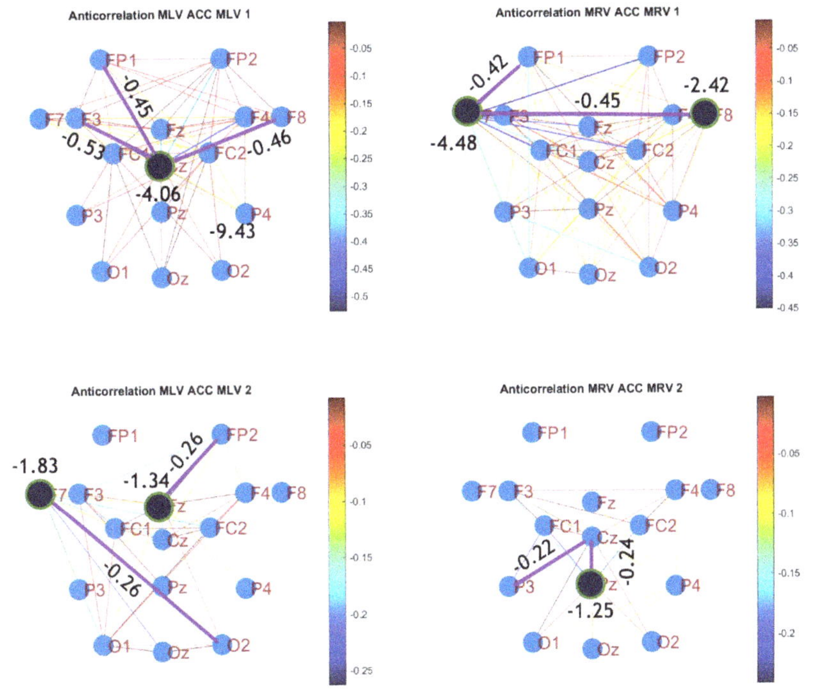

Figure 8. Disconnectivity graphs based on anticorrelation in the (**upper raw**) f_1 and (**lower raw**) f_2 frequency regions, associated with (**left column**) left-cube attention ($\Delta Corr_L < 0$) and (**right column**) right-cube attention ($\Delta Corr_R < 0$). Bold lines highlight the links with the largest negative cross-correlation differences (Min $\Delta Corr$). Violet dots denote the nodes with the lowest sums of their link weights, indicated next to these nodes.

By comparing Figures 5 and 6 with Figures 7 and 8, we observe a strong similarity between the connectivity patterns derived from cross-correlation and coherence measures. Specifically, the links showing the maximum increases and decreases in cross-correlation closely match those for coherence. This agreement between connectivity measures, based on both temporal and spectral analyses, provides strong validation for the robustness of our findings.

Table 2 summarizes the results of the brain connectivity analysis based on cross-correlation. The table presents the two largest positive and negative correlation differences (link weights) along with the corresponding brain regions.

Table 2. Maxima and minima cross-correlation differences.

Measure	Value	Lobes	Value	Lobes
Max $\Delta Corr_L(f_1)$	0.40	F7–Pz	0.37	F7–P3
Max $\Delta Corr_L(f_2)$	0.49	F8–Pz	0.46	F8–P4
Max $\Delta Corr_R(f_1)$	0.28	P3–Cz	0.28	Pz–Cz
Max $\Delta Corr_R(f_2)$	0.51	Oz–FP1	0.51	O2–FP1
Min $\Delta Corr_L(f_1)$	−0.53	Cz–F3	−0.45	Cz–FP1
Min $\Delta Corr_L(f_2)$	−0.26	Fz–FP2	−0.26	O2–F7
Min $\Delta Corr_R(f_1)$	−0.45	F7–F8	−0.42	F7–FP1
Min $\Delta Corr_R(f_2)$	−0.24	Pz–Cz	−0.22	P3–Cz

5.3. Connectivity Graphs Based on Mutual Information

The third connectivity measure explored in this study, mutual information, was calculated using Equation (11). The differences in mutual information induced by attention were computed using Equation (14). The resulting graphs depicting the attention-induced changes in mutual information are shown in Figures 9 and 10. Additionally, the corresponding mutual information matrices are provided in Appendix B, specifically in Figures A5 and A6, offering a detailed numerical representation of these connectivity changes.

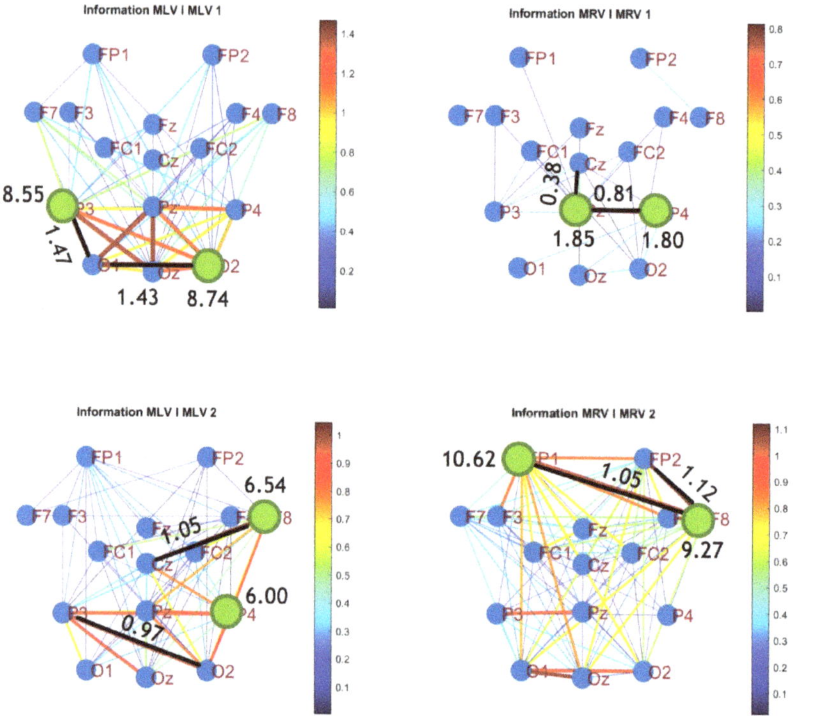

Figure 9. Connectivity graphs based on mutual information in the (**upper raw**) f_1 and (**lower raw**) f_2 frequency regions, associated with (**left column**) left-cube attention ($\Delta Inf_L > 0$) and (**right column**) right-cube attention ($\Delta Inf_R > 0$). Bold lines highlight the links with the largest mutual information differences (Max ΔInf), which are displayed next to the corresponding links. Orange dots denote the nodes with the highest sums of their link weights, indicated next to these nodes.

Table 3 summarizes the brain connectivity analysis results based on mutual information. It highlights the two largest positive and negative mutual information differences (link weights) along with their corresponding brain regions, providing insight into the most significant connectivity changes induced by attention shifts.

A strong correspondence can be observed among the connectivity graphs based on coherence, cross-correlation, and mutual information. Although some variations exist in the links with the highest weights, the nodes with the maximum and minimum labels largely overlap across all connectivity measures. This consistency further validates the robustness of our approach in capturing attention-related connectivity changes in the brain.

The shared features across different connectivity measures can be explored using hypergraphs. In the following section, we describe the process of constructing attention hypergraphs.

Table 3. Maxima and minima mutual information differences.

Measure	Value	Lobes	Value	Lobes
Max $\Delta Inf_L(f_1)$	1.41	O1–P3	1.43	O1– O2
Max $\Delta Inf_L(f_2)$	1.05	Cz–F3	0.97	O2–P3
Max $\Delta Inf_R(f_1)$	0.81	P4–Pz	0.38	Pz–Cz
Max $\Delta Inf_R(f_2)$	1.12	F8–FP2	1.05	F8–FP1
Min $\Delta Inf_L(f_1)$	−1.26	Cz–Fz	−1.18	Cz–F4
Min $\Delta Inf_L(f_2)$	−1.06	Cz–Fz	−0.81	Cz–FC2
Min $\Delta Inf_R(f_1)$	−1.07	F7–F4	−0.92	F7–FP1
Min $\Delta Inf_R(f_2)$	−0.57	Pz–Cz	−0.22	Pz–FC1

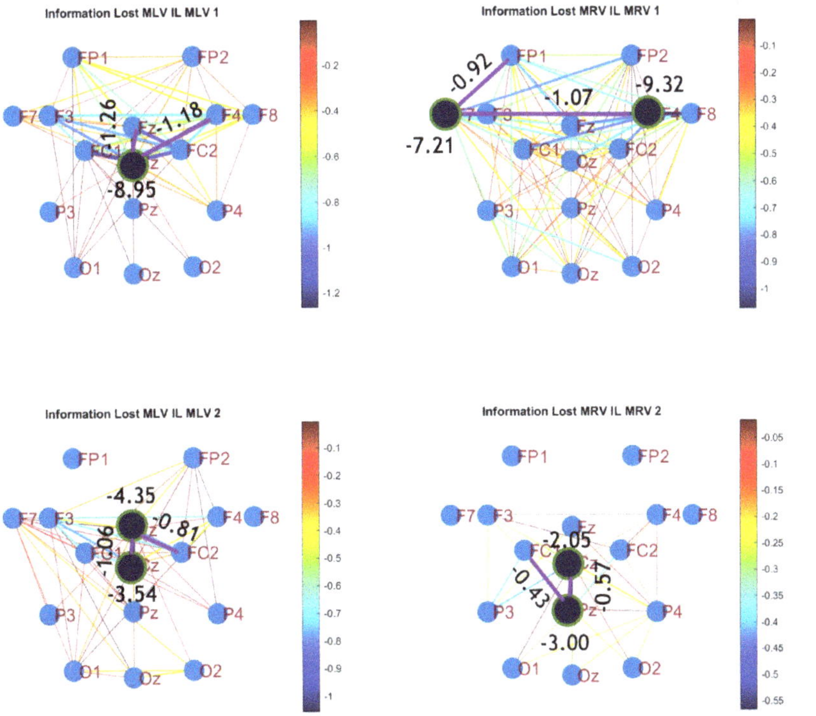

Figure 10. Disconnectivity graphs based on mutual information loss in the (**upper raw**) f_1 and (**lower raw**) f_2 frequency regions, associated with (**left column**) left-cube attention ($\Delta Inf_L < 0$) and (**right column**) right-cube attention ($\Delta Inf_R < 0$). Bold lines highlight the links with the largest negative mutual information differences (Min ΔInf). Violet dots denote the nodes with the lowest sums of their link weights, indicated next to these nodes.

6. Hypergraph Construction

Hypergraph analysis provides a simplified and more generalized representation of the brain network. Instead of relying on the twelve individual graphs presented in the previous section, we can construct four hypergraphs using the summarization technique described in Section 2.2.

To construct the hypergraphs, we follow these steps:

(1) Compute the label of every node i for each connectivity measure ΔC using Equation (2) (Lab $C_i = \sum_i^n \Delta C$).

(2) From the probability distribution of all n labels for each connectivity measure found the mean $M = \frac{1}{n}\sum_i^n \text{Lab}\, C_i$ and standard deviation σ.

(3) Calculate threshold value $C_{th} = M \pm \sigma$ for each connectivity measure ("+" for $M > 0$, "−" for $M < 0$).
(4) Using the threshold value, determine whether a change in connectivity is significant or not. The change is significant if the label $|\text{Lab}\, C_i| > |C_{th}|$.
(5) Using the summarization technique described in Section 2.2 remove insignificant nodes whose labels are smaller than the threshold value. Weak changes in connectivity are considered as brain noise and are not related to attention.
(6) Construct incident matrices using Equation (5).

The probability distributions for each connectivity measure are presented in Figures A7–A9 in Appendix C. One can also find there the values of means (M), standard deviations (σ), and thresholds (C_{th}), which are summarized in Table A1.

6.1. Simple Hypergraphs

Using the algorithm described above, we constructed four hypergraphs, denoted as $\mathcal{H}_{1-4}$. Two of these hypergraphs ($\mathcal{H}_{1,2}$) correspond to attention directed toward the left-cube orientation, while the other two ($\mathcal{H}_{3,4}$) correspond to attention directed toward the right-cube orientation. Hypergraphs $\mathcal{H}_1$ and $\mathcal{H}_3$ are constructed for frequency region f_1, while $\mathcal{H}_2$ and $\mathcal{H}_4$ are constructed for f_2.

Each hypergraph contains six hyperedges ($m = 6$): e_1 and e_2 represent coherence and anticoherence, e_3 and e_4 represent correlation and anticorrelation, and e_5 and e_6 represent mutual information and information loss, respectively.

The constructed hypergraphs $\mathcal{H}_{1-4}$ are visualized in Figures 11–14 in three distinct forms: as a bipartite graph (left panels), a Venn diagram (middle panels), and a brain map (right panels). The correspondence between node numbers and EEG channels is provided in Table 4.

Table 4. Correspondence of node numbers to EEG channels.

1	2	3	4	5	6	7	8	9	10	11	12	13	14	15	16
FP1	FP2	F7	Fz	F8	F3	FC1	Cz	FC2	F4	P3	Pz	P4	O1	Oz	O2

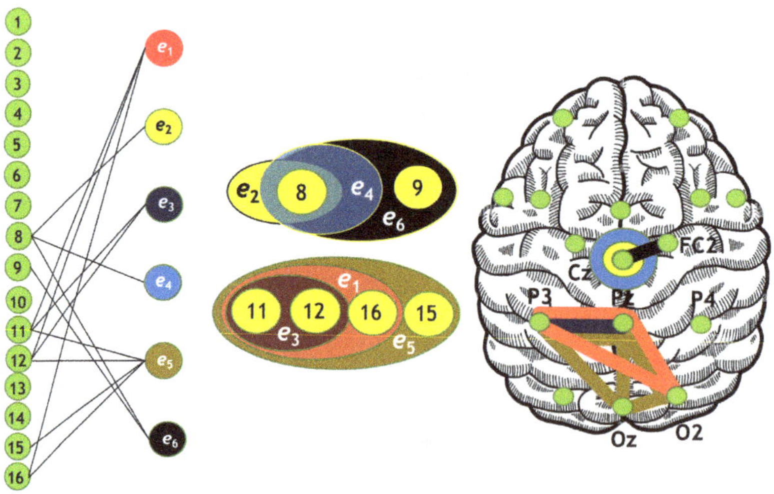

Figure 11. Hypergraph $\mathcal{H}_1$ associated with left-cube attention for f_1, presented as a bipartite graph (**left**), a Venn diagram (**middle**), and a brain map (**right**). The upper Venn diagram (e_2, e_4, e_6) indicates deactivation of central area, while the lower one (e_1, e_3, e_5) depicts activation of occipital-parietal area.

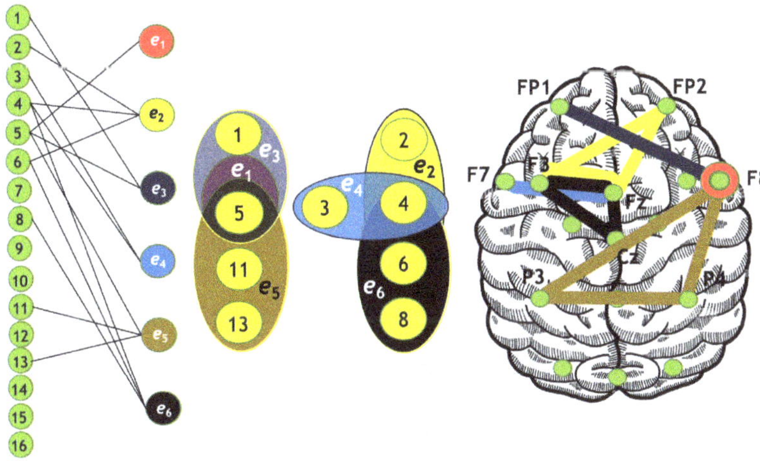

Figure 12. Hypergraph $\mathcal{H}_2$ associated with left-cube attention for f_2, presented as a bipartite graph (**left**), a Venn diagram (**middle**), and a brain map (**right**). The right Venn diagram (e_2, e_4, e_6) indicates deactivation of left frontal and right prefrontal lobes, while the left one (e_1, e_3, e_5) depicts activation of parietal, right anterior, and left prefrontal lobes.

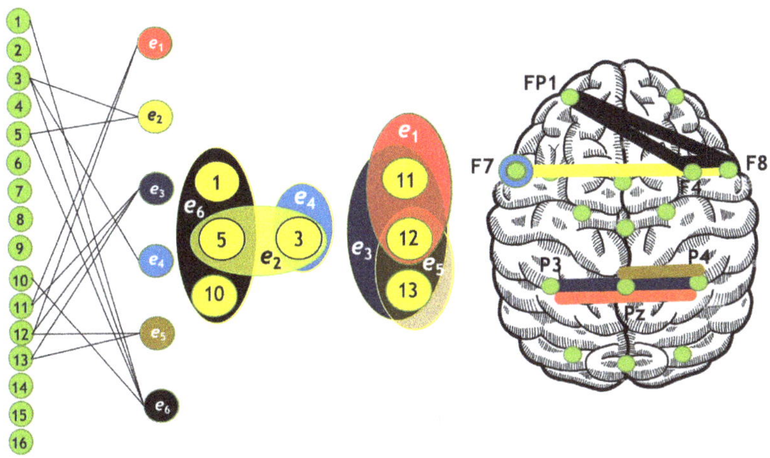

Figure 13. Hypergraph $\mathcal{H}_3$ associated with right-cube attention for f_1, presented as a bipartite graph (**left**), a Venn diagram (**middle**), and a brain map (**right**). The left Venn diagram (e_2, e_4, e_6) indicates deactivation of frontal cortex, while the right one (e_1, e_3, e_5) depicts activation of parietal area.

A hypergraph size is the number of hyperedges (in our case $m = 6$), and the order of a hypergraph is the number of vertices. The degrees of hyperedges and the orders of the hypergraphs are summarized in Table 5.

Table 5. Degrees of hyperedges and orders of hypergraphs.

Measure	Hyperedge	$\mathcal{H}_1$ (Left f_1)	$\mathcal{H}_2$ (Left f_2)	$\mathcal{H}_3$ (Right f_1)	$\mathcal{H}_4$ (Right f_2)
Coherence	e_1	3	1	3	4
Anticoherence	e_2	1	3	2	4
Correlation	e_3	2	2	3	4
Anticorrelation	e_4	1	2	1	3
Mut. Inform.	e_5	4	3	1	3
Inform. Loss	e_6	2	3	2	2
Order		6	9	7	8

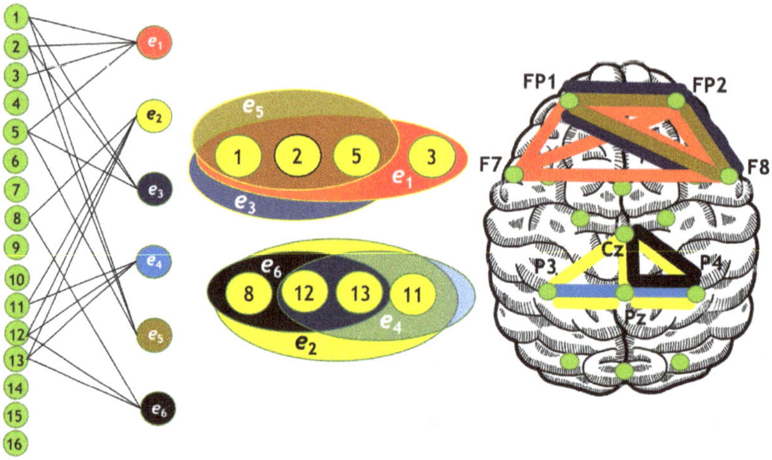

Figure 14. Hypergraph $\mathcal{H}_4$ associated with right-cube attention for f_2, presented as a bipartite graph (**left**), a Venn diagram (**middle**), and a brain map (**right**). The lower Venn diagram (e_2, e_4, e_6) indicates deactivation of central-parietal area, while the upper one (e_1, e_3, e_5) depicts activation of frontal cortex.

6.2. Multilayer Hypergraphs

The hypergraphs of figurative attention can be represented as multilayer hypergraphs, where each layer corresponds to a hypergraph capturing increased or decreased connectivity at a specific frequency. This approach allowed us to consolidate four separate hypergraphs into two four-layer hypergraphs, one for left-cube attention and another one for right-cube attention. The resulting four-layer hypergraphs are presented in Figures 15 and 16.

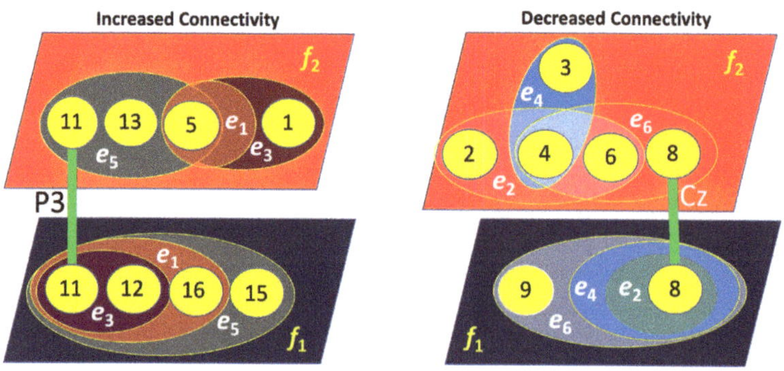

Figure 15. Four-layer hypergraphs representing brain connectivity (**left layers**) and disconnectivity (**right layers**) associated with left-cube attention. The lines indicate inter-layer connections, with the corresponding channels responsible for these links labeled near the lines.

As shown in Figure 15, for the left-cube attention, the layers exhibiting increased and decreased connectivity are not interconnected, meaning they do not share common nodes. Notably, the layers associated with increased connectivity at f_1 and f_2 are primarily linked through the left parietal lobe (P3). Conversely, the layers related to decreased connectivity at f_1 and f_2 are mainly connected via the central lobe (Cz). These nodes demonstrate the most significant increase and decrease in connectivity, respectively.

In contrast, Figure 16 reveals that for the right-cube attention, the layers of increased and decreased connectivity are strongly interconnected. Specifically, the layer with increased connectivity at f_1 is linked through three nodes (FP1, F7, and F8) to the layer of decreased connectivity at f_2. Simultaneously, the layer of increased connectivity at f_2 is connected via three nodes (Pz, P3, and P4) to the layer of decreased connectivity at f_1. This indicates a reciprocal relationship, where an increase in connectivity at f_2 is accompanied by a decrease at f_1, and vice versa.

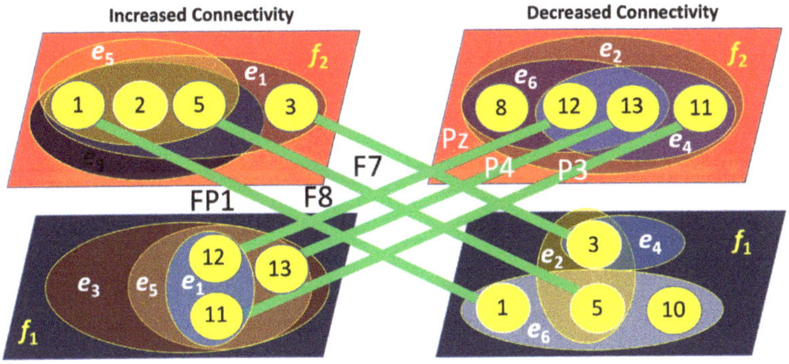

Figure 16. Four-layer hypergraphs representing brain connectivity (**left layers**) and disconnectivity (**right layers**) associated with right-cube attention. The lines indicate inter-layer connections, with the corresponding channels responsible for these links labeled near the lines.

6.3. Dual Clique Expansions

The high-order interactions shown as the hypergraphs in Figures 11–14 can also be presented in the form of dual clique expansions shown in Figures 17 and 18. As was defined in Section 2.1, a clique expansion is a transformation where each hyperedge is replaced by a clique connecting all vertices within that hyperedge. This is commonly used to approximate hypergraphs using graphs. To construct dual clique expansions of $\mathcal{H}_{1-4}$, we first created dual hypergraphs $\mathcal{H}^*_{1-4}$ and then performed clique expansions on $\mathcal{H}^*_{1-4}$, meaning that each hyperedge (which corresponds to a vertex in $\mathcal{H}^*$) is expanded into a clique. As a result, we created new hypergraphs $\mathcal{G}_{1-4}$ where hyperedges correspond to cliques in $\mathcal{H}^*_{1-4}$.

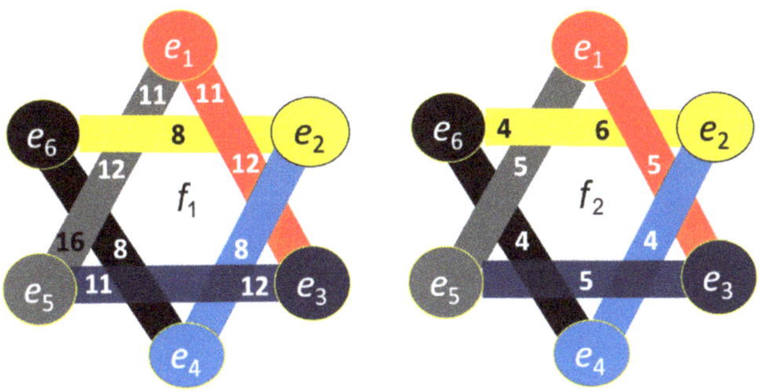

Figure 17. Dual clique expansions $\mathcal{G}_{1,2}$ of hypergraphs $\mathcal{H}_{1,2}$ associated with left-cube attention at (**left**) f_1 and (**right**) f_2.

The analysis of the dual clique expansions $\mathcal{G}_{1-4}$, presented in Figures 17 and 18, reveals distinct connectivity patterns among brain regions. Specifically, all connectivity measures—coherence, correlation, and mutual information—are mutually connected through the nodes $e_{1,3,5}$. Conversely, all disconnectivity measures—anticoherence, anticorrelation, and information loss—also form a mutually connected structure. In other words, these results highlight the brain regions (edges in these graphs) that predominantly contribute to both increasing and decreasing connectivity associated with figurative attention.

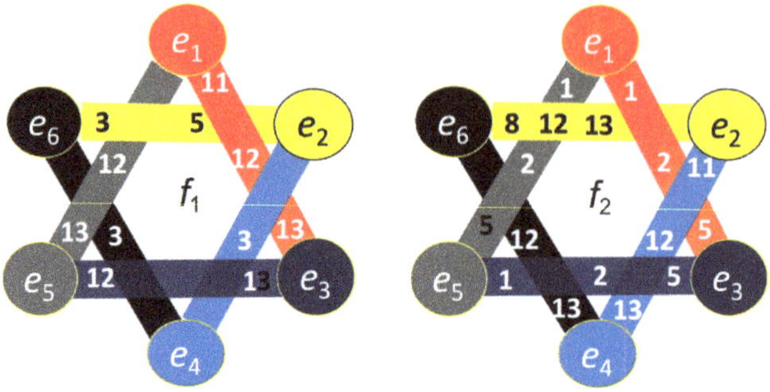

Figure 18. Dual clique expansions $\mathcal{G}_{3,4}$ of hypergraphs $\mathcal{H}_{3,4}$ associated with right-cube attention at (**left**) f_1 and (**right**) f_2.

For both left- and right-cube attention, the primary contributors to increased connectivity are the parietal region (P3, Pz, P4) at f_1 and the refrontal and frontal regions (FP1, FP5, F8) at f_2. In contrast, decreased connectivity is primarily driven by the centro-frontal region (Cz, F7, F8) at f_1 and the fronto-parietal region (Fz, F3, Cz, P3, Pz, P4) at f_2.

Additionally, an inverse relationship emerges between different frequency bands and cube percepts. The parietal cortex (P3, Pz, P4), which enhances connectivity at f_1, becomes disconnected at f_2 during right-cube attention. Similarly, the right frontal region (F8), which supports increased connectivity during left-cube attention task (MLV) at f_2, disconnects at f_1 during the right-cube attention task (MRV). This suggests an antiphase activation-deactivation dynamics, where increased connectivity for one percept at a specific frequency coincides with decreased connectivity for the alternative percept at another frequency.

7. Discussion

Hypergraphs have been extensively studied and applied in network analysis, including node classification [71], community detection [72], and link prediction [73]. In neuroscience, hypergraphs, as a generalization of the graph theory, provide a more sophisticated framework for modeling brain connectivity by capturing high-order interactions among multiple brain regions [74]. Recent studies [21] have constructed hypergraphs using functional and structural connectivity as hyperedges. In contrast, we created hypergraphs based on three connectivity measures (coherence, cross-correlation, and mutual information) and three disconnectivity measures (anticoherence, anticorrelation, and information loss). Additionally, we employed sparsification techniques to remove less significant nodes, reducing noise and enhancing functional connectivity interpretation.

By incorporating hypergraph theory into EEG-based functional connectivity analysis, our method provides a novel perspective on understanding attention and its neural correlates. Hypergraph-based models enable a subject-level representation of high-order relationships, facilitating a more refined analysis of individual differences in cognitive function. This has significant implications for identifying biomarkers of attention-related disorders and advancing personalized cognitive assessments. Recent advancements in deep learning have further amplified the potential of hypergraph-based models. Ji et al. [19] introduced the hypergraph attention network (FC-HAT), while Bi et al. [75] developed the hypergraph structural information aggregation generative adversarial network (HSIA-GAN) model for functional connectivity analysis. These models dynamically construct functional connectivity networks, extract abnormal connectivity patterns, and identify critical biomarkers for neurodevelopmental and neurodegenerative disorders.

Our hypergraph analysis reveals that hyperedges associated with increased and decreased connectivity are spatially distinct and localized in different brain regions, depending on both the cognitive task and frequency. Since attention to the left-cube orientation is linked to increased spectral energy at f_1, the hypergraph in Figure 11 highlights the key nodes involved in this cognitive process. Notably, the occipital-parietal cortex emerges as a central hub for functional connectivity, whereas the central cortex plays a predominant role in disconnectivity.

An intriguing observation is the frequency-dependent nature of brain region activity: nodes that exhibit increased connectivity at one frequency tend to show decreased connectivity at another. For instance, the left parietal lobe (P3), which is highly active at f_1 (Figure 11), demonstrates reduced activity at f_2 (Figure 12), suggesting a dynamic reconfiguration of neural networks in response to different attentional states.

A similar pattern is evident during right-cube attention. The hypergraph in Figure 14 indicates that the frontal cortex plays a central role in connectivity, while the central-parietal cortex is predominantly associated with disconnectivity. This observation aligns with the fact that right-cube attention corresponds to increased spectral energy at f_2. Once again, the same brain regions exhibit frequency-dependent activation and deactivation, reinforcing the notion that attentional shifts are accompanied by distinct neural reorganization patterns across different frequency bands. Specifically, for right-cube attention, the frontal and prefrontal cortices demonstrate increased activity at f_2 (Figure 14) but reduced activity at f_1 (Figure 13). In contrast, the parietal cortices show the opposite trend, increasing their activity at f_1 (Figure 13) while decreasing at f_2 (Figure 14).

Our study also confirms the crucial role of the prefrontal and parietal regions in attentional processes, consistent with findings from previous research. Numerous studies have investigated the neural correlates of attention using EEG-based approaches with varying methodological frameworks and analytical techniques. While these studies have provided valuable insights, their reliance on traditional graph-theoretical and spectral-based methods limits their ability to capture the full complexity of brain connectivity dynamics. In contrast, hypergraph-based approaches offer a richer framework that better represents the multi-dimensional and frequency-specific interactions underlying attentional mechanisms.

Early studies on EEG-based attention research [76–78] examined delta- and alpha-wave amplitudes as indicators of attention. While these studies laid the groundwork for EEG-based attentional analysis, they relied on a univariate spectral approach. Moreover, their interpretations—linking delta and alpha activity primarily to low awareness—may be overly simplistic. Fahimi et al. [79] explored EEG-based neural markers of attention in elderly participants, correlating frontal EEG features with reaction times. Although their work underscores the potential of single-channel EEG analysis, it overlooks the distributed and interconnected nature of attentional networks. Similarly, Wan et al. [80] computed spectral power ratios (β/γ, β/α, $\beta/(\alpha+\theta)$) for attention recognition—an approach that, while common, treats frequency bands as isolated rather than interdependent components of a complex system. In contrast, hypergraph-based analysis offers a more nuanced perspective by revealing interactions between different connectivity measures at specific frequencies. It addresses these limitations by capturing higher-order connectivity patterns across multiple regions and frequencies simultaneously.

Machine learning has also been widely applied to EEG-based attention detection. Several studies [81–83] used various classification techniques (e.g., Support Vector Machine (SVM), Linear Discriminant Analysis (LDA), Naive Bayes) to distinguish attentional states. While these methods are powerful for predictive modeling, they often lack interpretability regarding the underlying neural mechanisms. A hypergraph representation provides a more interpretable structure by modeling attention states as overlapping neural communities rather than treating EEG features as independent inputs for classification. Moreover, passive brain–computer interface (BCI) studies, such as those by Aci et al. [84], aimed to classify mental states (e.g., passive attention, disengagement, drowsiness) using EEG. However, most BCI-based attention-monitoring approaches rely on traditional feature extraction methods that fail to capture the full spectrum of interactions within attentional networks. Hypergraphs enhance these approaches by integrating multi-scale and multi-frequency relationships, leading to more robust and generalizable attention-monitoring systems.

While prior studies have made significant contributions to EEG-based attention research, they are often constrained by their methodological frameworks, which rely on simplistic spectral analyses, pairwise connectivity measures, and black-box machine learning models. By contrast, hypergraph-based approaches offer a more comprehensive representation of attentional processes, accounting for higher-order interactions, multi-frequency dynamics, and network-wide connectivity patterns.

It is essential to highlight that while the various hypergraph representations explored in this study offer similar insights into brain connectivity, each captures distinct aspects of neural dynamics. These differences contribute to a more comprehensive understanding of the mechanisms underlying figurative attention. Our findings suggest that figurative attention involves a dynamic redistribution of neural resources, with specific brain regions alternating between connectivity and disconnectivity based on cognitive demands and frequency. This reorganization reflects the brain's ability to adapt its functional networks, ensuring efficient information processing across different attentional states.

The observed frequency-dependent connectivity shifts underscore the importance of considering frequency-specific interactions in functional brain network analysis. Traditional graph-based methods may overlook these nuances, whereas hypergraph-based approaches provide a powerful framework for capturing the complex, higher-order dependencies that shape attentional processes. By integrating multiple connectivity measures within a hypergraph structure, this method offers a richer, more holistic representation of neural interactions.

Future research could extend this approach to other cognitive domains, such as working memory or decision-making, to further investigate the intricate dynamics of brain connectivity across different tasks and states.

8. Conclusions

This study presents a comprehensive hypergraph-based analysis of functional brain connectivity during figurative attention to an ambiguous visual stimulus, integrating coherence, cross-correlation, and mutual information from EEG data. Our findings support the hypothesis that figurative attention engages cortico–cortical interactions, with hypergraph representations revealing distinct frequency-dependent activation and deactivation patterns across brain regions. Notably, the parietal, frontal, and prefrontal lobes play a key role in integrating information, highlighting their functional specialization and dynamic reconfiguration in attentional processing. However, while these regions facilitate efficient cognitive functions, some information may be lost in the central cortex, underscoring the complexity of neural information processing.

By leveraging a frequency-tagging approach in the context of the Necker cube illusion, we identified distinct connectivity patterns corresponding to different cube orientations. Our results emphasize the role of bilateral cortico–cortical interactions and suggest the existence of integrated processing hubs that coordinate visual attention. Hypergraph analysis, extending beyond traditional graph-based methods, provided novel insights into higher-order relationships between multiple brain regions, offering a more comprehensive understanding of dynamic neural interactions.

The methodology developed in this study—incorporating three connectivity and three disconnectivity measures—enabled the identification of higher-order relationships among brain regions, facilitating a more precise characterization of functional connectivity networks underlying figurative attention. This multivariate approach allowed us to construct connectivity difference matrices, revealing both new connections induced by sustained attention and disconnections due to decreased connectivity. Additionally, by applying a sparsification method based on statistical thresholds, we filtered out spurious or noisy connections, enhancing the robustness and interpretability of our functional connectivity network.

Beyond its theoretical implications, this methodological framework holds promise for practical applications in cognitive neuroscience, attention monitoring, and the clinical assessment of attention-related disorders. While this proof-of-concept study is based on a single subject, it establishes a foundation for future large-scale investigations into the neural mechanisms underlying ambiguous visual perception and attentional control.

Although previous studies have explored frequency correlations and proposed biophysical models of neural interactions, our findings highlight the need for further studies investigating the mechanisms linking different frequency-dependent processes. A more refined hypergraph analysis, focusing on smaller neural ensembles or specific regions of interest, could provide deeper insights into the intricate dynamics of brain connectivity. This study lays essential groundwork for advancing our understanding of brain network dynamics and their role in cognitive function.

Hypergraphs provide a powerful framework for modeling high-order interactions in EEG-based functional connectivity networks. Our findings emphasize the importance of incorporating connectivity and disconnectivity measures in hypergraph construction, enabling a more comprehensive understanding of individual differences in attention and cognitive processing. Future research should focus on further refining hypergraph learning algorithms, integrating dynamic hypergraph representations, and expanding applications to clinical populations for biomarker discovery and personalized interventions in cognitive and psychiatric disorders.

While our findings demonstrate the potential of hypergraph-based models for capturing the complex neural dynamics underlying attention, several challenges must be addressed before this approach can be fully integrated into clinical practice. Key obstacles include the need for larger and more diverse datasets, enhanced interpretability of hypergraph representations, and rigorous validation across a broader range of cognitive tasks and populations. Overcoming these challenges will be essential for ensuring the reliability and applicability of hypergraph-based methods in both research and real-world settings.

Despite these hurdles, our study represents a significant step toward more precise and individualized assessments of brain function. By leveraging a hypergraph-based connectivity analysis, we provide a novel framework that moves beyond traditional graph-theoretic approaches, offering a richer and more nuanced understanding of attentional processes. Future research should focus on refining these models, integrating them into EEG analysis pipelines, and exploring their potential for real-time applications in cognitive neuroscience and clinical diagnostics. Ultimately, this work lays the foundation for more advanced and interpretable assessments of brain connectivity, with implications for both fundamental research and applied neurotechnology.

Author Contributions: A.N.P.: conceptualization, methodology, investigation, formal analysis, validation, visualization, writing—original draft preparation, supervision; N.P.S.: investigation, investigation, formal analysis, visualization; W.E.P.d.l.V.: investigation, data curation, software, methodology, formal analysis, writing; R.J.-R.: investigation, validation, review and editing. All authors have read and agreed to the published version of the manuscript.

Funding: This research received no external funding.

Institutional Review Board Statement: The study was conducted in accordance with the Declaration of Helsinki, and approved by the Ethics Committee of the Universidad Politécnica de Madrid (Approval Code: 2020-096).

Informed Consent Statement: Informed consent was obtained from the subject involved in the study.

Data Availability Statement: The data presented in this study are openly available in Google Drive at https://drive.google.com/drive/folders/1_PHVe2rsr4ohWufccE7oYNKNAEOB37BA (accessed on 5 February 2025).

Conflicts of Interest: The authors declare no conflicts of interest.

Appendix A. Data Acquisition Details

EEG signal acquisition followed a semi-automatic protocol that integrated manual event activation with automated processing via BrainVision Recorder 1.27 software. The procedure included the following steps:

(1) Recording (triggers): Triggers were configured within the BrainVision Recorder program, manually activated by the researcher through keypresses at the start and end of each recording. These triggers were labeled in the raw EEG signals, allowing for later identification of the analysis window.
(2) Start and End Margins: The recording included a 3 s margin before the start and 3 s after the end, resulting in a total window of 36 s, of which 30 s corresponded to the experimental period of interest.
(3) Justification for Temporal Margins: The implementation of these temporal margins aimed to minimize artifacts associated with the initiation and termination instructions of the task. The participant took approximately 3 s to process the start instruction, which could introduce transient artifacts in the signal. Similarly, the recording automatically stopped at 36 s with the trigger, reducing the variability associated with manual intervention in the recording termination.
(4) Preprocessing Segmentation: During preprocessing, the 30 s window was extracted, discarding the first and last 3 s to eliminate transient artifacts and ensure that the analysis was performed on stationary data, free from transient influences.

This methodological approach reduced variability and increased data reliability without requiring additional hardware for event activation.

Appendix B. Connectivity Matrices

The matrices of the explored connectivity measures are presented in Figures A1–A6.

Examining Figures A1, A3 and A5, one can observe a significant increase in connectivity between occipital and frontal regions during attention. Specifically, connectivity is enhanced at f_1 when the subject perceives the cube as left-oriented and at f_2 when the cube is perceived as right-oriented. This is evident from the warm colors in the lower-left corner of the matrices in the upper-left and lower-right panels.

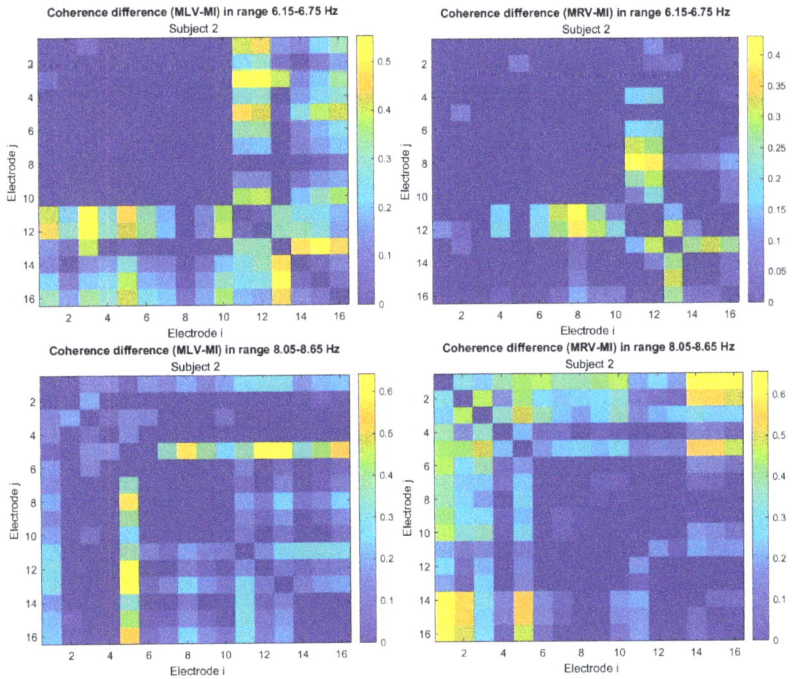

Figure A1. Increasing coherence ($\Delta Coh > 0$) associated with figurative attention to (**left panels**) left cube orientation and (**right panels**) at frequencies (**upper raw**) f_1 and (**lower raw**) f_2.

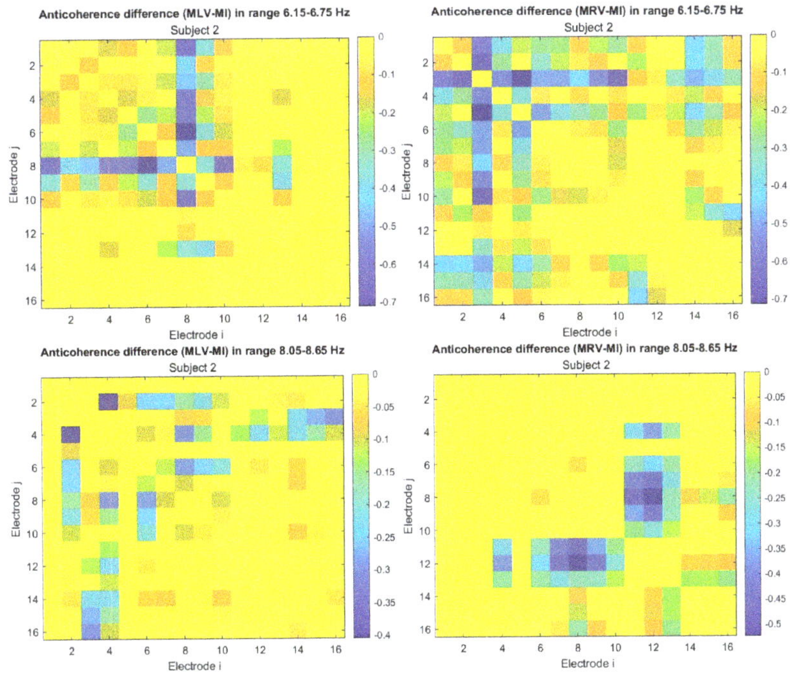

Figure A2. Decreasing coherence ($\Delta Coh < 0$) associated with figurative attention to (**left panels**) left cube orientation and (**right panels**) at frequencies (**upper raw**) f_1 and (**lower raw**) f_2.

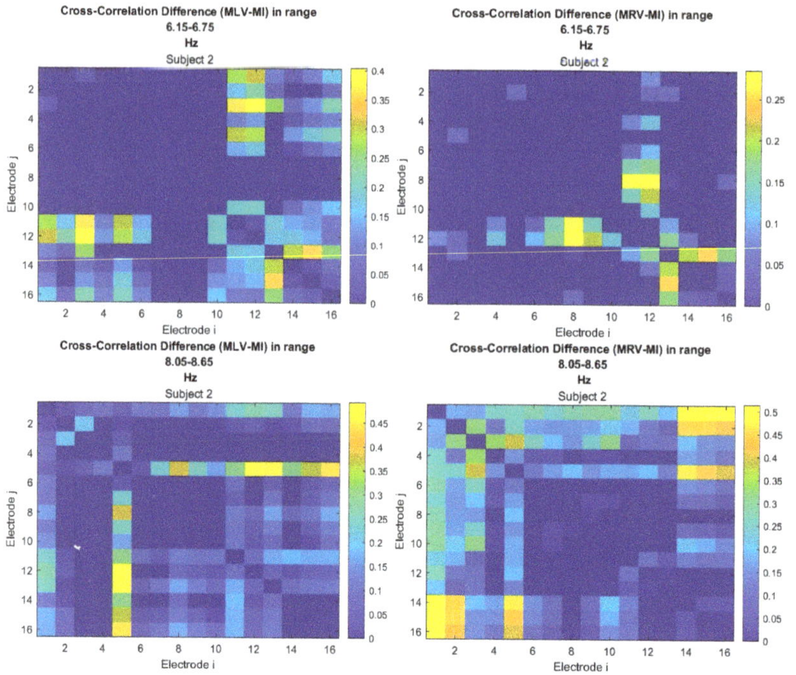

Figure A3. Increasing cross-correlation ($\Delta Corr > 0$) associated with figurative attention to (**left panels**) left cube orientation and (**right panels**) at frequencies (**upper raw**) f_1 and (**lower raw**) f_2.

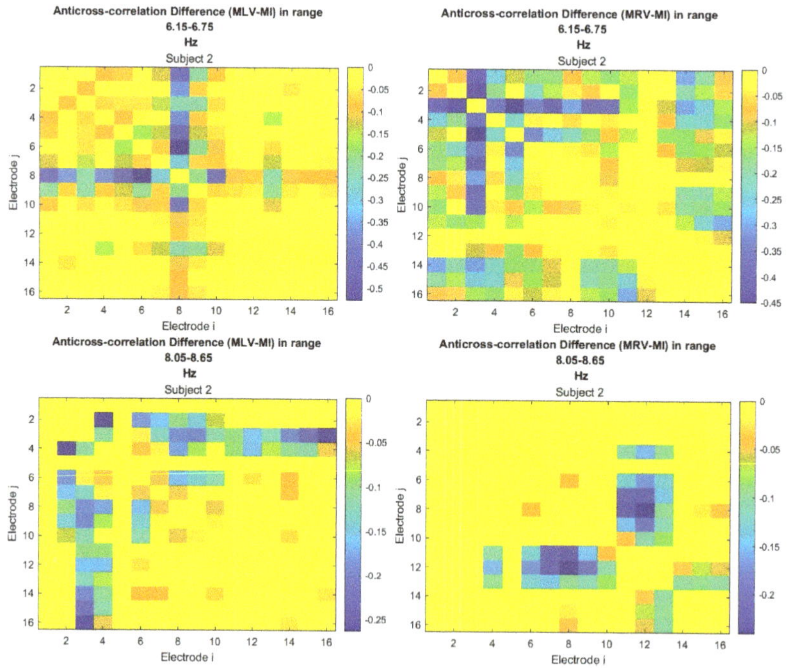

Figure A4. Decreasing cross-correlation ($\Delta Corr < 0$) associated with figurative attention to (**left panels**) left cube orientation and (**right panels**) at frequencies (**upper raw**) f_1 and (**lower raw**) f_2.

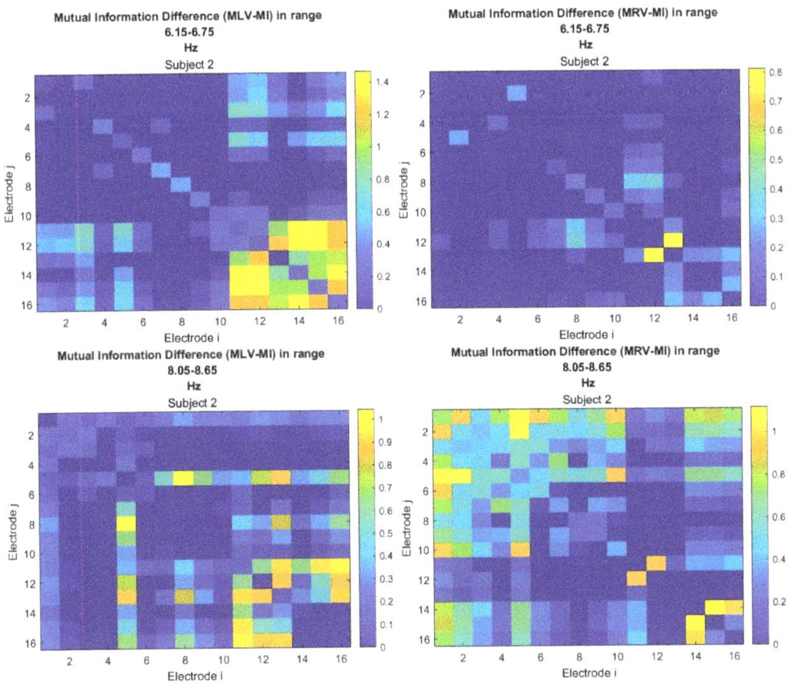

Figure A5. Increasing mutual information ($\Delta Inf > 0$) associated with figurative attention to (**left panels**) left cube orientation and (**right panels**) at frequencies (**upper raw**) f_1 and (**lower raw**) f_2.

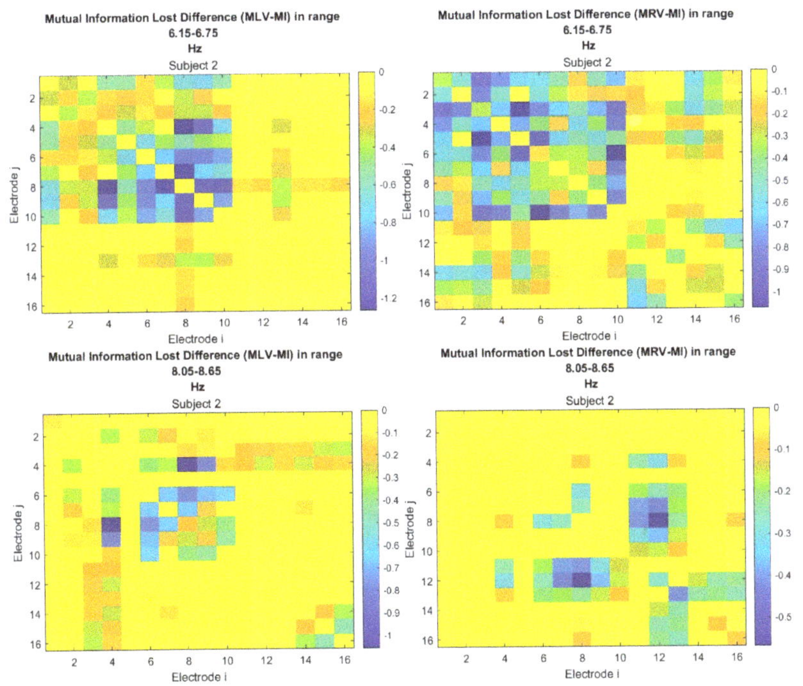

Figure A6. Decreasing mutual information ($\Delta Inf < 0$) associated with figurative attention to (**left panels**) left cube orientation and (**right panels**) at frequencies (**upper raw**) f_1 and (**lower raw**) f_2.

Appendix C. Probability Measures

The linear probability plots for explored connectivity measures are presented in Figures A7–A9.

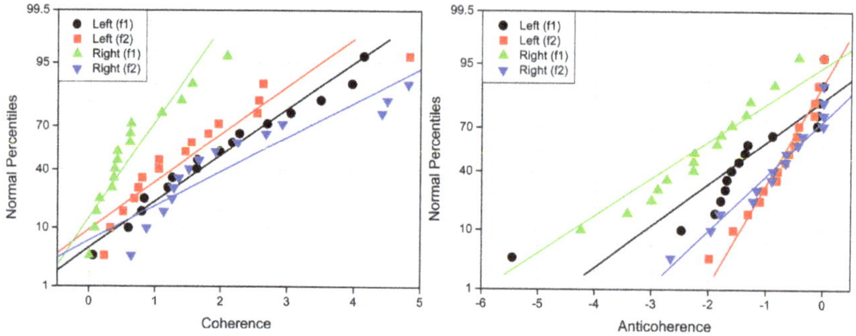

Figure A7. Normal probability plots of (**left**) coherence (Lab $Coh_n > 0$) and (**right**) anticoherence (Lab $Coh_n < 0$) labels. The reference lines have the same colors as percentiles.

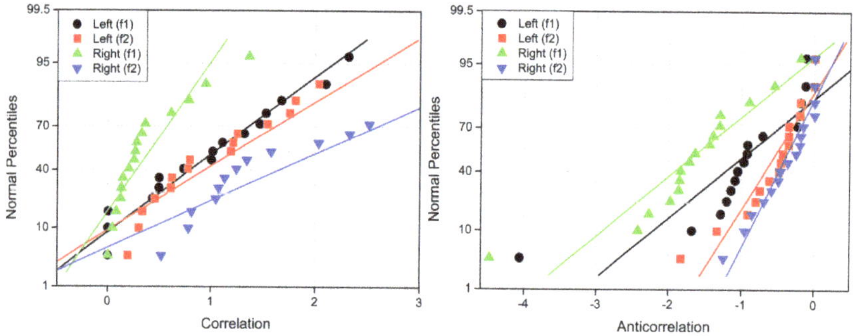

Figure A8. Normal probability plots of (**left**) correlation (Lab $Corr_n > 0$) and (**right**) anticoherence (Lab $Corr < 0$) labels. The reference lines have the same colors as percentiles.

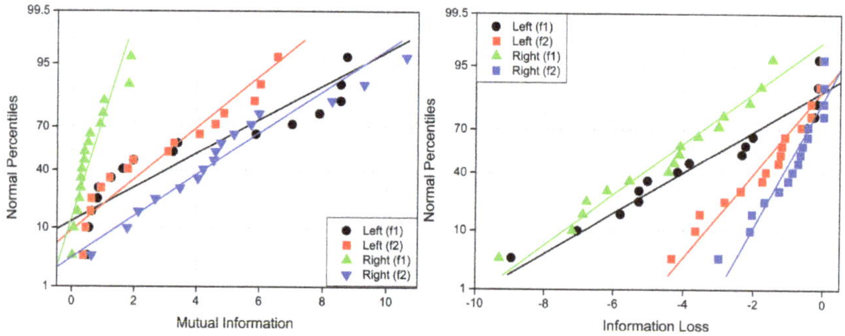

Figure A9. Normal probability plots of (**left**) mutual information (Lab $Inf_n > 0$) and (**right**) information loss (Lab $Inf_n < 0$) labels. The reference lines have the same colors as percentiles.

Table A1 presents the means and standard deviations derived from the normal probability plots.

Table A1. Mean (M), standard deviation (σ), and threshold (C_{th}) values for the following connectivity measures: coherence (Coh), anticoherence (ACoh), correlation (Corr), anticorrelation (ACorr), mutual information (MInf), and information loss (ILoss).

Measure	Left f_1			Left f_2			Right f_1			Right f_2		
	M	σ	C_{th}	m	σ	C_{th}	M	σ	C_{th}	M	σ	C_{th}
Coh	1.99	1.18	3.17	1.54	1.14	2.69	0.64	0.57	1.21	2.45	1.59	4.04
ACoh	−1.37	1.31	−2.68	−0.73	0.55	−1.28	−2.45	1.48	−3.92	−0.73	0.55	−1.28
Corr	0.98	0.70	1.68	1.15	0.85	2.00	0.37	0.36	0.73	1.95	1.16	3.11
ACorr	−0.99	0.92	−1.91	−0.57	0.47	−1.04	−1.70	0.91	−2.62	−0.40	0.37	−0.78
MInf	3.84	3.21	7.05	2.88	2.11	4.99	0.60	0.55	1.15	4.86	2.62	7.48
ILoss	−3.33	2.67	−6.01	−1.66	1.30	−2.96	−4.63	2.14	−6.77	−0.94	0.86	−1.80

References

1. Towle, V.L.; Hunter, J.D.; Edgar, J.C.; Chkhenkeli, S.A.; Castelle, M.C.; Frim, D.M.; Kohrman, M.; Hecox, K. Frequency domain analysis of human subdural recordings. *J. Clin. Neurophysiol.* **2007**, *24*, 205–213. [CrossRef] [PubMed]
2. Hramov, A.E.; Frolov, N.S.; Maksimenko, V.A.; Kurkin, S.A.; Kazantsev, V.B.; Pisarchik, A.N. Functional networks of the brain: from connectivity restoration to dynamic integration. *Phys. Uspekhi* **2021**, *64*, 584–616.
3. Boccaletti, S.; De Lellis, P.; del Genio, C.; Alfaro-Bittner, K.; Criado, R.; Jalan, S.; Romance, M. The structure and dynamics of networks with higher order interactions. *Phys. Rep.* **2023**, *1018*, 1–64.
4. Bullmore, E.; Sporns, O. Complex brain networks: Graph theoretical analysis of structural and functional systems. *Nat. Rev. Neurosci.* **2009**, *10*, 186–198. [CrossRef]
5. Friston, K.J. Functional and effective connectivity: A review. *Brain Connect.* **2011**, *1*, 13–36.
6. Tavor, I.; Jones, O.P.; Mars, R.; Smith, S.; Behrens, T.; Jbabdi, S. Task-free MRI predicts individual differences in brain activity during task performance. *Science* **2016**, *352*, 216–220.
7. Greicius, M. Resting-state functional connectivity in neuropsychiatric disorders. *Curr. Opin. Neurol.* **2008**, *21*, 424–430. [CrossRef]
8. Zhang D, R.M. Disease and the brain's dark energy. *Nat. Rev. Neurol.* **2010**, *6*, 15–28.
9. Dennis, E.L.; Thompson, P.M. Functional brain connectivity using fMRI in aging and Alzheimer's disease. *Neuropsychol. Rev.* **2014**, *24*, 49–62.
10. Tomasi, D.; Volkow, N.D. Aging and functional brain networks. *Mol. Psychiatry* **2012**, *17*, 549–558.
11. Sala-Llonch, R.; Bartrés-Faz, D.; Junqué, C. Reorganization of brain networks in aging: A review of functional connectivity studies. *Front. Psychol.* **2015**, *6*, 663.
12. Contreras, J.A.; Goñi, J.; Risacher, S.L.; Sporns, O.; Saykin, A.J. The structural and functional connectome and prediction of risk for cognitive impairment in older adults. *Curr. Behav. Neurosci. Rep.* **2015**, *2*, 234–245. [PubMed]
13. Davison, E.N.; Turner, B.O.; Schlesinger, K.J.; Miller, M.B.; Grafton, S.T.; Bassett, D.S.; Carlson, J.M. Individual differences in dynamic functional brain connectivity across the human lifespan. *PLoS Comput. Biol.* **2016**, *12*, e1005178.
14. Zuo, X.N.; Ehmke, R.; Mennes, M.; Imperati, B.; Castellanos, F.X.; Sporns, O.; Milham, M.P. Network centrality in the human functional connectome. *Cereb. Cortex* **2012**, *22*, 1862–1875.
15. Greene, A.S.; Gao, S.; Scheinost, D.; Constable, R.T. Task-induced brain state manipulation improves prediction of individual traits. *Nat. Commun.* **2018**, *9*, 2807.
16. Gu, S.; Yang, M.; Medaglia, J.D.; Gur, R.C.; Gur, R.E.; Satterthwaite, T.D.; Bassett, D.S. Functional hypergraph uncovers novel covariant structures over neurodevelopment. *Hum. Brain Mapp.* **2017**, *38*, 3823–3835.
17. Xiao, L.; Wang, J.; Kassani, P.H.; Zhang, Y.; Bai, Y.; Stephen, J.M.; Wilson, T.W.; Calhoun, V.D.; Wang, Y.P. Multi-hypergraph learning based brain functional connectivity analysis in fMRI data. *IEEE Trans. Med. Imaging* **2020**, *39*, 1746–1758.
18. Ji, Y.; Zhang, Y.; Shi, H.; Jiao, Z.; Wang, S.H.; Wang, C. Constructing dynamic brain functional networks via hyper-graph manifold regularization for mild cognitive impairment classification. *Front. Neurosci.* **2021**, *15*, 669345.
19. Ji, J.; Ren, Y.; Lei, M. FC–HAT: Hypergraph attention network for functional brain network classification. *Inf. Sci.* **2022**, *608*, 1301–1316.
20. Wu, K.; Song, X.; Chai, L. Graph metrics based brain hypergraph learning in schizophrenia patients. In Proceedings of the 2022 China Automation Congress (CAC), Xiamen, China, 25–27 November 2022; pp. 3251–3256.
21. Xi, Z.; Liu, T.; Shi, H.; Jiao, Z. Hypergraph representation of multimodal brain networks for patients with end-stage renal disease associated with mild cognitive impairment. *Math. Biosci. Eng.* **2023**, *20*, 1882–1902.
22. Wang, J.; Li, H.; Qu, G.; Cecil, K.M.; Dillman, J.R.; Parikh, N.A.; He, L. Dynamic weighted hypergraph convolutional network for brain functional connectome analysist. *Med. Image Anal.* **2023**, *87*, 102828. [CrossRef] [PubMed]

23. Niua, J.; Dua, Y. Applications of hypergraph-based methods in classifying and subtyping psychiatric disorders: A survey. *Radiol. Sci.* **2023**, *2*, 83–95. [CrossRef]
24. Yang, J.; Hu, M.; Hu, Y.; Zhang, Z.; Zhong, J. Diagnosis of autism spectrum disorder (ASD) using recursive feature elimination-graph neural network (RFE-GNN) and phenotypic feature extractor (PFE). *Sensors* **2023**, *23*, 9647. [CrossRef] [PubMed]
25. Hao, X.; Li, J.; Ma, M.; Qin, J.; Zhang, D.; Liu, F.; for Alzheimer's Disease Neuroimaging Initiative. Hypergraph convolutional network for longitudinal data analysis in Alzheimer's disease. *Comput. Biol. Med.* **2024**, *168*, 107765. [CrossRef]
26. Gao, X.; Zhu, Y.; Yang, Y.; Zhang, F.; Zhou, F.; Tian, X.; Xu, K.; Chen, Y. A seizure detection method based on hypergraph features and machine learning. *Biomed. Signal Process. Control* **2022**, *77*, 103769. [CrossRef]
27. Li, M.; Qiu, M.; Zhu, L.; Kong, W. Feature hypergraph representation learning on spatial-temporal correlations for EEG emotion recognition. *Cogn. Neurodyn.* **2023**, *17*, 1271–1281. [CrossRef]
28. Zhu, L.; Liu, Y.; Liu, R.; Peng, Y.; Cao, J.; Li, J.; Kong, W. Decoding multi-brain motor imagery from EEG using coupling feature extraction and few-shot learning. *IEEE Trans. Neural Syst. Rehabil. Eng.* **2023**, *31*, 4683–4692. [CrossRef]
29. Liu, M.; Zhang, J.; Yap, P.T.; Shen, D. View-aligned hypergraph learning for Alzheimer's disease diagnosis with incomplete multi-modality data. *Med. Image Anal.* **2017**, *36*, 123–134.
30. Peña Serrano, N.; Jaimes-Reátegui, R.; Pisarchik, A.N. Hypergraph of functional connectivity based on event-related coherence: Magnetoencephalography data analysis. *Appl. Sci.* **2024**, *14*, 2343. [CrossRef]
31. Bretto, A. *Hypergraph Theory: An Introduction*; Mathematical Engineering Series; Springer: Cham, Switzerland, 2013.
32. Dai, Q.; Gao, Y. *Hypergraph Computation*; Springer: Singapore, 2023.
33. Wang, Z.; Liu, J.; Zhong, N.; Qin, Y.; Zhou, H.; Yang, J.; Li, K., A naïve hypergraph model of brain networks. In *Brain Informatics. BI 2012. Lecture Notes in Computer Science*; Zanzotto, F.M., Tsumoto, S., Taatgen, N., Yao, Y., Eds.; Springer: Berlin/Heidelberg, Germany, 2012; Volume 7670, pp. 119–129.
34. Caetano e Souza, R.H.; Martins Naves, E.L. Attention detection in virtual environments using EEG signals: A Scoping review. *Front. Physiol.* **2021**, *12*, 727840.
35. Esqueda-Elizondo, J.J.; Juárez-Ramírez, R.; López-Bonilla, O.R.; García-Guerrero, E.E.; Galindo-Aldana, G.M.; Jiménez-Beristáin, L.; Serrano-Trujillo, A.; Tlelo-Cuautle, E.; Inzunza-González, E. Attention measurement of an autism spectrum disorder user using EEG signals: A case study. *Math. Comput. Appl.* **2022**, *27*, 21. [CrossRef]
36. Mateos, D.M.; Krumm, G.; Arán Filippetti, V.; Gutierrez, M. Power spectrum and connectivity analysis in EEG recording during attention and creativity performance in children. *NeuroSci* **2022**, *3*, 347–365. [CrossRef]
37. Al-Nafjan, A.; Aldayel, M. Predict students' attention in online learning using EEG data. *Sustainability* **2022**, *14*, 6553. [CrossRef]
38. Escalante Puente de la Vega, W.; Imbert Paredes, R.; Villalba Mora, E.; Pisarchik, A.N. Figurative attention: EEG-based assessment using ambiguous stimuli. *Prepr. (Res. Sq.)* **2025**. [CrossRef]
39. Bowyer, S.M. Coherence a measure of the brain networks: Past and present. *Neuropsychiatr. Electrophysiol.* **2016**, *2*, 1.
40. Chiarion, G.; Sparacino, L.; Antonacci, Y.; Faes, L.; Mesin, L. Connectivity analysis in EEG data: A tutorial review of the state of the art and emerging trends. *Bioengineering* **2023**, *10*, 372. [CrossRef]
41. Jeong, J.; Gore, J.C.; Peterson, B.S. Mutual information analysis of the EEG in patients with Alzheimer's disease. *Clin. Neurophysiol.* **2001**, *112*, 827–835. [CrossRef]
42. Zien, J.Y.; Schlag, M.D.F.; Chan, P.K. Multilevel spectral hypergraph partitioning with arbitrary vertex sizes. *IEEE Trans. Comput.-Aided Des. Integr.* **1999**, *18*, 1389–1399.
43. Chholak, P.; Maksimenko, V.A.; Hramov, A.E.; Pisarchik, A.N. Voluntary and involuntary attention in bistable visual perception: A MEG study. *Front. Hum. Neurosci.* **2020**, *14*, 597895. [CrossRef]
44. Liu, Y.; Safavi, T.; Dighe, A.; Koutra, D. Graph summarization methods and applications: A survey. *ACM Comput. Surv.* **2018**, *51*, 62.
45. Maccioni, A.; Abadi, D.J. Scalable pattern matching over compressed graphs via dedensification. In Proceedings of the ACM Conference on Knowledge Discovery and Data Mining (KDD'16), San Francisco, CA, USA, 13–17 August 2016; ACM Press: Cambridge, MA, USA, 2016; pp. 1755–1764.
46. Andrew, C.; Pfurtscheller, G. Event-related coherence as a tool for studying dynamic interaction of brain regions. *Electroencephalogr. Clin. Neurophysiol.* **1996**, *98*, 144–148. [PubMed]
47. Pisarchik, A.; Hramov, A. Coherence resonance in neural networks: Theory and experiments. *Phys. Rep.* **2023**, *1000*, 1–57.
48. Gross, J.; Schmitz, F.; Schnitzler, I.; Kessler, K.; Shapiro, K.; Hommel, B.; Schnitzler, A. Modulation of long-range neural synchrony reflects temporal limitations of visual attention in humans. *Proc. Natl. Acad. Sci. USA* **2004**, *101*, 13050–13055. [PubMed]
49. Guggisberg, A.G.; Honma, S.M.; Findlay, A.M.; Dalal, S.S.; Kirsch, H.E.; Berger, M.S.; Nagarajan, S.S. Mapping functional connectivity in patients with brain lesions. *Ann. Neurol.* **2008**, *63*, 193–203.
50. Belardinelli, P.; Ciancetta, L.; Staudt, M.; Pizzella, V.; Londei, A.; Birbaumer, N.; Romani, G.L.; Braun, C. Cerebro-muscular and cerebro-cerebral coherence in patients with pre- and perinatally acquired unilateral brain lesions. *NeuroImage* **2007**, *37*, 1301–1314.

51. de Pasquale, F.; Della Penna, S.; Snyder, A.Z.; Lewis, C.; Mantini, D.; Marzetti, L.; Belardinelli, P.; Ciancetta, L.; Pizzella, V.; Romani, G.L.; et al. Temporal dynamics of spontaneous MEG activity in brain networks. *Proc. Natl. Acad. Sci. USA* **2010**, *107*, 6040–6045. [CrossRef]
52. Kim, J.S.; Shin, K.S.; Jung, W.H.; Kim, S.N.; Kwon, J.S.; Chung, C.K. Power spectral aspects of the default mode network in schizophrenia: An MEG study. *BMC Neurosci.* **2014**, *15*, 104.
53. Bowyer, S.M.; Gjini, K.; Zhu, X.; Kim, L.; Moran, J.E.; Rizvi, S.U.; Gumenyuk, N.T.; Tepley, N.; Boutros, N.N. Potential biomarkers of schizophrenia from MEG resting-state functional connectivity networks: Preliminary data. *J. Behav. Brain Sci.* **2015**, *5*, 1.
54. Boutros, N.N.; Galloway, M.P.; Ghosh, S.; Gjini, K.; Bowyer, S.M. Abnormal coherence imaging in panic disorder: A magnetoencephalography investigation. *Neuroreport* **2013**, *24*, 487–491.
55. Chholak, P.; Niso, G.; Maksimenko, V.A.; Kurkin, S.A.; Frolov, N.S.; Pitsik, E.N.; Hramov, A.E.; Pisarchik, A.N. Visual and kinesthetic modes affect motor imagery classification in untrained subjects. *Sci. Rep.* **2019**, *9*, 9838.
56. Pisarchik, A.N.; Chholak, P.; Hramov, A.E. Brain noise estimation from MEG response to flickering visual stimulation. *Chaos Solitons Fractals X* **2019**, *1*, 100005. [CrossRef]
57. Baldassarre, A.; Lewis, C.M.; Committeri, G.; Snyder, A.Z.; Luca Romani, G.; Corbetta, M. Individual variability in functional connectivity predicts performance of a perceptual task. *Proc. Natl. Acad. Sci. USA* **2012**, *109*, 3516–3521. [CrossRef] [PubMed]
58. Kashyap, R.; Kong, R.; Bhattacharjee, S.; Li, J.; Zhou, J.; Thomas Yeo, B.T. Individual-specific fMRI-Subspaces improve functional connectivity prediction of behavior. *NeuroImage* **2019**, *189*, 804–812.
59. Shannon, C.E.; Weaver, W. *The Mathematical Theory of Communication*; University of Illinois Press: Urbana, IL, USA, 1949.
60. Cover, T.M.; Thomas, J.A. *Elements of Information Theory*; John Wiley & Sons: New York, NY, USA, 1991.
61. Tadel, F.; Baillet, S.; Mosher, J.C.; Pantazis, D.; Leahy, R.M. Brainstorm: A user-friendly application for MEG/EEG analysis. *Comput. Intell. Neurosci.* **2011**, *2011*, 879716. [CrossRef]
62. GitLab.com. Available online: https://gitlab.com/bci_project/visual_stimulation_codes/datas_eeg (accessed on 10 January 2025).
63. Hennighausen, E.; Heil, M.; Rösler, F. A correction method for DC drift artifacts. *Electroencephalogr. Clin. Neurophysiol.* **1993**, *86*, 199–204. [CrossRef]
64. Kim, H.; Luo, J.; Chu, S.; Cannard, C.; Hoffmann, S.; Miyakoshi, M. ICA's bug: How ghost ICs emerge from effective rank deficiency caused by EEG electrode interpolation and incorrect re-referencing. *Front. Signal Process.* **2023**, *3*, 1064138. [CrossRef]
65. Isoglu-Alkaç, U.; Basar-Eroglu, C.; Ademoglu, A.; Demiralp, T.; Miener, M.; Stadler, M. Alpha activity decreases during the perception of necker cube reversals: An application of wavelet transform. *Biol. Cybern.* **2000**, *82*, 313–320. [CrossRef]
66. Adeli, H.; Zhou, Z.; Dadmehr, N. Analysis of EEG records in an epileptic patient using wavelet transform. *J. Neurosci. Methods* **2003**, *123*, 69–87. [CrossRef]
67. Ghuman, A.S.; McDaniel, J.R.; Martin, A. A wavelet-based method for measuring the oscillatory dynamics of resting-state functional connectivity in MEG. *Neuroimage* **2011**, *56*, 69–77. [CrossRef]
68. Maksimenko, V.A.; Hramov, A.E.; Grubov, V.V.; Nedaivozov, V.O.; Makarov, V.V.; Pisarchik, A.N. Nonlinear effect of biological feedback on brain attentional state. *Nonlin. Dyn.* **2019**, *95*, 1923–1939. [CrossRef]
69. Maksimenko, V.A.; Frolov, N.S.; Hramov, A.E.; Runnova, A.E.; Grubov, V.V.; Kurths, J.; Pisarchik, A.N. Neural interactions in a spatially-distributed cortical network during attentional tasks. *Front. Behav. Neurosci.* **2019**, *13*, 220. [CrossRef] [PubMed]
70. Khramova, M.V.; Kuc, A.K.; Maksimenko, V.A.; Frolov, N.S.; Grubov, V.V.; Kurkin, S.A.; Pisarchik, A.N.; Shusharina, N.N.; Fedorov, A.A.; Hramov, A.E. Monitoring the cortical activity of children and adults during cognitive task completion. *Sensors* **2021**, *21*, 6021. [CrossRef] [PubMed]
71. Fang, Q.; Sang, J.; Xu, C.; Rui, Y. Topic-sensitive influencer mining in interest-based social media networks via hypergraph learning. *IEEE Trans. Multimed.* **2014**, *16*, 796–812.
72. Martinet, L.E.; Kramer, M.; Viles, W.; Perkins, L.; Spencer, E.; Chu, C.J.; Cash, S.S.; Kolaczyk, E.D. Robust dynamic community detection with applications to human brain functional networks. *Nat. Commun.* **2020**, *11*, 2785. [CrossRef]
73. Zhao, Z.; Yang, K.; Guo, J. Link prediction with hypergraphs via network embedding. *Appl. Sci.* **2023**, *13*, 523. [CrossRef]
74. Berge, C. *Graphs and Hypergraphs*; North-Holland: Amsterdam, The Netherlands, 1976.
75. Bi, X.A.; Wang, Y.; Luo, S.; Chen, K.; Xing, Z.; Xu, L. Hypergraph structural information aggregation generative adversarial networks for diagnosis and pathogenetic factors identification of Alzheimer's disease with imaging genetic data. *IEEE Trans. Neural Netw. Learn. Syst.* **2024**, *35*, 7420–7434.
76. Harmony, T.; Fernández, T.; Silva, J.; Bernal, J.; Díaz-Comas, L.; Reyes, A.; Marosi, E.; Rodríguez, M.; Rodríguez, M. EEG delta activity an indicator of attention to internal processing during performance of mental tasks. *Int. J. Psychophysiol.* **1996**, *24*, 161–171.
77. Li, X.; Hu, B.; Zhu, T.; Yan, J.; Zheng, F. Towards affective learning with an EEG feedback approach. In Proceedings of the First ACM International Workshop on Multimedia Technologies for Distance Learning, Beijing, China, 23 October 2009; pp. 33–38.
78. Zhou, S.; Gao, T. Brain activity recognition method based on attention-based RNN mode. *Appl. Sci.* **2011**, *11*, 425.

79. Fahimi, F.; Goh, W.B.; Lee, T.S.; Guan, C. Neural indexes of attention extracted from EEG correlate with elderly reaction time in response to an attentional task. In Proceedings of the 3rd International Conference on Crowd Science and Engineering, Singapore, 28–31 July 2018; pp. 1–6.
80. Wan, W.; Cui, X.; Gao, Z.; Gu, Z. Frontal EEG-based multi-level attention states recognition using dynamical complexity and extreme gradient boosting. *Int. J. Psychophysiol.* **2021**, *15*, 673955.
81. Myrden, A.; Chau, T. A passive EEG-BCI for single-trial detection of changes in mental state. *IEEE Trans. Neural Syst. Rehabil. Eng.* **2017**, *25*, 345–356.
82. Alirezaei, M.; Hajipour Sardouie, S. Detection of human attention using EEG signals. In Proceedings of the 24th Iranian Conference on Biomedical Engineering and 2nd International Iranian Conference on Biomedical Engineering, Tehran, Iran, 30 November–1 December 2017; pp. 1–5.
83. Nuamah, J.K.; Seong, Y. Support vector machine (SVM) classification of cognitive tasks based on electroencephalography (EEG) engagement index. *Brain-Comput. Interfaces* **2018**, *5*, 1–12.
84. Aci, c.I.; Kaya, M.; Mishchenko, Y. Distinguishing mental attention states of humans via an EEG-based passive BCI using machine learning methods. *Expert Syst. Appl.* **2019**, *134*, 153–166. [CrossRef]

applied sciences

Article

Predicting Attachment Class Using Coherence Graphs: Insights from EEG Studies on the Secretary Problem

Dor Mizrahi *, Ilan Laufer and Inon Zuckerman

Department of Industrial Engineering and Management, Ariel University, Ariel 4070000, Israel; ilanl@ariel.ac.il (I.L.); inonzu@ariel.ac.il (I.Z.)
* Correspondence: dor.mizrahi1@msmail.ariel.ac.il

Abstract

Attachment styles, rooted in Bowlby's Attachment Theory, significantly influence our romantic relationships, workplace behavior, and decision-making processes. Traditional methods like self-report questionnaires often have biases, so we aimed to develop a predictive model using *objective* physiological data. In our study, participants engaged in the Secretary problem, a sequential decision-making task, while their brain activity was recorded with a 16-electrode EEG device. We transformed this data into coherence graphs and used Node2Vec and PCA to convert these graphs into feature vectors. These vectors were then used to train a machine learning model, XGBoost, to predict attachment styles. Using participant-level nested 5-fold cross-validation, our first model achieved 80% accuracy for Secure and 88% for Fearful-avoidant styles but had difficulty distinguishing between Avoidant and Anxious styles. Analysis of the first three principal components showed these two groups overlapped in coherence space, explaining the confusion. To address this, we created a second model that categorized participants as Secure, Insecure, or Extremely Insecure, improving the overall accuracy to about 92%. Together, the results highlight (i) large-scale EEG connectivity as a viable biomarker of attachment, and (ii) the empirical similarity between Anxious and Avoidant profiles when measured electrophysiologically. This method shows promise in using EEG data and machine learning to understand attachment styles. Our findings suggest that future research should include larger and more diverse samples to refine these models. If validated in multi-site cohorts, such graph-based EEG markers could guide personalised interventions by objectively assessing attachment-related vulnerabilities. This study demonstrates the potential for using EEG data to classify attachment styles, which could have important implications for both research and therapeutic practices.

Keywords: sequential decision-making; attachment theory; coherence graphs

1. Introduction

The main idea behind Bowlby's Attachment Theory is that people vary in forming emotional bonds with others [1]. Typically, the social and behavioral sciences identified four types of attachment styles: secure, anxious, avoidant, and fearful-avoidant styles. The primary distinction in attachment styles is between secure and insecure attachment. Securely attached individuals exhibit mutual trust, support, emotional stability during conflicts, and the ability to set healthy boundaries. Insecure attachment styles are broadly categorized into anxious and avoidant types. Anxiously attached individuals fear abandonment, need constant validation, and rely heavily on their partner for self-worth. Avoidantly

https://doi.org/10.3390/app15169009

attached individuals shun intimacy and vulnerability, struggle with commitment, and tend to be guarded and emotionally distant. Some people display both high anxiety and high avoidance, known as fearful-avoidant (or disorganized) attachment. These individuals have a strong fear of rejection, difficulty trusting and depending on partners, and low self-esteem.

Attachment style is typically formed in the early childhood years of zero to three; it is now well known that it affects not only our adult romantic relationships [2] but also our behavior in the work environment and strategic decision-making in general [3–5]. For example [6], it was shown that attachment style is related to adolescents' autonomy in decision-making. Another study [7] shows that career decision-making of final-year high school students was correlated to the student attachment style. Finally, ref. [8] showed that the students' attachment styles significantly predict decision self-esteem, decision-making styles, and personality traits.

Today, self-report questionnaires and narratives are two primary methods for measuring adult attachment style [9]. Various self-report questionnaires exist, each with unique attributes; some classify individuals into one of the four attachment styles, while others assess the degree to which attachment dimensions (anxiety or avoidance) are present [10]. For instance, the ECR-R (Experiences in Close Relationships—Revisited) questionnaire [11] includes 36 items. It is a reliable and valid self-report tool that quantifies the anxiety and avoidance dimensions on a scale from 1 to 7.

While self-report questionnaires and narratives are valuable for measuring adult attachment styles, they come with biases and limitations. These methods depend on individuals' self-awareness and honesty, which can lead to inaccuracies. Due to these shortcomings, there is increasing interest in finding objective physiological measures that provide more reliable and unbiased data. Exploring psychophysiological responses, such as EEG signals, offers a promising way to develop predictive models that classify attachment styles accurately without relying on self-reported information.

This shift towards objective measures is supported by advancements in understanding the psychophysiological aspects of attachment styles.

Studies have examined cardiovascular activity, galvanic skin response, and adrenocortical output to explore how attachment style influences physiological regulation, especially under stress [9]. These findings suggest that attachment is not only a psychological construct but also expressed through measurable biological responses [12]. More recently, EEG research has provided valuable insights into how attachment patterns affect brain function. For example, Verbeke et al. [13] showed that individuals with anxious attachment display heightened alpha, beta, and theta power in social situations, indicating increased cortical arousal. Sloan et al. [14] linked attachment anxiety to atypical alpha activity during sleep, suggesting a persistent state of physiological vigilance. Rognoni et al. [15] associated adult attachment styles with frontal EEG asymmetry, a marker often tied to emotional regulation strategies [9]. Zuckerman et al. [16] further demonstrated that attachment style modulates early and late ERP components during emotionally relevant decision-making tasks, particularly under conditions of cognitive conflict. Together, these studies show that attachment-related differences are reflected in neural activity across a range of contexts. However, most of this research has focused on average group-level comparisons rather than on building predictive models that classify attachment style at the individual level. Addressing this gap could lead to more precise tools for both research and clinical use.

Although previous studies have explored the connection between attachment and physiological responses, one important area remains underexplored: using objective physiological data to predict attachment styles at the individual level. As noted in a recent study on attachment and depression, there is a growing need to move beyond self-report questionnaires, which are vulnerable to bias, and to include physiological markers as a more objective alternative [12]. Most research continues to depend on self-report tools, which can be influenced by memory limitations or the desire to present oneself favorably. Physiological measures such as EEG provide a more direct way to observe how attachment tendencies are reflected in brain activity. EEG data can be analyzed using a range of methods, including time- and frequency-based approaches, as well as more advanced techniques like wavelet transforms, principal component analysis (PCA), and independent component analysis (ICA).

Several recent studies demonstrate their potential in emotion and attachment-related contexts. For example, a two-step PCA–ICA pipeline was successfully used to isolate EEG components associated with social acceptance and rejection in adolescent romantic couples, reflecting both spatial and temporal aspects of emotional processing [17]. In patients with depression, Morlet wavelet transformation revealed elevated gamma activity in response to attachment-related stimuli, especially among those with higher emotional reactivity [18]. In the stress domain, PCA, ICA, and wavelet-based DCT were used in EEG-based stress detection from the DEAP dataset, showing that these preprocessing methods contribute meaningfully to emotional state classification [19]. Similarly, EEG responses to music-based attention modulation were clarified using ICA and wavelet transforms, which revealed differences in alpha and beta band activity depending on stimulus type [20]. These findings highlight the versatility of such techniques in detecting neural patterns associated with psychological states. Even so, this approach remains underused in attachment research compared to its growing application in other domains of psychophysiology.

In our previous studies, we explored the Secretary Problem by analyzing neurophysiological data, focusing on changes in Theta and Beta band power and Event-Related Potentials (ERPs) to understand shifts in cognitive load during sequential decision-making [21,22]. These investigations provided a detailed look at how mental resources vary during decision-making. The Secretary Problem is an ideal test-bed for attachment research because it forces an irrevocable, high-stakes choice: once an offer is rejected, it cannot be revisited. This "no-return" rule elevates both cognitive demand (continuous rule-updating) and situational anxiety over missing the optimum [23]. Attachment theory predicts divergent reactions: anxious profiles typically hyper-monitor loss and show heightened arousal, whereas avoidant profiles down-regulate threat and invest less cognitive effort [24]. EEG work confirms that anxious individuals display stronger fronto-parietal coherence under stress, whereas avoidant individuals show weaker long-range coupling [25]. Analyzing such coherence graphs links our paradigm to broader personality-neuroscience research that relates graph metrics to individual traits [26]. Thus, the Secretary Task simultaneously taxes cognition and triggers attachment-relevant affect, offering a theoretically grounded context for distinguishing attachment styles. The current study, however, broadens this scope by combining the Secretary Problem with attachment theory, aiming to predict attachment styles using coherence graphs derived from EEG data. This study addresses the current research gap by integrating the Secretary Problem with attachment theory to predict attachment styles using coherence graphs from EEG data. By developing a new predictive model, we can observe how brain regions interact during decision-making tasks. Coherence graphs help identify patterns in brain activity, which can then be linked to attachment styles, offering a more objective assessment based on physiological data. This approach could deepen our understanding of how attachment styles influence decision-making, directly connecting psychological traits with brain activity.

To explore these connections further, we utilize the node2vec algorithm as part of our analytical approach, which brings several benefits [27]. Node2vec creates continuous feature representations for nodes in a graph by simulating random walks, which helps capture both local and global structural information. Understanding the complex relationships within EEG coherence data is important. Traditional methods for analyzing EEG data often depend on predefined features, which might miss the deeper connections within the data. In contrast, node2vec explores the graph through random walks, allowing it to capture both local neighborhoods and broader structural patterns.

This dual ability is especially useful for EEG coherence graphs, where the relationships between electrodes can show complex patterns at various scales. Node2vec's flexibility lets us adjust the random walks to favor either breadth-first or depth-first search strategies, tailoring the process to capture the most relevant patterns in our EEG data. Additionally, the algorithm is efficient and scalable, making it suitable for handling the large and detailed coherence graphs derived from EEG data. By embedding the graph into a continuous vector space, node2vec maintains the essential relationships between nodes, resulting in more informative and distinctive features for our machine learning models. This innovative method allows us to build a more accurate predictive model, advancing the field of attachment style classification based on objective physiological data [27,28].

To implement this approach, we collected EEG data from participants while they engaged in a sequential decision-making task. Specifically, participants played an instance of the Secretary problem [29] while the electrical brain signals on their scalp were recorded using a 16-electrodes EEG device. The EEG data was used to construct a coherence graph (e.g., [30,31]) describing the relationship between the electrodes. The coherence graph was transformed into a feature vector using a virtual node embedding technique and the Node2vec algorithm [27,32]. The feature vectors were input into the XGBoost [33] decision-learning tree algorithm to construct a predictive model.

Our results show great success in predicting secure and fearful-avoidant attachment styles. The secure cluster was correctly predicted with around 80% accuracy, and the fearful-avoidant group was correctly predicted with around 88% accuracy. However, the two other attachment clusters, anxious and avoidant, were harder to predict, and the model could not differentiate them. Some of the anxious individuals were classified as avoidants and the other way around. Next, we constructed a second model to classify three clusters: Secure, Insecure, and Extreme Insecure, and we were able to achieve an improved correct classification of around 92%. Further research is needed to understand why and utilize more data or other techniques to construct more robust models.

2. Methods

The study consisted of two primary phases. First, a questionnaire was conducted to determine the participants' attachment styles, and then an EEG study was conducted to record brain patterns. The experiment received approval from the institution's Institutional Review Board (IRB) committee, and all participants signed a formal agreement form before participating.

2.1. First Phase—ECR-R Questionnaire

In the first phase of the experiments, we asked 96 participants, fourth-year senior engineering students (average age of 24.25 with $\sigma = 2.0673$), to fill out an ECR-R questionnaire (with 36 items). In each item in the ECR-R questionnaire, one must determine how much she agrees or disagrees with the presented statement on a scale from 1 to 7. A value of 1 means "Strongly disagree", and 7 means "Strongly agree". The questionnaire items include questions such as: "I'm afraid that I will lose my partner's love", "I prefer not to show my partner how I feel deep down", and "I am very comfortable being close to romantic partners".

The ECR-R questionnaire outputs two values as numbers from 1 to 7. One value is the quantified degree of anxiety, and the other is the quantified degree of avoidance. When these two values are low, the person is said to have a secure attachment style. When these two values are high, the person is said to have a fearful-avoidant attachment style. When one of these values is high, and the other is low, this person is said to have an anxious or avoidance attachment based on the higher values of these two. Next, we employed the k-means clustering algorithm (with k = 4) to classify the samples into the four clusters of attachment styles [34]. Figure 1 depicts the results of the k-means algorithm. It is worth noting that the sizes of the different clusters were very similar to the common percentages available in the psychological literature. Most studies on different populations show that roughly 50% of the population exhibits a secure attachment style, around 20% exhibit an anxious and avoidant attachment style, and around 10% have a fearful avoidant attachment style [3,35].

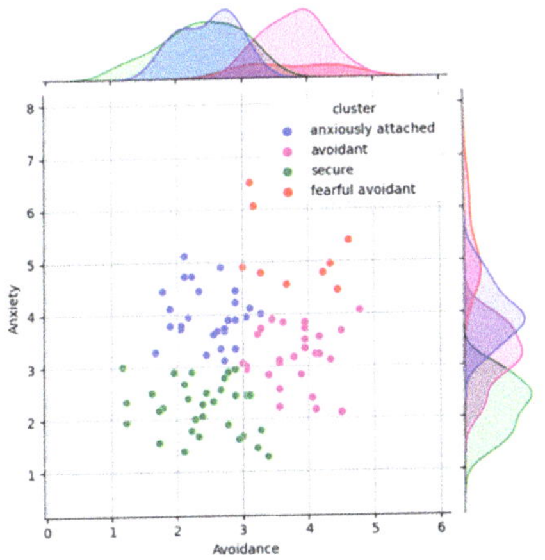

Figure 1. Grouped attachment outcomes based on the ECR-R questionnaire [36].

Following the initial assessment of attachment style using the ECR-R questionnaire, a subset of participants was invited to take part in the EEG-based decision-making experiment described below. The following Sections 2.2 and 2.3 provide full details on participant distribution and experimental design, while Section 2.4 outlines the EEG data acquisition and preprocessing procedures.

2.2. Second Phase—EEG Recording in the Secretary Game

In the second phase of the experiment, we issued invitations to participate in the EEG laboratory part. Each session was one hour long, and participants were paid for their efforts. We employed a proportional allocation method to ensure a representative distribution across attachment clusters. The secure group consisted of six participants; nine were anxiously attached, seven were avoidants, and five participants had a fearful avoidant attachment. Figure 2 shows the 27 individuals who participated in the EEG recording part of the experiment. The blue dots represent participants who completed only the ECR-R questionnaire, while the orange dots represent those who participated in both the ECR-R questionnaire and the EEG session. This scatter plot maps the participants based on their levels of Anxiety and Avoidance, which are key dimensions in attachment theory. From the figure, it is clear that the participants in the EEG session (orange dots) provide a relatively uniform sample compared to the broader population (blue dots) surveyed in the initial stage. This consistency ensures that the EEG sample accurately reflects the broader attachment style distribution observed in the larger group.

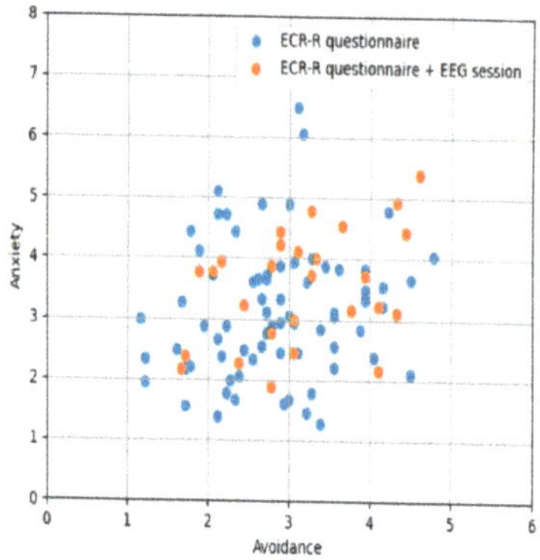

Figure 2. The sample of individuals who participated in the 2nd stage of the experiment, the EEG session.

While the scatter plot does not display distinct clusters for each attachment style, it demonstrates the continuous nature of Anxiety and Avoidance. Individuals are distributed across a range of values for both dimensions, showing variability in how these traits are expressed. This distribution supports the idea that Avoidant and Anxious attachment styles can coexist within individuals, as no clear boundaries separate these traits. Instead, individuals may exhibit varying levels of both Anxiety and Avoidance, rather than fitting neatly into discrete categories. This visual representation underscores the complexity and variability of attachment styles, highlighting the importance of using detailed methods like EEG coherence graphs (see Section 3 below) to gain deeper insights into the underlying neurophysiological mechanisms of attachment behaviors.

In the second phase, participants engaged in six blocks of the secretary problem task, each consisting of 20 monetary offers. These offers, represented as sums of money displayed on-screen, required participants to accept (pressing the 'Y' key) or reject (pressing the 'N' key) each one. Simulating the role of apartment sellers, participants knew they could neither see future offers nor revisit past ones. The primary objective was to select the highest offer in each block to maximize potential earnings, where their actual monetary compensation for participating in the experiment was a function of their success in the Secretary game. It is important to note that if a participant did not choose any offer within a block, it was considered as if they had accepted the last offer. The stream of offers and the experimental setting are adapted from the study conducted by Hsiao and Kemp [37]. However, we increased the original bids tenfold in blocks 2, 4, and 6 of our experiments to see if this increased value changes one's decision-making process, either behaviorally or in his brain patterns as recorded by the EEG.

Crucially, each of the six blocks generated an independent EEG coherence graph, so that the 27 participants contributed a total of $27 \times 6 = 162$ graph-level observations. These 162 feature vectors form the input dataset for our machine-learning pipeline, allowing the model to learn from multiple decision episodes per participant and thereby reducing the risk that idiosyncratic traits of any single individual drive the results.

2.3. Experiment Procedure

As stated above, each participant played six blocks of the secretary problem. Figure 3 depicts graphically the construction of a single game block. The participant first sees a welcome screen with some instructions on the upcoming task. Next, a waiting screen is displayed for a randomized duration of 200 to 500 milliseconds. The offer screen follows with a single amount of money in the center of the screen. This screen will remain active (without time restrictions) until a decision is made by the participant (either accept or reject). If the offer was rejected and the session had not yet concluded, the procedure reverted to the waiting screen, maintaining the decision-making cycle.

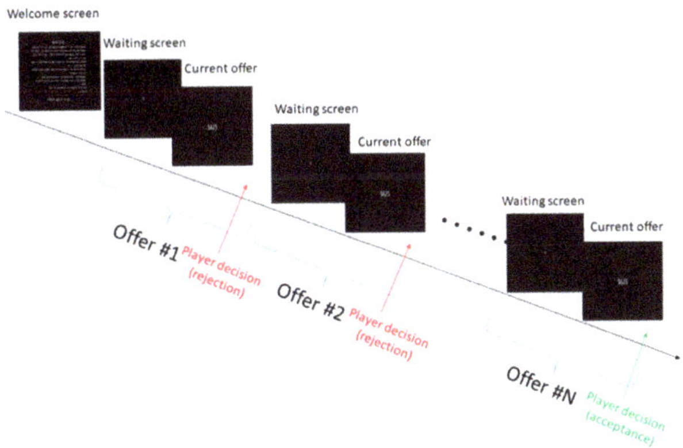

Figure 3. The structure of one secretary game block.

Before beginning data collection, participants underwent a thorough briefing and were equipped with an EEG cap. Two training sessions were conducted to ensure they were comfortable with the task and the experimental environment. These sessions, held without EEG recording, were important for familiarizing participants with the task.

EEG signals were recorded using a 16-channel active EEG amplifier (e.g., USBAMP by g.tec, Schiedlberg, Austria) operating at a sampling rate of 512 Hz, following the 10–20 international system for electrode placement. The electrode impedance was kept below 5 Kohm, monitored through OpenVibe software (version 3.6.0). For data preprocessing, we applied a 1–30 Hz FIR bandpass filter and used Independent Component Analysis (ICA) to separate neural signals from artifacts. The EEGLAB software's (version 2023.1) algorithm, described by Delorme and Makeig (2004) [38] facilitated this process. We identified and eliminated artifacts such as eye movements, blinks, muscle activity from facial and neck regions, and electrical noise from the environment. Specifically, a notch FIR filter was used to remove the 50 Hz line noise often introduced by power lines.

2.4. Data Processing and Analysis

In our analysis, we focused on a fixed 2-s "decision-making" window for each round of the secretary game. This window spans from 1 s before the participant's decision (pressing "Y" or "N") to 1 s after the decision. Each participant completed 6 game blocks (rounds), yielding a total of 162 decision-window epochs (27 participants $\times$ 6 blocks). Because each game round could end at a different offer (out of 20 possible offers) depending on when the participant made their final decision, the length of each block (and the exact timing of the decision within it) varied across trials.

The end-to-end pipeline comprises five stages:

1. Compute coherence between EEG channels and construct a 16 $\times$ 16 brain network (coherence graph) for each epoch.
2. Add a virtual central node to the coherence graph and apply the node2vec algorithm to embed the graph into a feature vector representation [27].

3. Reduce the dimensionality of the resulting feature vectors using Principal Component Analysis (PCA) to retain the most important features [39].
4. Use the resulting features as input to train an XGBoost ensemble decision-tree classification model to predict attachment class [33].
5. Evaluate the model using nested 5-fold cross-validation at the participant level to tune hyperparameters and assess performance while preventing overfitting.

2.4.1. EEG Coherence

In signal processing, coherence is a statistical measure of the relationship between two signals. It is defined for two signals x(t) and y(t) by the formula:

$$C_{xy}(f) = \frac{|G_{xy}(f)|^2}{G_{xx}(f) * G_{yy}(f)}$$

where $G_{xy}(f)$ is the cross-spectral density between x and y, and $G_{xx}(f)$ and $G_{yy}(f)$ are the auto-spectral densities of x(t) and y(t), respectively. The cross-spectral density can be calculated as $G_{xy}(f) = X(f)Y^*(f)$, where $X(f)$ and $Y^*(f)$ are the Fourier transforms of x(t) and the complex conjugate of y(t). Coherence has been used to efficiently quantify cortical connectivity in EEG signals [40]. EEG coherence indicates the level of synchronization between two brain regions of the same person or the similarity of brain activity in the same region between two different people [40,41]. There is a direct relationship between the coherence index and brain synchronization: a higher coherence value signifies stronger synchronization. Moreover, synchronization in specific EEG frequency bands (e.g., alpha, theta, beta) has been linked to different cognitive processes [42].

The coherence value ranges from -1 to 1 (0 = no correlation, analogous to a correlation coefficient). For each 2-s decision epoch (1 s before to 1 s after the key-press), we constructed a 16×16 coherence graph: nodes are EEG electrodes; edge weights are the corresponding coherence values. Figure 4 shows an example graph in which edges are displayed only when |coherence| > 0.5. Nodes follow the 10–20 system.

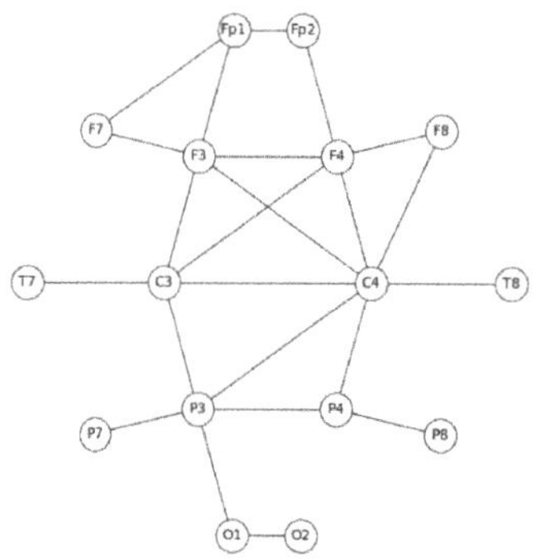

Figure 4. Example coherence graph for one 2-s decision epoch (edges discretized on a 0.5 threshold).

We chose the absolute-coherence threshold of 0.5 for three practical reasons: (i) values below ≈ 0.4 are often indistinguishable from noise in short EEG epochs [43]; (ii) a 0.5 cut-off yields a sparse yet fully connected graph, which stabilizes node2vec random-walk statistics and prevents the embedding from over-fitting small, noisy edge weights; and (iii) in a pilot sweep (θ = 0.3–0.7) the downstream XGBoost accuracy peaked near θ = 0.5, indicating that this level balances information retention with model generalizability in our limited sample.

While we used a fixed threshold of 0.5 to binarize the coherence matrices, this choice followed an informal evaluation of values ranging from 0.3 to 0.7 in increments of 0.1. We found that 0.5 offered the most stable and interpretable network properties. However, we recognize that using a static threshold has limitations. Such fixed values can introduce arbitrary cutoffs that may distort the underlying network topology, especially in graphs that are sparse or include a relatively small number of nodes. Although this method is simple, commonly used in studies with similarly sized graphs, and easy to interpret, it does not rest on a formal theoretical foundation. That said, comparative analyses have shown that the fixed threshold method performs robustly across a range of thresholds and yields similar classification power to weighted approaches in distinguishing clinical populations, such as in Alzheimer's disease [44]. Recent studies have also proposed more rigorous alternatives, such as percolation-based thresholding, which adaptively select thresholds that preserve meaningful topological features across different graph densities. These methods help maintain both global connectivity and local specificity (e.g., [45–47]). Looking ahead, future work should consider using such data-driven strategies to more accurately capture the fine-grained modular and functional organization of brain networks, particularly when studying conditions where weaker or local connections are disrupted.

2.4.2. Transforming the Coherence Graph into a Feature Vector

To convert each 16×16 coherence graph into a fixed-length vector, we first added a virtual hub node that connects to every electrode with unit-weight edges, following the virtual-node embedding scheme of [32]. This hub keeps the graph connected even after thresholding ($|\text{coherence}| > 0.5$), propagates global context, and stabilizes random-walk statistics—an important consideration with a limited participant pool.

We then embedded the augmented graph using the node2vec algorithm [27]. Guided walks of length 10 (256 walks per node) were generated with hyper-parameters $p = 0.5$ (local bias) and $q = 2$ (global bias), values selected in the inner fold of the nested cross-validation (2.4.4). The embedding dimension was set to 64, capturing both fine-scale synchrony and broader network topology mediated by the virtual node (Figure 5). Although 64 dimensions are modest, pilot tests showed that removing noisy or redundant axes improved generalizability. We therefore applied Principal Component Analysis [39] to the 64-D embeddings.

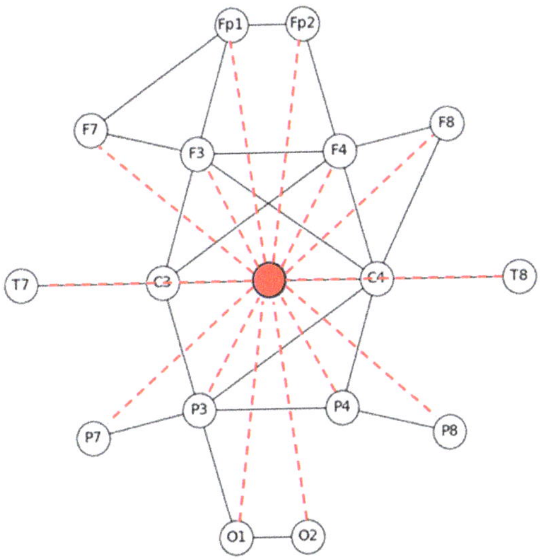

Figure 5. Coherence graph with virtual hub node (red dashed edges) prior to node2vec embedding.

The scree eigenvalue graph (Figure 6) shows a clear knee at component 28; beyond this point, each additional component contributes less than 2% of the cumulative variance. Keeping the first 28 components preserves ≈ 90% of the variance while reducing the risk of overfitting. These 28-element feature vectors are used as input to the XGBoost classifier in Section 2.4.3.

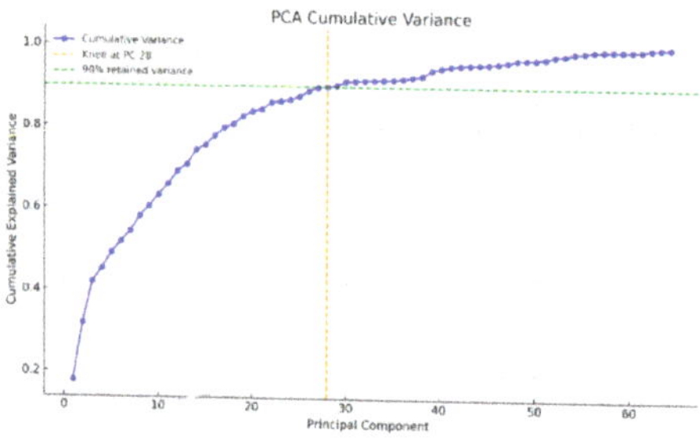

Figure 6. PCA scree plot for the 64-D node2vec embeddings.

2.4.3. Classification Model (XGBoost)

We employed XGBoost [33] to build an ensemble of gradient-boosted decision trees, chosen for its strong regularization [33,48] and documented success on high-dimensional biomedical data [49–52]. Built-in L1/L2 penalties limit overfitting [53]; tree pruning, parallelized training, and sparse-aware splits further enhance performance [54].

2.4.4. Model Training and Validation

To obtain an unbiased performance estimate we used nested 5-fold cross-validation at the participant level. This protocol is widely recommended for small-sample neuroimaging studies [55] and guards against optimistic bias [56].

- Outer loop: In each of five folds, data from ~5–6 participants were held out as the external test set; the remaining participants formed the training set.
- Inner loop: A second 5-fold split on the training participants tuned XGBoost hyperparameters (learning rate, tree depth, subsample ratio.

Because all six epochs from any participant reside wholly in either the training or the test set, person-level data leakage is eliminated. Testing on each unseen group in turn forces the model to generalize to new individuals—crucial with only 27 subjects. Final metrics are averaged across the five outer folds.

In addition to the XGBoost classifier described above, we also tested support vector machines (SVM) with both linear and non-linear kernels, as well as multivariate linear regression models. These alternative classifiers consistently yielded substantially lower classification performance and were therefore excluded from the final analysis.

3. Results

We can see in Figure 6 a graphical representation of the first three principal components that maintain 42% of the variance.

Figure 7 illustrates the first three principal components, capturing 42% of the variance in our data. This graph shows the distribution of the four attachment styles—Secure, Avoidant, Anxiously Attached, and Fearful Avoidant—across these components. Each point represents an individual's feature vector reduced to three dimensions, making it easier to see how well our model distinguishes between the different attachment styles. Notably, the Secure group (blue) forms a distinct cluster, reflecting the high accuracy of our model in predicting this style. The Fearful Avoidant group (yellow) also clusters together, indicating our model's success in identifying this style. However, the Anxiously Attached (green) and Avoidant (red) groups overlap more, demonstrating the model's challenges in distinguishing between these two attachment styles.

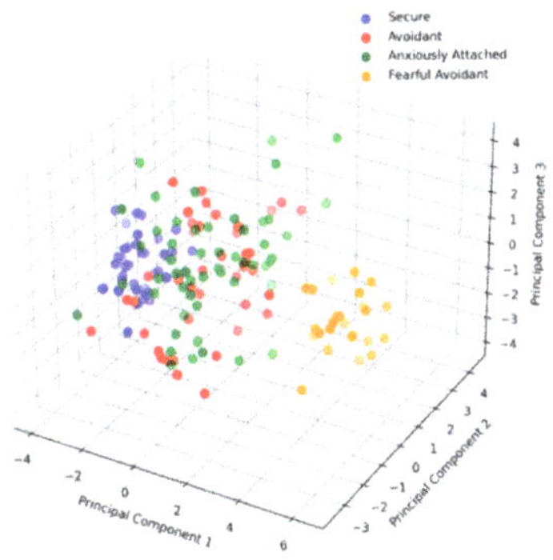

Figure 7. Graphic representation of the first three Principal components.

This visualization underscores the value of using principal component analysis to simplify the complexity of EEG coherence data. By focusing on the most significant components, we can more easily identify patterns and relationships within the data. The graph highlights the effectiveness of our approach in capturing the underlying structure of EEG data and points to areas where further refinement is needed, particularly in differentiating between the Anxiously Attached and Avoidant styles.

Table 1 presents the classification results. The model accurately identified Secure and Fearful-avoidant categories with around 80% and 88% accuracy, respectively. However, it found it more challenging to differentiate Avoidant and Anxious categories, with accuracy between 57% and 65%. The most frequent misclassifications were between these two categories: 11 Avoidant instances were incorrectly classified as Anxious, and 16 Anxious instances were classified as Avoidant. Figure 7, which plots the first three principal components, reinforces this point: Avoidant and Anxious observations form a single, largely inseparable cluster in the EEG-coherence space, whereas Secure and Fearful-avoidant remain distinct.

Table 1. Results of 4-class classification.

		Predicted				
		Secure	**Avoidant**	**Anxious**	**Fearful Avoidant**	**Recall**
Real	Secure	29	3	4	0	80.55%
	Avoidant	2	28	11	1	53.70%
	Anxious	5	16	31	2	57.41%
	Fearful avoidant	0	2	1	22	88.00%
Precision		80.55%	57.14%	65.95%	88.00%	Accuracy 68%

The difficulty in distinguishing between the Avoidant and Anxious categories may stem from how these attachment styles are divided. The ECR-R questionnaire scores between 1 and 7 for avoidance and anxiety. The k-means algorithm then sets thresholds to define attachment styles. For instance, someone with an anxiety score of 3.25 might be classified as Secure, while someone with a score of 3.75 might be classified as Anxious. This arbitrary separation creates challenges for the classification algorithm. Additionally, Avoidant and Anxious categories rely mainly on one dimension—either avoidance or anxiety—whereas Secure and Fearful-avoidant categories consider both dimensions. This likely contributes to the model's higher accuracy for these latter categories. In light of the overlap observed here—and in keeping with prior work showing that Anxious and Avoidant participants share a similar position on the R_attachment continuum [36]—we judged it methodologically prudent to treat them as a single Insecure category in the present analysis, while still reporting the original four-class results for full transparency.

To further investigate, we developed a new classification model focusing on Secure, Insecure, and Extremely Insecure attachments. Secure attachments exhibit low anxiety and low avoidance, Extremely Insecure attachments show high anxiety and high avoidance, and Insecure attachments have high anxiety or high avoidance, but not both. We trained a new XGBoost model on the same data and parameters to classify these three categories. We grouped the Avoidant and Anxious categories into one Insecure category due to the significant overlap and difficulty in distinguishing between them. Both categories exhibit high emotional dysregulation, manifesting as either high anxiety or high avoidance [57].

We chose to classify Avoidant and Anxious as "Insecure" rather than "Extreme Insecure" because, while both show high emotional dysregulation, they are characterized by high scores in only one dimension (either anxiety or avoidance). "Extreme Insecure" includes those with high scores in both dimensions, making the "Insecure" category more appropriate for those with significant distress in one area.

As shown in Table 2, the new model achieved around 88% overall accuracy, with approximately 93% accuracy for the Insecure group. Our EEG coherence graph technique effectively classified individuals with insecure attachment styles. These results suggest two potential improvements: using a larger dataset and employing a regression model. Our current dataset included 27 participants, with 6 in the Secure category, 7 in Avoidant, 9 in Anxious, and 5 in Fearful-avoidant. A larger, more balanced dataset could improve learning. Alternatively, a regression model predicting exact ECR-R scores for avoidance and anxiety might provide more precise classifications than the traditional four-category approach.

Table 2. Classification results on 3 attachment classes.

		Predicted			
		Secure	**Insecure**	**Extreme Insecure**	**Recall**
Real	Secure	32	4	0	88.88%
	Insecure	5	89	2	92.71%
	Extreme Insecure	0	3	22	88.00%
Precision		86.49%	92.71%	91.66%	Accuracy 88.27%

4. Discussion

Attachment style is a key psychological construct that shapes behavior in relationships, work, and other life domains [58–60]. In this study, we set out to classify participants' attachment styles using EEG data recorded during a sequential decision-making task: the secretary problem. Twenty-seven participants, representing the four main attachment categories, completed six task sessions each while EEG data were collected. These signals were segmented into 2-s windows and transformed into coherence graphs representing connectivity across EEG channels. Using Node2Vec, we embedded each graph into a 64-dimensional feature space, then applied Principal Component Analysis (PCA) to reduce these to 28 key components. These features served as input for an XGBoost classifier aimed at predicting attachment style. We also note that alternative models such as SVM and multivariate regression were tested but yielded lower accuracy, supporting our choice of classifier for the main analysis.

The model classified Secure and Fearful-avoidant attachment styles with high accuracy (80% and 88%, respectively), but was less accurate for Anxious and Avoidant styles (57–65%). It reliably identified individuals with either low (Secure) or high (Fearful-avoidant) levels of both anxiety and avoidance. However, it struggled with cases where anxiety and avoidance were asymmetrical, with one being high and the other low. This aligns with prior work noting significant behavioral and emotional overlap between Anxious and Avoidant styles [61]. Both reflect underlying insecurity, though their coping strategies differ markedly.

To address this, we combined Anxious and Avoidant types into a single Insecure category, which improved model accuracy to around 92%. This consolidation is consistent with findings that anxiety and avoidance often co-exist and may reflect a shared emotional core. While Anxious individuals typically seek proximity and reassurance, Avoidant individuals distance themselves emotionally, yet both responses stem from relational insecurity [24,62,63]. This conceptual shift supports growing evidence that attachment should be viewed dimensionally rather than categorically.

Our results also challenge the validity of self-report questionnaires that impose rigid classifications. The fluidity reflected in our EEG-based model suggests that attachment behaviors lie along a continuum. For example, taxometric analyses have shown that anxiety and avoidance vary along continuous gradients [64]. Research using both categorical and dimensional approaches have further demonstrated that insecure attachment patterns often overlap in ways that defy binary labeling [24,65].

Node2Vec generates graph embeddings by simulating biased random walks, which allows it to capture both local and global connectivity patterns across the network [27]. In our context, this property is useful for modeling EEG coherence graphs that may include both short-range synchrony and longer-range inter-regional interactions. We tuned key node2vec parameters, including walk length, number of walks, and return/in-out biases, to better reflect the topology of our small-scale graphs. In addition, we introduced a virtual hub node to preserve full connectivity and prevent walk fragmentation. This strategy is consistent with recent findings showing that embeddings, even in small graphs, can recover informative structural patterns when appropriately configured [66–68]. PCA then filtered these representations to retain the most informative features. This contributed to the model's success in identifying Secure and Fearful-avoidant profiles [69].

Recent studies have demonstrated the effectiveness of graph-based and embedding techniques in brain network research. For example, vector embeddings have been used to map structure-function relationships in connectomes [70], EEG coherence networks have been shown to exhibit modular structure amenable to graph analysis [71], and graph-theoretical features derived from EEG connectivity have been applied to classification tasks involving clinical populations [72].

Despite our model's success, several limitations remain. The small sample of 27 participants may not fully represent the diversity of attachment styles in the broader population. Although each participant contributed six Secretary Problem rounds (162 coherence graphs in total), external validity is still limited by the number of distinct individuals. The imbalance in participant numbers across attachment styles may also have affected performance, particularly for the Avoidant and Anxious categories, because class sizes were unequal. We mitigated some small sample bias by using participant-level nested 5-fold cross-validation, which prevents data leakage between training and testing. Nevertheless, the results should still be interpreted with caution given the modest cohort. Another limitation is the reliance on discrete attachment groups, which may oversimplify the continuous nature of attachment-related anxiety and avoidance. In addition, although we applied symmetric 1-s pre/post response windows and excluded frequencies above 30 Hz to reduce muscle-related artifacts, we acknowledge that beta-band activity may still partially reflect motor processes. Since all participants performed the same brief button press across all trials, any residual motor contamination is expected to be uniformly distributed and unlikely to bias the results across conditions. Nevertheless, this remains a potential limitation of our approach and should be examined more thoroughly in future studies, for example by varying analysis windows or isolating motor components. Finally, our use of a fixed |coherence| > 0.5 threshold, while empirically justified in short epoch EEG studies, imposes a binary view of functional connectivity and could restrict the model's sensitivity to subtler network variations.

Our study used the Secretary Problem task to probe how attachment styles shape sequential decision-making. Future research should recruit substantially larger and more evenly balanced cohorts, ideally through multi-site sampling, where two or more independent laboratories or clinics collect data using a harmonized protocol. Such multi-site designs expand the participant pool, broaden the range of demographics and equipment settings represented, and reduce the risk that findings are influenced by site-specific factors [73,74]. Because attachment-related anxiety and avoidance lie on continuous dimensions, upcoming models should move beyond discrete classes and adopt dimensional or multi-label regression that directly predicts each participant's anxiety and avoidance scores [75,76]. Methodologically, recording high-density EEG (64–128 channels) and applying source localization methods, such as sLORETA or beamforming, can provide finer neuroanatomical precision [77,78], enabling researchers to map coherence patterns to specific cortical generators. Expanding the paradigm to additional real-life decision contexts will further clarify how attachment influences behavior across situations. Finally, integrating objective physiological measures (e.g., EEG-based coherence graphs) with traditional questionnaires may yield hybrid tools that improve the assessment of attachment and inform personalized therapeutic interventions.

Several key takeaways emerged from our work. First, modeling EEG signals as coherence graphs for classification models is feasible. While EEG data collection can be costly and time-consuming, our research showed that even a small dataset could be effective. Second, our findings suggest that graph embedding methods like Node2Vec may help reveal latent topological patterns that contribute to classification. While these embeddings operate in a space that limits direct interpretability, their utility is reflected in the model's ability to generalize across participants. This generalization was evaluated using participant-level nested cross-validation, where all data from held-out individuals were excluded from training. The classifier's success in predicting unseen participants' attachment styles suggests that the embedding captures transferable features of brain connectivity. Future studies could complement this approach with ablation analyses or interpretable, handcrafted features to further clarify the contribution of specific graph components. Third, while Secure and Fearful-avoidant styles were accurately classified, distinguishing between Anxious and Avoidant styles remains challenging due to their overlapping characteristics.

Another important design choice was our use of EEG coherence to construct functional connectivity graphs. Coherence was selected because it captures the statistical coupling between signals from different brain regions, offering a measure of large-scale neural coordination that aligns with our network-level hypothesis. Unlike raw EEG or PCA-reduced time series data, which emphasize variance within individual electrodes, coherence preserves inter-regional relationships that are central to understanding distributed processing. While future work could benchmark raw EEG features, we selected coherence because it supports graph-theoretical modeling and reflects biologically grounded inter-regional connectivity. It has been widely used to study functional brain networks, especially when large-scale coordination is central to the cognitive process [79].

In conclusion, EEG coherence graphs and machine learning can classify attachment styles effectively. By addressing the fluidity within insecure attachment categories [64] and using advanced graph embedding techniques, we can develop better models for understanding human behavior. Even though EEG data collection is complex and resource-intensive, our research suggests that coherence graphs can reveal valuable patterns even in small datasets. Using techniques like Node2Vec and XGBoost helped us classify Secure and Fearful-avoidant styles accurately, though it was challenging to distinguish between Anxious and Avoidant styles due to their overlapping characteristics. XGBoost enhanced the model's generalizability, particularly by handling variation across EEG signals through its built-in regularization mechanisms [80]. Nonetheless, the model's difficulties with Anxious and Avoidant cases underscore the complexity of classifying individuals with overlapping emotional and behavioral traits [81].

Although recent large-sample EEG studies have applied deep CNN or graph neural architectures successfully when thousands of trials are available [82,83], the present dataset with 162 coherence graphs from 27 participants favors a lighter pipeline. Node2Vec embeddings coupled with XGBoost strike a practical balance between model capacity and sample size in this context. This method provided detailed insights into the neurophysiological patterns of different attachment behaviors. However, since node2vec is commonly applied to larger graphs, we acknowledge that using this technique on a 16-node network, as in our case, presents certain limitations. Small graphs restrict the diversity of random walk paths and reduce the expressiveness of the resulting embeddings. To address this, we tuned key parameters, including the number and length of walks, as well as the return and in-out biases, to capture as much structural information as possible within the network's limited topology. We also introduced a virtual hub node to maintain full connectivity and prevent walk fragmentation. This design aligns with recent findings that highlight the importance of walk modeling precision, particularly when sampling space is constrained [84].

Despite its common application in larger graphs, our results indicate that, even in small graphs, node2vec can extract informative structural patterns when parameters are tuned to the graph's scale. Importantly, our goal was not to identify 'hubs', but to preserve and extract broader patterns of functional connectivity. This approach is supported by interpretability frameworks showing that embeddings, even in small graphs, can capture meaningful subgraph structures [Towards Interpretation of Node Embeddings]. In addition, the use of standardized pipelines for evaluating graph construction methods helps ensure reproducibility and methodological rigor in low-resolution settings [85]. Finally, meaningful structural recovery remains feasible even in small or noisy graphs when combined with appropriate constraints and preprocessing [86].

We emphasize that this is a proof-of-concept, demonstrating node2vec's complementary value alongside traditional graph metrics. We plan to explore alternative approaches in the future, such as spectral methods, simpler graph statistics, or CNNs, particularly in settings where higher node counts or more granular data are available. Future studies should explore attachment behaviors in varied contexts and incorporate broader and more diverse samples to extend the reach and validity of these methods.

Author Contributions: Conceptualization, D.M., I.L. and I.Z.; Methodology, D.M., I.L. and I.Z.; Software, D.M.; Validation, D.M., I.L. and I.Z.; Formal analysis, D.M., I.L. and I.Z.; Investigation, D.M., I.L. and I.Z.; Data curation, D.M., I.L. and I.Z.; Writing—original draft, D.M., I.L. and I.Z.; Writing—review & editing, D.M., I.L. and I.Z.; Visualization, D.M. All authors have read and agreed to the published version of the manuscript.

Funding: This research received no external funding.

Institutional Review Board Statement: The experimental protocols used in this work were evaluated and approved by the Ethics Committee of Ariel University (confirmation number: AU-ENG-IZ-20220404, 4 April 2022). Permission to perform the electrophysiological recordings in the experiment was granted for the period from 4 February 2022, to 4 February 2023. All methods were carried out in accordance with relevant guidelines and regulations.

Informed Consent Statement: Written informed consent was obtained from all subjects and/or their legal guardian(s) involved in the study. It is important to note that no minors were involved in this study.

Data Availability Statement: All the experimental data, which include the players' electrophysiological recordings and the corresponding secretary problem game logs, are stored on the servers of Ariel University. The data can be obtained by request from The IRB member, Chen Hajaj (chenha@ariel.ac.il) or from one of the authors (Dor Mizrahi—dor.mizrahi1@msmail.ariel.ac.il, Ilan Laufer—ilanl@ariel.ac.il, Inon Zuckerman—inonzu@ariel.ac.il).

Conflicts of Interest: The authors declare no conflict of interest.

References

1. Bretherton, I. Attachment theory: Retrospect and prospect. *Monogr. Soc. Res. Child Dev.* **1985**, *50*, 3–35. [CrossRef]
2. Hazan, C.; Shaver, P. Romantic love conceptualized as an attachment process. *J. Personal. Soc. Psychol.* **1987**, *52*, 511. [CrossRef]
3. Fearon, R.P.; Roisman, G.I. Attachment theory: Progress and future directions. *Curr. Opin. Psychol.* **2017**, *15*, 131–136. [CrossRef]
4. Mikulincer, M.; Shaver, P.R. Applications of attachment theory and research: The blossoming of relationship science. In *Applications of Social Psychology*; Routledge: Abingdon, UK, 2020; pp. 187–206.
5. Sutton, T.E. Review of attachment theory: Familial predictors, continuity and change, and intrapersonal and relational outcomes. *Marriage Fam. Rev.* **2019**, *55*, 1–22. [CrossRef]
6. Van Petegem, S.; Beyers, W.; Brenning, K.; Vansteenkiste, M. Exploring the association between insecure attachment styles and adolescent autonomy in family decision making: A differentiated approach. *J. Youth Adolesc.* **2013**, *42*, 1837–1846. [CrossRef]
7. Bolat, N.; Odacı, H. High school final year students' career decision-making self-efficacy, attachment styles and gender role orientations. *Curr. Psychol.* **2017**, *36*, 252–259. [CrossRef]
8. Deniz, M. An investigation of decision making styles and the five-factor personality traits with respect to attachment styles. *Educ. Sci. Theory Pract.* **2011**, *11*, 105–113.
9. Gander, M.; Buchheim, A. Attachment classification, psychophysiology and frontal EEG asymmetry across the lifespan: A review. *Front. Hum. Neurosci.* **2015**, *9*, 79. [CrossRef]
10. Ravitz, P.; Maunder, R.; Hunter, J.; Sthankiya, B.; Lancee, W. Adult attachment measures: A 25-year review. *J. Psychosom. Res.* **2010**, *69*, 419–432. [CrossRef]
11. Sibley, C.G.; Fischer, R.; Liu, J.H. Reliability and validity of the revised experiences in close relationships (ECR-R) self-report measure of adult romantic attachment. *Personal. Soc. Psychol. Bull.* **2005**, *31*, 1524–1536. [CrossRef]
12. Cho, Y.J.; Ryu, J.S.; Seok, J.-H.; Kim, E.; Oh, J.; Kim, B.-H. Machine Learning Prediction of Attachment Type From Bio-Psychological Factors in Patients With Depression. *Psychiatry Investig.* **2025**, *22*, 412. [CrossRef]
13. Verbeke, W.J.M.I.; Pozharliev, R.; Strien, J.W.V.; Belschak, F.; Bagozzi, R.P. "I am resting but rest less well with you." The moderating effect of anxious attachment style on alpha power during EEG resting state in a social context. *Front. Hum. Neurosci.* **2014**, *8*, 486. [CrossRef]
14. Sloan, E.P.; Maunder, R.G.; Hunter, J.J.; Moldofsky, H. Insecure attachment is associated with the α-EEG anomaly during sleep. *Biopsychosoc. Med.* **2007**, *1*, 20. [CrossRef]
15. Rognoni, E.; Galati, D.; Costa, T.; Crini, M. Relationship between adult attachment patterns, emotional experience and EEG frontal asymmetry. *Personal. Individ. Differ.* **2008**, *44*, 909–920. [CrossRef]
16. Zuckerman, I.; Mizrahi, D.; Laufer, I. Attachment style, emotional feedback, and neural processing: Investigating the influence of attachment on the P200 and P400 components of event-related potentials. *Front. Hum. Neurosci.* **2023**, *17*, 1249978. [CrossRef]

17. Kuo, C.-C.; Ha, T.; Ebbert, A.M.; Tucker, D.M.; Dishion, T.J. Dynamic Responses in Brain Networks to Social Feedback: A Dual EEG Acquisition Study in Adolescent Couples. *Front. Comput. Neurosci.* **2017**, *11*, 46. [CrossRef]
18. Labek, K.; Karch, S.; Taubner, S.; Kessler, H.; Kächele, H.; Cierpka, M.; Roth, G.; Chrobok, A.; Pogarell, O.; Buchheim, A. EPA-1756-EEG activity in the gamma band and late positive potential (LPP) during processing attachment-related stimuli. *Eur. Psychiatry* **2014**, *29*, 1. [CrossRef]
19. Dhake, D.; Angal, Y. A Comparative Analysis of EEG-based Stress Detection Utilizing Machine Learning and Deep Learning Classifiers with a Critical Literature Review. *Int. J. Recent Innov. Trends Comput. Commun.* **2023**, *11*, 61–73. [CrossRef]
20. Xu, Y.; Xu, X.; Deng, L. EEG research based on the influence of different music effects. *J. Phys. Conf. Ser.* **2020**, *1631*, 012147. [CrossRef]
21. Mizrahi, D.; Zuckerman, I.; Laufer, I. Exploring Cognitive Load in the Secretary Problem Using EEG Signals. In Proceedings of the Intelligent Systems Conference, Amsterdam, The Netherlands, 5–6 September 2024; Springer Nature: Cham, Switzerland, 2024; pp. 664–674.
22. Mizrahi, D.; Laufer, I.; Zuckerman, I. Neurophysiological insights into sequential decision-making: Exploring the secretary problem through ERPs and TBR dynamics. *BMC Psychol.* **2024**, *12*, 245. [CrossRef]
23. Seale, D.A.; Rapoport, A. Sequential decision making with relative ranks: An experimental investigation of the "secretary problem". *Organ. Behav. Hum. Decis. Process.* **1997**, *69*, 221–236. [CrossRef]
24. Mikulincer, M.; Shaver, P.R. *Attachment in Adulthood: Structure, Dynamics, and Change*; Guilford Publications: New York, NY, USA, 2010.
25. Vrtička, P.; Andersson, F.; Grandjean, D.; Sander, D.; Vuilleumier, P. Individual attachment style modulates human amygdala and striatum activation during social appraisal. *PLoS ONE* **2008**, *3*, e2868. [CrossRef]
26. Finn, E.S.; Shen, X.; Scheinost, D.; Rosenberg, M.D.; Huang, J.; Chun, M.M.; Papademetris, X.; Constable, R.T. Functional connectome fingerprinting: Identifying individuals using patterns of brain connectivity. *Nat. Neurosci.* **2015**, *18*, 1664–1671. [CrossRef]
27. Grover, A.; Leskovec, J. node2vec: Scalable feature learning for networks. In Proceedings of the 22nd ACM SIGKDD International Conference on Knowledge Discovery and Data Mining, San Francisco, CA, USA, 13–17 August 2016.
28. Li, S.; Zaidi, N.A.; Du, M.; Zhou, Z.; Zhang, H.; Li, G. Property graph representation learning for node classification. *Knowl. Inf. Syst.* **2024**, *66*, 237–265. [CrossRef]
29. Freeman, P.R. The secretary problem and its extensions: A review. *Int. Stat. Rev. Int. Stat.* **1983**, *51*, 189–206. [CrossRef]
30. Quiroz, G.; Espinoza-Valdez, A.; Salido-Ruiz, R.A.; Mercado, L. Coherence analysis of EEG in locomotion using graphs. *Rev. Mex. Ing. Biomédica* **2017**, *38*, 235–246. [CrossRef]
31. Zuckerman, I.; Mizrahi, D.; Laufer, I. EEG Pattern Classification of Picking and Coordination Using Anonymous Random Walks. *Algorithms* **2022**, *15*, 114. [CrossRef]
32. Li, Y.; Tarlow, D.; Brockschmidt, M.; Zemel, R. Gated Graph Sequence Neural Networks. In Proceedings of the International Conference on Learning Representations, San Juan, Puerto Rico, 2–4 May 2016.
33. Chen, T.; Guestrin, C. Xgboost: A scalable tree boosting system. In Proceedings of the 22nd ACM SIGKDD International Conference on Knowledge Discovery and Data Mining, San Francisco, CA, USA, 13–17 August 2016; pp. 785–794.
34. Xu, R.; Wunsch, D. Survey of clustering algorithms. *IEEE Trans. Neural Netw.* **2005**, *16*, 645–678. [CrossRef]
35. Waters, E.; Merrick, S.; Treboux, D.; Crowell, J.; Albersheim, L. Attachment security in infancy and early adulthood: A twenty-year longitudinal study. *Child Dev.* **2000**, *71*, 684–689. [CrossRef]
36. Laufer, I.; Mizrahi, D.; Zuckerman, I. Enhancing EEG-Based Attachment Style Prediction: Unveiling the Impact of Feature Domains. *Front. Psychol.* **2024**, *15*, 1326791. [CrossRef]
37. Hsiao, Y.-C.; Kemp, S. The effect of incentive structure on search in the secretary problem. *Judgm. Decis. Mak.* **2020**, *15*, 182–192. [CrossRef]
38. Delorme, A.; Makeig, S. EEGLAB: An open source toolbox for analysis of single-trial EEG dynamics including independent component analysis. *J. Neurosci. Methods* **2004**, *134*, 9–21. [CrossRef]
39. Jolliffe, I.T.; Cadima, J. Principal component analysis: A review and recent developments. *Philos. Trans. R. Soc. A Math. Phys. Eng. Sci.* **2016**, *374*, 20150202. [CrossRef] [PubMed]
40. Murias, M.; Webb, S.J.; Greenson, J.; Dawson, G. Resting state cortical connectivity reflected in EEG coherence in individuals with autism. *Biol. Psychiatry* **2007**, *62*, 270–273. [CrossRef]
41. Basharpoor, S.; Heidar, F.; Molavi, P. EEG coherence in theta, alpha, and beta bands in frontal regions and executive functions. *Appl. Neuropsychol. Adult* **2021**, *28*, 310–317. [CrossRef]
42. Rippon, G.; Brock, J.; Brown, C.; Boucher, J. Disordered connectivity in the autistic brain: Challenges for the 'new psychophysiology'. *Int. J. Psychophysiol.* **2007**, *63*, 164–172. [CrossRef]
43. Bocci, T.; Moretto, C.; Tognazzi, S.; Briscese, L.; Naraci, M.; Leocani, L.; Mosca, F.; Ferrari, M.; Sartucci, F. How does a surgeon's brain buzz? An EEG coherence study on the interaction between humans and robot. *Behav. Brain Funct.* **2013**, *9*, 14. [CrossRef]

44. Ahmadi, H.; Fatemizadeh, E.; Nasrabadi, A.M. A comparative study of the effect of weighted or binary functional brain networks in fMRI data analysis. *Front. Biomed. Technol.* **2020**, *7*. [CrossRef]
45. Nicolini, C.; Bordier, C.; Bifone, A. Community detection in weighted brain connectivity networks beyond the resolution limit. *Neuroimage* **2017**, *146*, 28–39. [CrossRef]
46. Bardella, G.; Pani, P.; Brunamonti, E.; Giarrocco, F.; Ferraina, S. The small scale functional topology of movement control: Hierarchical organization of local activity anticipates movement generation in the premotor cortex of primates. *Neuroimage* **2020**, *207*, 116354. [CrossRef]
47. Bardella, G.; Giuffrida, V.; Giarrocco, F.; Brunamonti, E.; Pani, P.; Ferraina, S. Response inhibition in premotor cortex corresponds to a complex reshuffle of the mesoscopic information network. *Netw. Neurosci.* **2024**, *8*, 597–622. [CrossRef]
48. Wang, F.; Tian, Y.-C.; Zhang, X.; Hu, F. An ensemble of Xgboost models for detecting disorders of consciousness in brain injuries through EEG connectivity. *Expert Syst. Appl.* **2022**, *198*, 116778. [CrossRef]
49. Elgart, M.; Lyons, G.; Romero-Brufau, S.; Kurniansyah, N.; Brody, J.A.; Guo, X.; Lin, H.J.; Raffield, L.; Gao, Y.; Chen, H.; et al. Non-linear machine learning models incorporating SNPs and PRS improve polygenic prediction in diverse human populations. *Commun. Biol.* **2022**, *5*, 256. [CrossRef]
50. Tiwari, A.; Chaturvedi, A. A multiclass EEG signal classification model using spatial feature extraction and XGBoost algorithm. In Proceedings of the IEEE/RSJ International Conference on Intelligent Robots and Systems (IROS), Macau, China, 4–8 November 2019.
51. Deng, X.; Li, M.; Deng, S.; Wang, L. Hybrid gene selection approach using XGBoost and multi-objective genetic algorithm for cancer classification. *Med. Biol. Eng. Comput.* **2022**, *60*, 663–681. [CrossRef]
52. Lundberg, S.M.; Erion, G.; Chen, H.; DeGrave, A.; Prutkin, J.M.; Nair, B.; Katz, R.; Himmelfarb, J.; Bansal, N.; Lee, S.-I. Explainable AI for trees: From local explanations to global understanding. *arXiv* **2019**, arXiv:1905.04610. [CrossRef]
53. Zhai, W.; Wang, Y. Lasso Ridge based XGBoost and Deep_LSTM Help Tennis Players Perform better. *arXiv* **2024**, arXiv:2405.07030. [CrossRef]
54. Mitchell, R.; Frank, E. Accelerating the XGBoost algorithm using GPU computing. *PeerJ Comput. Sci.* **2017**, *3*, e127. [CrossRef]
55. Varma, S.; Simon, R. Bias in error estimation when using cross-validation for model selection. *BMC Bioinform.* **2006**, *7*, 91. [CrossRef]
56. Cawley, G.C.; Talbot, N.L. On over-fitting in model selection and subsequent selection bias in performance evaluation. *J. Mach. Learn. Res.* **2010**, *11*, 2079–2107.
57. Bartholomew, K.; Horowitz, L.M. Attachment styles among young adults: A test of a four-category model. *J. Personal. Soc. Psychol.* **1991**, *61*, 226. [CrossRef] [PubMed]
58. Sweet, J. The Association Between Attachment Style and Decision Making Style. Honors Thesis, Union College, Schenectady, NY, USA, 2021.
59. Blalock, D.V.; Franzese, A.T.; Machell, K.A.; Strauman, T.J. Attachment style and self-regulation: How our patterns in relationships reflect broader motivational styles. *Personal. Individ. Differ.* **2015**, *87*, 90–98. [CrossRef]
60. Harms, P.D. Adult attachment styles in the workplace. *Hum. Resour. Manag. Rev.* **2011**, *21*, 285–296. [CrossRef]
61. Mikulincer, M.; Orbach, I. Attachment styles and repressive defensiveness: The accessibility and architecture of affective memories. *J. Personal. Soc. Psychol.* **1995**, *68*, 917. [CrossRef]
62. Fraley, R.C.; Waller, N.G.; Brennan, K.A. An item response theory analysis of self-report measures of adult attachment. *J. Personal. Soc. Psychol.* **2000**, *78*, 350. [CrossRef]
63. Brennan, K.A.; Clark, C.L.; Shaver, P.R. Self-report measurement of adult attachment: An integrative overview. *Attach. Theory Close Relatsh.* **1998**, *46*, 70–100.
64. Fraley, R.C.; Hudson, N.W.; Heffernan, M.E.; Segal, N. Are adult attachment styles categorical or dimensional? A taxometric analysis of general and relationship-specific attachment orientations. *J. Personal. Soc. Psychol.* **2015**, *109*, 354. [CrossRef]
65. Levy, K.N.; Blatt, S.J.; Shaver, P.R. Attachment styles and parental representations. *J. Personal. Soc. Psychol.* **1998**, *74*, 407. [CrossRef]
66. Hacker, C.; Rieck, B. On the surprising behaviour of node2vec. *arXiv* **2022**, arXiv:2206.08252. [CrossRef]
67. Dalmia, A.; Gupta, M. Towards interpretation of node embeddings. In Proceedings of the Companion Proceedings of the The Web Conference 2018, Lyon, France, 23–27 April 2018; pp. 945–952.
68. Knežević, D.; Babić, J.; Savić, M.; Radovanović, M. Evaluation of LID-aware graph embedding methods for node clustering. In Proceedings of the International Conference on Similarity Search and Applications, Bologna, Italy, 5–7 October 2022; Springer International Publishing: Cham, Switzerland, 2022; pp. 222–233.
69. Gan, A.; Gong, A.; Ding, P.; Yuan, X.; Chen, M.; Fu, Y.; Cheng, Y. Computer-aided diagnosis of schizophrenia based on node2vec and Transformer. *J. Neurosci. Methods* **2023**, *389*, 109824. [CrossRef]

70. Rosenthal, G.; Váša, F.; Griffa, A.; Hagmann, P.; Amico, E.; Goñi, J.; Avidan, G.; Sporns, O. Mapping higher-order relations between brain structure and function with embedded vector representations of connectomes. *Nat. Commun.* **2018**, *9*, 2178. [CrossRef]
71. Cattai, T.; Caporali, C.; Corsi, M.-C.; Colonnese, S. Introducing the modularity graph: An application to brain functional networks. In Proceedings of the 2024 32nd European Signal Processing Conference (EUSIPCO), Lyon, France, 26–30 August 2024; IEEE: New York, NY, USA, 2024; pp. 1611–1615.
72. Kim, J.-G.; Kim, H.; Hwang, J.; Kang, S.H.; Lee, C.-N.; Woo, J.; Kim, C.; Han, K.; Kim, J.B.; Park, K.-W. Differentiating amnestic from non-amnestic mild cognitive impairment subtypes using graph theoretical measures of electroencephalography. *Sci. Rep.* **2022**, *12*, 6219. [CrossRef]
73. Paap, K.R.; Anders-Jefferson, R.T.; Balakrishnan, N.; Majoubi, J.B. The many foibles of Likert scales challenge claims that self-report measures of self-control are better than performance-based measures. *Behav. Res. Methods* **2024**, *56*, 908–933. [CrossRef]
74. McBride, C.; Atkinson, L.; Quilty, L.C.; Bagby, R.M. Attachment as moderator of treatment outcome in major depression: A randomized control trial of interpersonal psychotherapy versus cognitive behavior therapy. *J. Consult. Clin. Psychol.* **2006**, *74*, 1041. [CrossRef]
75. Fraley, R.C.; Waller, N.G. Adult attachment patterns: A test of the typological model. In *Attachment Theory and Close Relationships*; The Guilford Press: New York, NY, USA, 1998; pp. 77–114.
76. Gillath, O.; Karantzas, G. Attachment security priming: A systematic review. *Curr. Opin. Psychol.* **2019**, *25*, 86–98. [CrossRef]
77. Pascual-Marqui, R.D. Standardized low-resolution brain electromagnetic tomography (sLORETA): Technical details. *Methods Find. Exp. Clin. Pharmacol.* **2002**, *24*, 5–12.
78. Van Veen, B.; van Drongelen, W.; Yuchtman, M.; Suzuki, A. Localization of brain electrical activity via linearly constrained minimum variance spatial filtering. *IEEE Trans. Biomed. Eng.* **1997**, *44*, 867–880. [CrossRef]
79. Decker, S.; Fillmore, P.T.; Roberts, A. Coherence: The measurement and application of brain connectivity. *NeuroRegulation* **2017**, *4*, 3. [CrossRef]
80. Suresh, G.V.; Reddy, S. Uncertain data analysis with regularized XGBoost. *Webology* **2022**, *19*, 3722–3740. [CrossRef]
81. Dark-Freudeman, A.; Pond, R.S.J.; Paschall, R.E.; Greskovich, L. Attachment style in adulthood: Attachment style moderates the impact of social support on depressive symptoms. *J. Soc. Pers. Relatsh.* **2020**, *37*, 2871–2889. [CrossRef]
82. Rasool, A.; Aslam, S.; Xu, Y.; Wang, Y.; Pan, Y.; Chen, W. Deep neurocomputational fusion for ASD diagnosis using multi-domain EEG analysis. *Neurocomputing* **2025**, *641*, 130353. [CrossRef]
83. Bunterngchit, C.; Wang, J.; Su, J.; Wang, Y.; Liu, S.; Hou, Z.-G. Temporal attention fusion network with custom loss function for EEG–fNIRS classification. *J. Neural Eng.* **2024**, *21*, 066016. [CrossRef]
84. Liu, R.; Hirn, M.; Krishnan, A. Accurately modeling biased random walks on weighted networks using node2vec+. *Bioinformatics* **2023**, *39*, btad047. [CrossRef]
85. Zhao, W.; Zhou, D.; Qiu, X.; Jiang, W. A pipeline for fair comparison of graph neural networks in node classification tasks. *arXiv* **2020**, arXiv:2012.10619. [CrossRef]
86. Zhang, Y.; Tang, M. Exact recovery of community structures using DeepWalk and node2vec. *arXiv* **2021**, arXiv:2101.07354.

applied sciences

MDPI

Article

Slower Ageing of Cross-Frequency Coupling Mechanisms Across Resting-State Networks Is Associated with Better Cognitive Performance in the Picture Priming Task

Vasily A. Vakorin [1,2,*], Taha Liaqat [3], Hayyan Liaqat [4], Sam M. Doesburg [1], George Medvedev [2] and Sylvain Moreno [4]

1 Department of Biomedical Physiology and Kinesiology, Simon Fraser University, Burnaby, BC V5A 1S6, Canada
2 Royal Columbian Hospital, Fraser Health Authority, New Westminster, BC V3L 0A2, Canada
3 School of Public Health Sciences, University of Waterloo, Waterloo, ON N2L 3G1, Canada
4 School of Interactive Arts and Technology, Simon Fraser University, Surrey, BC V3T 0A3, Canada
* Correspondence: vasily_vakorin@sfu.ca

Abstract: The brain age gap (BAG), the divergence of an individual's neurobiologically predicted brain age from their chronological age, is a key indicator of brain health. While BAG can be derived from diverse brain metrics, its interpretation often polarizes between early-life trait influences and current state-dependent factors like cognitive decline. Here, we propose an integrative framework that moves beyond single summary statistics by considering the full distribution of brain metrics across regions or time. We distinguish between a neural system's "baseline" (typical values, e.g., mean) and its "capacity" (extreme values, e.g., maximum) within these distributions. To test this, we analyzed resting-state magnetoencephalography (MEG) from the Cam-CAN adult cohort, focusing on cross-frequency coupling (CFC) within functional MRI-defined networks. We derived network-specific CFC baseline (mean) and capacity (maximum) measures. Separate brain age prediction models were trained for each measure. The resulting BAGs (baseline-BAG and capacity-BAG) for each network were then correlated with cognitive performance on a picture priming task. Both baseline-BAG and capacity-BAG profiles showed associations with cognitive scores, with younger predicted brain age correlating with better performance. However, capacity-BAG exhibited more conclusive relationships, suggesting that metrics reflecting a neural system's peak operational ability (capacity) may better capture an individual's current cognitive state. These findings indicate that brain age models emphasizing neural capacity, rather than just baseline activity, could offer a more sensitive lens for understanding the state-dependent aspects of brain ageing.

Keywords: brain age; brain age gap; cross-frequency coupling; phase–amplitude coupling; magnetoencephalography; resting-state networks; brain–cognition associations; Cam-CAN study

1. Introduction

The concept of "brain age" has emerged as a valuable tool in neuroscience to quantify an individual's brain health relative to their chronological age [1]. Typically estimated from neuroimaging or neurophysiological data, predicted brain age serves as a potential biomarker for cognitive status, susceptibility to neurodegenerative diseases, and various psychiatric conditions [2]. The discrepancy between this predicted brain age and an individual's chronological age is termed the "brain age gap" (BAG) or delta. While BAG offers a compelling index, its interpretation remains a significant challenge, as it may encapsulate both enduring, early-life neurodevelopmental influences and more transient effects of current health or clinical deterioration [3].

Two primary, though not mutually exclusive, perspectives frame the interpretation of BAG: the "state" hypothesis and the "trait" hypothesis [4]. The state hypothesis posits that BAG reflects dynamic, potentially reversible brain function or structure changes linked to an individual's current physiological or pathological condition. Conversely, the trait hypothesis suggests that BAG is a more stable, enduring characteristic, possibly established early in life and reflecting long-term developmental trajectories or inherent predispositions.

The state hypothesis suggests that an increased brain age gap may indicate brain pathology. For instance, studies have shown that individuals with schizophrenia exhibit a greater brain age gap, which correlates with negative symptom severity, suggesting that the brain age gap could serve as a biomarker for psychosis conversion in high-risk individuals [5]. Alzheimer's disease (AD) and mild cognitive impairment (MCI) are associated with an increased BAG, indicating accelerated brain ageing. This gap is linked to cognitive impairment and is a predictive marker for disease progression [6]. MRI models have identified higher than control-based brain age for diabetes, multiple sclerosis, mild cognitive impairment, and Alzheimer's dementia [7]. In another MRI study, neurologically older brains were associated with heavy episodic drinking, higher blood pressure, and higher blood glucose [8].

The trait hypothesis proposes that the brain age gap is a stable, individual characteristic rather than a reflection of transient physiological states [4]. Several studies support this hypothesis, suggesting that the brain age gap primarily reflects inherent individual differences rather than current health conditions or temporary influences [3,4,9].

EEG-based research provides evidence for the trait hypothesis. One study found that the brain age gap does not significantly correlate with pathological EEG changes, suggesting acute neurological abnormalities do not drive it [4]. Another study reported no association between cross-sectional brain age and longitudinal ageing rates [3]. Instead, brain age in adulthood was linked to congenital factors such as birth weight and polygenic scores for brain age, indicating a lifelong influence on brain structure originating in early development.

Additional EEG evidence comes from a study comparing variability in brain age prediction between young and older adults using age prediction models based on middle-aged adults [9]. Greater variability in older adults would be expected due to stochastic accumulation of pathologies, leading to poorer prediction and thus supporting the state hypothesis. However, the study found no significant differences in prediction accuracy between age groups, reinforcing the trait hypothesis.

Studies highlight the overlap between state and trait components in age-related changes in neural irregularity [10]. This irregularity can capture trait-like differences and state-like fluctuations, indicating that brain age may encompass stable and transient elements [11]. Research on functional connectivity shows that state and trait components shape individual differences in brain function. This suggests that while some aspects of brain age are stable, others may vary with mental states [10].

A potential avenue for reconciling the trait and state hypotheses of brain age lies in acknowledging the inherently skewed nature of many neurophysiological parameters. Whether assessed longitudinally or spatially across different regions, brain metrics frequently exhibit distributions more akin to log-normal than Gaussian patterns [12]. This prevalence of skewed, heavy-tailed distributions is not unique to neuroscience. Still, it is a common feature in complex biological systems, particularly for measures constrained to positive values, such as synaptic strengths, neuronal firing rates, or network connectivity [13].

The ubiquity of such distributions in brain organization implies that relying on a single summary statistic, like the mean, can be misleading. In skewed distributions, a small number of high-magnitude observations can disproportionately influence the mean, rendering it an incomplete or even biased descriptor of the system's central tendency or typical state. For instance, the average firing rate of a neuronal population can be heavily skewed by a minority of hyperactive neurons, even if most neurons fire at much lower rates [13]. Consequently, brain age estimates derived from metrics of local or coordinated neural activity might be unduly influenced by extreme values in a specific subset of brain regions or during particular epochs, rather than reflecting a homogenous process across the entire brain or consistently over time [12]. This suggests that different aspects of these skewed distributions, such as the typical value versus the extreme tail, might capture distinct facets of brain function relevant to the trait versus state interpretations of brain age.

We propose that the central mode and the tail of skewed distributions represent two complementary aspects of the same neural dynamics: a baseline reflecting typical values and an indicator of the system's overall capacity [14]. Functional parameters in biological systems, including neural networks in the brain, often exhibit a wide dynamic range, enabling adaptation to varying demands [15]. The most representative values, such as the mean or median of a parameter measured across time or brain regions, primarily capture the system's typical operation under normal conditions. However, these central tendencies may overlook critical differences that emerge under extreme conditions, such as peak cognitive demand, stress, or pathology [16].

The right tail of skewed distributions, representing the observed values of a brain parameter beyond the central tendency, may provide insight into the system's transient effects, which in turn may reflect reserve capacity [17]. In the context of neural activity, this could correspond to the maximal firing rates of a subset of neurons, the peak amplitude of evoked responses, or the highest levels of functional connectivity observed under demanding conditions [18]. These extreme values may represent the system's upper limits, revealing potential vulnerabilities or resilience in response to physiological or pathological challenges. One example demonstrating the idea of neural capacity is the concept of cognitive reserve.

We further propose that the distinction between the trait and state hypotheses can be reframed in terms of a neural network's baseline function and its capacity. Specifically, we propose that the baseline represents the system's typical operation under normal conditions, while extreme event dynamics reflect the dynamic repertoire of brain states [19]. We hypothesize that the baseline, characterized by the most representative values of neural activity, is primarily shaped by stable, individual-specific factors established early in life. In contrast, extreme neural dynamics, captured by the distribution's right tail, reflect the system's capacity to adapt to challenges, such as cognitive load, stress, or pathological disruptions. These fluctuations in peak activity or network efficiency may provide a more sensitive window into the current state of the neural system.

Our study illustrates how neural networks' baseline and capacity contribute to brain age prediction and their associations with cognitive performance. We specifically test whether the brain age gap, computed separately for these two aspects of skewed distributions, relates to cognitive function. To investigate this, we analyze a large cohort of adults spanning a wide age range from the Cam-CAN study. We use resting-state magnetoencephalography (MEG) data and cognitive performance across multiple domains. We analyze neural networks based on how they are defined in the literature on resting-state functional magnetic resonance imaging (rs-fMRI). For each brain region of interest (ROI) defined by a cortical parcellation as our primary brain metric, we compute cross-frequency coupling (CFC), a key mechanism for coordinating brain rhythms. We define the baseline function for each network as the typical CFC value across ROIs within the given network, representing the system's normal operational range.

In contrast, we define capacity as the maximum CFC value within the same network, reflecting its upper functional limits. We then predict biological age using both baseline and capacity measures of CFC and estimate the brain age gap for each approach. Finally, we examine correlations between these neural metrics and cognitive performance to determine whether baseline and capacity provide distinct insights into cognitive ageing.

2. Methods

2.1. Data Statement

We analyzed data from the Cambridge Centre for Ageing and Neuroscience (Cam-CAN) study [20,21]. This is a large-scale, multidisciplinary research initiative that examines the effects of ageing on brain structure, function, and cognition across the lifespan. Cam-CAN integrates behavioural, neuroimaging (MRI, MEG), and cognitive data from a population-based sample of adults aged 18 to over 88. For this study, we used magnetoencephalography (MEG), structural magnetic resonance imaging (MRI), and psychometric assessments. The psychometric evaluation included seven tasks to measure executive function, working memory, and semantic memory. Our final sample consisted of 447 participants between 18 and 87 years old, with complete data across all required modalities.

Our use of this de-identified MEG, structural MRI, and cognitive data for secondary analysis was approved by the Simon Fraser University Research Ethics Board (Protocol #30001248). This secondary use was determined to present minimal risk to participants as all data were fully anonymized prior to their release by Cam-CAN and their access by our research team.

2.2. Neurophysiological Recordings

Resting-state MEG was recorded with a VectorView MEG scanner (Elekta Neuromag, Helsinki, Finland) with 102 magnetometers and 204 gradiometers and a sampling rate of 1 kHz. The eyes-closed recordings were approximately eight minutes long in a magnetically shielded room. We analyzed the recordings preprocessed from Cam-CAN Release 005. Specifically, the MEG recordings were MaxFiltered (v2.2.12). The MaxFilter algorithm allowed them to remove noise from external sources, provide head movement compensation, and reconstruct noisy channels. For a complete description of the MEG acquisition protocol, refer to Taylor et al., 2017 [21].

To reconstruct neuromagnetic dynamics, we used structural MRI data. The MRI scans were recorded using a 3T Siemens TIM Trio Scanner (Erlangen, Germany) with a 32-channel head coil. The T1-weighted MRI images were obtained using the magnetization-prepared rapid gradient-echo (MP-RAGE) sequence with a repetition time (TR) of 2250 ms, an echo time (TE) of 2.99 ms, and a flip angle of 9°. The field of view was $256 \times 240 \times 192$ mm^3, with an isotropic voxel size of 1 mm in all directions. For further details on the MRI acquisition protocol, refer to Taylor et al., 2017 [21].

2.3. Psychometrics

From the initial Cam-CAN cohort of 652 participants with available MEG/MRI data, our preprocessing pipeline yielded successful source reconstructions for 605 individuals (296 females, 309 males). This group was then cross-referenced with available cognitive assessments. After excluding additional participants who lacked complete cognitive score data, our final sample for analysis comprised 447 participants (211 females, 236 males).

Participants completed cognitive tests assessing multiple domains, including executive function, short-term memory, semantic memory, and somatosensory function [20]. Each test included multiple metrics capturing different aspects of cognitive performance. To ensure data completeness, we selected only tests with available data for at least two-thirds of participants who also had neuroimaging data. This selection yielded 8 cognitive tests, comprising 21 total performance metrics, for 447 participants. The selected tasks included the colour wheel task, hotel task, picture priming task, reaction time choice task, simple reaction time task, tip-of-the-tongue task, proverbs task, and famous faces task. Table 1 provides a complete list and description of the test metrics.

Table 1. Cognitive test measurement names and short descriptions, as defined in the original Cam_CAN study [20,21].

Name of Measurement	Description
Tip-of-tongue—Ratio	The ratio of the total number of tip-of-tongue occurrences divided by the total number of faces the participants say they can recognize
Picture Priming—High semantic priming effect	Effect of high semantic priming on reaction time
Picture Priming—High phonetic priming effect	Effect of high phonetic priming on reaction time
Picture Priming—Low semantic priming difference	Change in naming accuracy between baseline and low semantic priming
Picture Priming—High semantic priming difference	Change in naming accuracy between baseline and high semantic priming
Picture Priming—Low phonetic priming difference	Change in naming accuracy between baseline and low phonological priming
Picture Priming—High phonetic priming difference	Change in naming accuracy between baseline and high phonological priming
Picture Priming—Number recall	Total number across all conditions where participant names the image correctly with a correct reaction time (>200 ms)
Simple Reaction Time—Median	The median reaction time of correct responses
Simple Reaction Time—Mean	The mean reaction time of correct responses
Simple Reaction Time—Accuracy %	The percentage of button presses where the participant correctly used their index finger
Reaction Time Choice—Median	The median reaction time of correct responses
Reaction Time Choice—Mean	The mean reaction time of correct responses
Reaction Time Choice—Accuracy %	The accuracy of pressing a button with the correct finger (each finger is on one of 4 buttons)
Proverbs—Score	Rating of participant's explanation of English proverbs (range of 0–2)
Hotel—Total delta time	The summed total deviation from the allotted 2 min to complete each one of 5 tasks
Hotel—Number tasks completed	The number of tasks completed out of 5
Colours—Colour recall	The accuracy of the colour selected by participants after being presented with a randomly determined number of color discs, ranging from 1 to 4.
Faces—Familiar recall %	The percentage of faces of public figures that are familiar with the number of expected responses
Faces—Occupation recall %	The percentage of faces of public figures where participants can recall occupation over the number of expected responses
Faces—Name recall %	The percentage of faces of public figures where participants can recall their name over the number of expected responses

For further analysis, we normalized each cognitive score to a mean of zero and a variance of one. Some metrics, such as reaction time, were inversely related to performance, with lower values indicating better outcomes. To ensure consistency in brain–behaviour correlations, we inverted these metrics by multiplying their values by -1, so that higher scores consistently reflected better performance.

2.4. MEG Preprocessing

The preparation of MEG data for further analysis involved several key stages, including preprocessing and reconstructing neuromagnetic dynamics in brain space. We preprocessed the Max-Filtered MEG recordings from the Cam-CAN dataset as follows. First, we removed the initial 30 s of each recording to exclude potential transient artefacts, then extracted a uniform one-minute segment for each participant. We followed the preprocessing workflow established in previous studies that analyzed the Cam-CAN data algorithms [22]. For MEG preprocessing and subsequent source reconstruction, we used algorithms implemented in the MNE-Python toolbox [23].

We bandpass filtered the signals between 1 Hz and 90 Hz and downsampled the data to 512 Hz. To remove electrooculography (EOG) and electrocardiography (ECG) artefacts, we applied independent component analysis (ICA). Using the 'Picard' method with 40 components, we detected eye blink and heartbeat interference by correlating the components with the EOG and ECG channels, respectively. The MNE-Python methods (Version 1.7), *find_bads_eog()* and *find_bads_ecg()*, were used to identify components strongly correlated with EOG and ECG signals based on default parameters for the correlation threshold [24].

2.5. Neuromagnetic Source Reconstruction

To estimate cortical source activity from sensor-level MEG data, we applied a linearly constrained minimum variance (LCMV) beamformer [25,26] implemented in MNE-Python. The reconstruction process involved several key steps.

First, the MRI and MEG coordinate systems were co-registered for each participant using anatomical landmarks to generate a transformation matrix. The individual's cortical surface was reconstructed from T1-weighted MR images using the FreeSurfer *recon-all* algorithm [27–29]. This process yielded a boundary element model (BEM) and defined the source space at an oct6 resolution.

Second, a forward solution, modelling the contribution of each cortical source to the MEG sensor signals, was computed using MNE-Python's *make_forward_solution()* function. This step integrated the co-registration transformation, the *oct6* surface mesh (serving as the source model), the BEM, and preprocessed MEG metadata. To ensure consistent source orientation across participants, the forward solution was subsequently converted to have fixed orientations normal to the cortical surface using *convert_forward_solution()* with the *surf_ori=True* and *force_fixed=True* parameters.

Third, the LCMV beamformer was then applied to the preprocessed MEG data using the *make_lcmv()* and *apply_lcmv_raw()* functions to reconstruct neuromagnetic source activity across the *oct6* cortical mesh. To obtain representative dynamics for functionally defined regions, we applied the Schaefer 400-region cortical parcellation atlas [30], which segments the cortex into regions of interest (ROIs), which are aligned with large-scale intrinsic connectivity networks identified from resting-state functional MRI [31] and assigns each of the 400 ROIs (200 per hemisphere) to one of 17 established functional networks. For each of these 400 ROIs, a single representative time course was extracted using MNE-Python's () *extract_label_time_course()* function with the *mode='mean_flip'* parameter. This method averages activity across vertices within an ROI after aligning their orientations by flipping the sign of time series at vertices whose orientation deviates by more than 90° from the dominant orientation within that ROI.

This pipeline resulted in estimating neuromagnetic dynamics for each participant at 400 functionally defined cortical locations, corresponding to the Schaefer atlas parcellation.

2.6. Cross-Frequency Coupling

We computed cross-frequency coupling (CFC) matrices for each participant to characterize the interaction between neural oscillations at different frequencies [32,33]. CFC facilitates information integration across brain regions and timescales and plays a key role in cognitive functions such as memory, perception, attention, and motor control [34]. Various methods exist to quantify CFC [32]. In our study, we focused on phase–amplitude coupling (PAC), which measures how the phase of a slow oscillation modulates the amplitude of a faster oscillation [35,36].

We computed a 20×30 PAC matrix for each of the 400 ROIs defined by the Schaefer parcellation. This matrix represents the interaction between 20 low-frequency oscillations (1–12 Hz) and 30 high-frequency oscillations (13–75 Hz). Using a time-frequency decomposition, we reconstructed single-source MEG dynamics at these frequency points evenly spaced on a logarithmic scale. We applied an eighth-order complex Gaussian derivative wavelet to reconstruct the analytic signal for each wavelet frequency [37]. We then computed the instantaneous phase of low-frequency signals and the amplitude envelope of high-frequency signals.

After calculating low-frequency phases and high-frequency amplitudes, we estimated the normalized modulation index (MI) for each phase–amplitude frequency pair. The normalized MI quantifies the coupling strength while accounting for signal amplitude, ensuring participant comparability.

As a result of this procedure, each participant was characterized by 400 ROIs, with each ROI associated with a 20×30 PAC matrix. Each ROI was assigned to one of 17 functional networks, separately for the left and right hemispheres, yielding a total of 34 network-based classifications.

2.7. Mean and Maximum Values of Cross-Frequency Coupling

We characterized each functional network using two parameters: baseline cross-frequency coupling (CFC) and CFC capacity. Each network consisted of multiple regions of interest (ROIs). We computed the median phase–amplitude coupling (PAC) across all ROIs within a given network to quantify baseline coordinated activity. We calculated the maximum PAC value across ROIs within the same network to assess CFC capacity.

This procedure resulted in 34 functional networks per participant, represented by two 20×30 PAC matrices. One matrix reflected the network's baseline CFC, while the other represented its capacity to sustain phase–amplitude coupling. We used these network-level metrics in subsequent analyses to predict participants' age.

2.8. Brain Age Gap Based on Cross-Frequency Coupling

We predicted brain age and estimated the brain age gap separately for each of the 34 functional networks, considering the baseline and capacity cross-frequency coupling (CFC) metrics separately. To predict brain age from cross-frequency phase–amplitude coupling (PAC) features, we employed Relevance Vector Regression (RVR), a sparse Bayesian machine learning method [38]. While sharing operational principles with Support Vector Regression (SVR), RVR incorporates probabilistic inference, typically yielding a more parsimonious model that uses fewer input features (or "relevance vectors") and is less prone to overfitting. In this study, the RVR model was trained using these CFC features derived from all participants, with chronological age as the target variable for prediction. The analysis was performed using the *sklearn-rvm* Python package, with the RVR configured to use a polynomial kernel of degree 3.

To estimate the brain age gap for each network and CFC metric, we used 10-fold cross-validation. We divided participants into ten equal groups, trained the model on 90% of the data, and tested it on the remaining 10%. This process was repeated ten times, ensuring that each participant's brain age was predicted without their data being included in the training set.

After obtaining brain age predictions, we retained only the networks that showed robust associations with chronological age. We computed a distance correlation t-test for each network between chronological and predicted age using the *distance_correlation_t_test()* function from the *dcor* Python package [39]. We corrected network-specific p-values for multiple comparisons using a false discovery rate (FDR) threshold of 5%. In subsequent analyses examining correlations between the brain age gap and psychometric performance, we included only those functional networks where predicted age showed a robust correlation with chronological age.

We computed the brain age gap (BAG) for each selected functional network as the deviation of a participant's predicted brain age from the population average. Specifically, we applied a quadratic regression for each CFC metric and functional network to model the relationship between chronological age and predicted brain age. We then calculated the BAG for each participant by subtracting the predicted brain age from the regression-based population average.

Positive BAG values indicated slower ageing, meaning that the participant's cross-frequency coupling profile resembled a younger individual's. Conversely, negative BAG values reflected accelerated ageing, where the participant's CFC profile appeared older than expected for their chronological age. Each participant was ultimately characterized by two sets of network-specific brain age gap estimates, one for each CFC metric. Notably, the same participant could exhibit both accelerated and slowed ageing effects across different functional networks.

2.9. Associations Between Brain Age Gap and Psychometrics

In the final analysis stage, we examined correlations between brain age gap (BAG) estimates and cognitive performance. We applied behavioural partial least squares (PLS), a multivariate technique commonly used in neuroimaging to assess brain–behaviour relationships [40–43]. This analysis was conducted separately for each of the two BAG profiles and each cognitive test, resulting in 16 Behavioural PLS analyses (2 BAG profiles × 8 cognitive tests).

Behavioural PLS decomposed the covariance between network-specific BAG estimates and cognitive scores into latent variables (LVs) using singular value decomposition (SVD). Each LV captured a distinct pattern of covariance between brain age gaps and cognitive performance across participants.

To assess the significance of the overall brain–cognition correlations, we performed a permutation test wherein we randomly shuffled data across participants and tested for stability of correlations based on the original data. This produced a vector of overall correlations, with dimensionality determined by the number of cognitive scores. We performed a bootstrap test to evaluate the robustness of individual networks' contributions to the identified overall correlations. In each bootstrap iteration, we resampled participants with replacements while preserving the correspondence between each participant's BAG and cognitive scores. Behavioural PLS then computed bootstrap ratio values, defined as the ratio of original network saliences to the standard error of the bootstrap distribution. These values are equivalent to z-scores. In our study, we used the terms of bootstrap ratio values and z-scores interchangeably. We applied 10,000 iterations for both permutation and bootstrap tests and focused on the first LV, which accounted for the largest proportion of variance in the brain–cognition covariances.

For each PLS analysis, we reported three key results: (i) a vector of overall correlations, with each value representing the association between BAG and a cognitive variable within a given test; (ii) a p-value assessing the statistical significance of these correlations across the entire dataset; and (iii) a vector of network-specific z-scores, indicating the robustness of networks' contribution to the observed brain–cognition overall correlations, as defined in (i). Larger absolute z-scores reflected more potent effects, and we focused on networks with z-scores greater than ± 2, approximately corresponding to a 95% confidence interval.

We note that PLS analysis yields several metrics for interpreting brain–behaviour relationships. First, it provides a global significance value for each identified latent variable (LV), which represents the strength of the overall multivariate correlation between the combined imaging features and behavioural scores. The individual correlation values between the cognitive scores and the LV's brain scores indicate the effect size concerning specific behavioural outcomes. For the imaging data, the contribution of individual fMRI networks to the LV is quantified using Z-scores, which are computed as bootstrap ratio values. These ratios are derived by dividing the original salience of each network by the standard deviation of its bootstrap distribution, thereby providing a measure of effect size for each network's involvement in the identified pattern.

Interpretation of individual correlations between BAG and cognitive scores required considering overall correlations and z-scores together. Positive brain–cognition associations were reflected either by both a positive overall correlation and z-score or by both values being negative. Conversely, negative brain–cognition associations occurred when one metric (correlation or z-score) was positive and the other was negative.

3. Results

For both profiles of network-specific brain age gaps, we identified 24 networks, which expressed robust associations between predicted and chronological age. 20 networks were common for both profiles: 'DorsAttnB_LH', 'ContA_LH', 'VisPeri_RH', 'SomMotA_RH', 'DefaultA_RH', 'SalVentAttnB_RH', 'DefaultA_LH', 'SalVentAttnB_LH', 'DorsAttnB_RH', 'SomMotA_LH', 'SomMotB_LH', 'DorsAttnA_RH', 'SalVentAttnA_LH', 'DefaultB_LH', 'SalVentAttnA_RH', 'SomMotB_RH', 'DorsAttnA_LH', 'LimbicA_RH', 'DefaultC_LH', 'TempPar_RH'. The labels' LH' and 'RH' indicate the reference to the left and right hemisphere, respectively. Four networks were unique for the baseline-related metric: 'VisPeri_LH', 'ContB_LH', 'VisCent_LH', 'TempPar_LH'. Also, four networks were unique for the capacity-based metric: 'DefaultB_RH', 'DefaultC_RH', 'LimbicA_LH', 'LimbicA_LH'. We applied a series of multivariate analyses (behavioural PLS) to explore correlations between brain age gap and cognitive test performance.

3.1. Picture Priming Task

Partial least squares (PLS) analysis revealed a significant overall relationship between brain age gap (BAG) and cognitive performance on one specific cognitive test, the Picture Priming Task (overall $p < 0.05$). Figure 1 displays the results of two separate PLS analyses for this task, one for BAG derived from baseline cross-frequency coupling (CFC) metrics (left column) and another for BAG derived from capacity CFC metrics (right column). In each column, the lower plot shows the overall correlation coefficients between BAG and individual cognitive scales from the Picture Priming Task, while the upper plot shows the corresponding bootstrap ratios (Z-scores) for each functional network's contribution to this relationship.

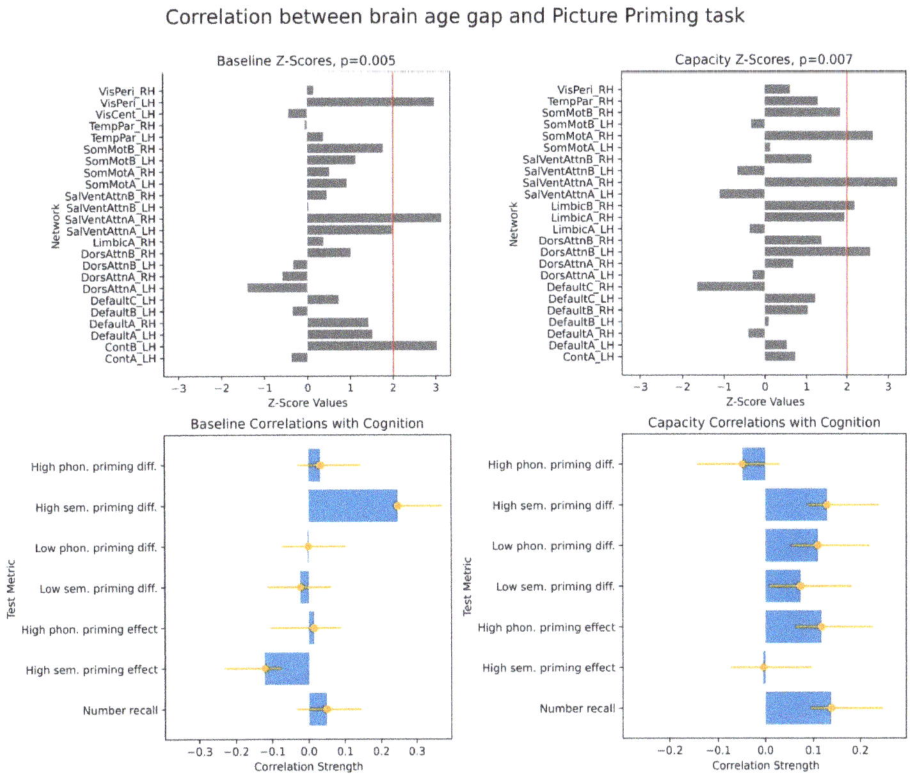

Figure 1. Partial least squares (PLS) Analysis of the brain age gap (BAG) and Picture Priming Task Performance. This figure presents results from two separate partial least squares (PLS) analyses examining the relationship between cognitive performance on the Picture Priming Task and brain age gap (BAG). The left column displays results for BAG derived from baseline cross-frequency coupling (CFC) metrics ("baseline-based BAG"). In contrast, the right column shows results for BAG derived from capacity CFC metrics ("capacity-based BAG"). The lower plot illustrates the overall correlation coefficients between individual cognitive scales and the BAG derived from all functional networks for each BAG type. The upper plot displays the corresponding bootstrap ratios (z-scores) for each functional network, indicating the reliability of each network's contribution to the observed brain–behaviour correlation. Positive Z-scores and positive correlations in the lower plots would suggest that a higher BAG (interpreted as relatively slower brain ageing) is associated with better cognitive performance, and vice versa.

For both baseline-based and capacity-based BAG analyses, the distributions of network Z-scores (Figure 1, upper plots) were predominantly positive, indicating a consistent directionality of network contributions to the overall PLS effect. We focused on networks with Z-scores exceeding 2, signifying robust contributions. Four such networks were identified for the baseline-CFC BAG, and four for the capacity-CFC BAG. The consistently positive Z-scores allow for a direct interpretation of the brain–behaviour relationships based on the sign of the overall correlation coefficients (Figure 1, lower plots).

Examining the baseline CFC-derived BAG (Figure 1, left column), the pattern of correlations with cognitive scales from the Picture Priming Task was mixed. For instance, BAG positively correlated with lower semantic priming difference scores but negatively correlated with lower phonetic priming difference scores. This heterogeneity in the direction of effects did not yield a clear, unidirectional relationship between baseline-BAG and cognitive performance, suggesting that both positive (slower ageing) and negative (accelerated ageing) BAG values could be associated with better performance on different aspects of the task.

In contrast, the capacity CFC-derived BAG (Figure 1, right column) exhibited a qualitatively distinct and more consistent pattern. Specifically, BAG derived from CFC capacity measures showed positive correlations with five out of the seven cognitive scales of the Picture Priming Task. Given that a positive BAG indicates a brain appearing younger than its chronological age (i.e., slower ageing) and positive values on these cognitive scales represent better performance, this pattern consistently suggests that slower brain ageing, as indexed by capacity-BAG, is associated with enhanced cognitive function on this task. This consistent relationship was supported by the predominantly positive network-level Z-scores.

3.2. Other Cognitive Tasks

No statistically significant overall relationships were found between the brain age gap (BAG) and cognitive scores for the remaining cognitive tasks included in our analyses. All p-values derived from the partial least squares (PLS) analyses for these tasks ranged between 0.15 and 0.81. The full results of these non-significant PLS analyses are presented in Figures A1–A7 for transparency and completeness. These appendix figures adhere to the same format as Figure 1, illustrating the overall correlation coefficients and network-specific Z-scores for each task, though none achieved statistical significance, as tested by a series of PLS analyses.

4. Discussion

In our study, we explored the utility of the brain age gap (BAG), which is the difference between an individual's neurobiologically predicted age and chronological age, as a lens to understand cognitive ageing. Conventionally, a positive BAG implies a brain appearing "younger" than its chronological age, whereas a negative BAG suggests accelerated brain ageing. While brain age can be derived from diverse neurophysiological measures, our approach centred on cross-frequency coupling (CFC). This choice was motivated by CFC's critical role in orchestrating neural oscillations across multiple timescales within brain regions, providing a sensitive measure of coordinated neural dynamics relevant to ageing and cognitive function.

We estimated brain age separately for a collection of neural networks defined a priori using a functional parcellation of resting-state fMRI networks [30,31]. Critically, we reconstructed two distinct profiles of brain age. In one approach, we predicted chronological age using CFC averaged across regions of interest (ROIs) within each network, representing the network's baseline activity. In the second approach, we used the maximum CFC value across ROIs within the same network, reflecting the network's capacity to generate strongly coupled temporal activity. These two profiles of network-specific brain age gap estimates capture different aspects of neural dynamics: baseline levels of functional coupling and the potential for extreme neural synchronization across frequencies.

Finally, we examined associations between these two brain age profiles and participants' cognitive performance, measured across eight tasks assessing working memory, executive function, and semantic memory. We found robust positive correlations between the capacity-based brain age gap and scores on the picture priming task, which reflects semantic memory. In contrast, results for the baseline-based brain age in the same task were inconclusive. Our findings align with the hypothesis that the brain age gap reflects an individual's current cognitive state, supporting the state hypothesis of brain ageing.

Our overall rationale for distinguishing two types of cross-frequency coupling was based on the observation that brain parameters commonly exhibit skewed distributions [44]. This pattern is prevalent across multiple levels of neural hierarchy and closely resembles log-normal distributions [13]. In our interpretation of the brain age gap, we specifically analyzed these distributions' central mode and right tail, focusing on variability within functional networks.

We computed the maximum cross-frequency coupling (CFC) within each network to characterize extreme values. While multiple approaches exist for quantifying distribution tails, statistical properties such as skewness and kurtosis, among other metrics, could also provide meaningful insights into cognitive and clinical states.

Skewness effectively differentiates cognitive states such as meditation, mathematical problem-solving, and open-eye conditions, making it a versatile metric for cognitive state monitoring [45]. Similarly, kurtosis has demonstrated clinical relevance across various domains. In pediatric epilepsy, high-frequency brain signals show significantly elevated kurtosis in epileptogenic zones compared to controls [46]. In neurodegenerative disorders, EEG studies of Alzheimer's patients report increased kurtosis values, which have contributed to the development of kurtosis-based denoising techniques that enhance diagnostic accuracy [47]. Additionally, brain–computer interface (BCI) research has leveraged kurtosis to classify motor imagery tasks in EEG data, achieving high accuracy in distinguishing between imagined left- and right-hand movements [48].

A study using resting-state MEG data from the Cam-CAN dataset further demonstrated the relevance of extreme values in characterizing age-related changes in neural activity [49]. By analyzing frequency-specific signal amplitudes in terms of skewness and kurtosis, the study described how these statistical features captured the variability of neural dynamics across the lifespan. Similarly, another study used resting-state EEG to predict Fugl-Meyer upper extremity motor scores in individuals with chronic stroke [50]. With functional connectivity estimated as the mean and maximum phase-locking value, the study found that maximum-based estimates achieved higher prediction accuracy [50]. These findings reinforce the importance of considering distribution tails when interpreting neural metrics.

Characterizing the tails of the skewed distributions and focusing on their maximum values, skewness, or kurtosis falls into the domain of extreme value theory (EVT). In neural networks, EVT can be used to evaluate robustness against adversarial examples [51]. The extreme events help assess neural networks' capacity to resist extreme perturbations, thereby defining their robustness and capacity. In a study exploring the interaction between olfactory and visual systems in vertebrates, EVT was applied to model cross-modal sensory information integration [14]. This integration was mediated by extreme responses rather than average cell responses, suggesting that neural capacity is influenced by the ability to handle extreme sensory inputs.

Cross-frequency coupling (CFC) in neural ensembles requires understanding how different oscillatory frequencies interact to enhance cognitive processes such as memory, perception, and attention [34]. CFC is a mechanism by which neural oscillations at different frequencies synchronize, facilitating information integration across neural populations [33]. This coupling is crucial for robust information storage and retrieval and for the modulation of cognitive functions [52].

Our study has limitations. We recognize that using the maximum value of cross-frequency coupling across brain regions to represent neural network capacity is a simplification. From a theoretical perspective, interpreting this peak, or the upper tail of its distribution, as a direct measure of neural capacity currently lacks strong theoretical grounding. CFC can occur through various mechanisms, including phase-phase and phase–

amplitude coupling, which are influenced by synaptic time scales, synaptic coupling, and neural population excitability [32]. Inhibitory neural mass models demonstrate that CFC can emerge from the collective dynamics of neural populations, with different dynamical regimes such as periodic and quasi-periodic oscillations [53]. Nonlinear interactions in neural fields, rather than specialized cells, can generate CFC, suggesting a universal mechanism across multi-scale systems [54].

Neuronal ensembles are considered modular units of neural circuits [55]. Neuronal ensembles represent and process information through coordinated activity patterns, which are influenced by factors such as the size of the ensemble, the diversity of neuronal responses, and the synaptic connections within the ensemble [55]. The capacity of a neural ensemble can be conceptualized in terms of its ability to encode, store, and retrieve patterns or information, which is influenced by the structure and dynamics of the neuronal ensemble. For example, cross-frequency coupling, particularly between theta and gamma oscillations, has been shown to increase the memory capacity of neural networks [56].

Other limitations should be acknowledged. First, although we analyzed a broad range of cognitive assessments available within the Cam-CAN dataset, our analyses did not incorporate all potentially influential variables. Factors such as detailed educational history, comprehensive health status indicators beyond gross clinical categories, and specific lifestyle variables (e.g., diet, physical activity) were not included. The inclusion of such covariates could further refine our models of brain–behaviour relationships and potentially reveal nuances in the observed effects or identify moderating influences.

Second, our investigation of brain network dynamics relied on specific, well-established cortical parcellations [30,31]. While these atlases are prevalent in functional MRI research, facilitating comparability with the vast body of existing literature, they represent only a subset of available parcellation schemes. These parcellation, while foundational, is now relatively dated [57]. Using additional or more contemporary atlases, potentially derived from different modalities or based on different organizational principles, could offer further insights into the robustness and generalizability of our findings across various network definitions.

Third, in our cross-frequency coupling (CFC) analysis, we focused on mean values and metrics reflecting the upper tail of the distribution (i.e., maximum values). This decision was motivated by the observation that neurophysiological measures often exhibit skewed distributions where the mean alone provides an incomplete representation, potentially obscuring functionally relevant extreme values [13]. While we interpret these as speculatively representing baseline and peak coupling capacity, respectively, this is a simplification. A more exhaustive characterization of the entire CFC distribution, or the exploration of alternative metrics beyond the mean and maximum, could provide a more nuanced understanding of CFC dynamics. Future studies could also benefit from explicitly modelling the skewness or other distributional properties [49].

The observed correlation between the brain age gap derived from the cross-frequency coupling (CFC) and performance specifically on the picture priming task, but not other cognitive tasks, warrants further consideration. While the precise reasons for this specificity remain fully elucidated, the neurophysiological underpinnings of picture priming offer plausible explanations. Picture priming tasks rely on the efficient processing and integration of visual object information, processes known to involve coordinated activity across several brain regions where CFC plays a crucial modulatory role. For instance, the fusiform gyrus, a key region for object recognition and consistently implicated in picture priming, exhibits enhanced theta/alpha band evoked power for repeated stimuli, suggesting that CFC within this region may facilitate more efficient neural processing during priming [58]. Similarly, the inferior temporal cortex, critical for visual object memory, shows enhanced theta/alpha-beta phase–amplitude coupling during visual working memory, indicating a potential role for CFC in maintaining primed representations [59]. Furthermore, successful priming often involves attentional modulation and decision-making processes supported by the prefrontal cortex, where theta–gamma CFC is implicated in coordinating large-scale network activity [59,60]. Finally, the parietal cortex, involved in attention and integrating sensory information with working memory, demonstrates theta–gamma phase synchronization during memory matching, a related cognitive function [61]. Therefore, the sensitivity of the CFC-derived brain age gap to picture priming performance might stem from its capacity to capture the integrity of these distributed, CFC-mediated neural interactions that are particularly critical for the rapid recognition and facilitated processing inherent to priming effects, potentially more so than for the cognitive domains assessed by our other tasks.

In conclusion, our study demonstrates associations between cross-frequency coupling (CFC) and semantic memory performance, framed within the context of brain age. By defining the brain age gap (BAG) relative to typical ageing trajectories within resting-state fMRI networks, we specifically investigated how different characteristics of network-specific CFC relate to cognition. Our novel approach distinguished between BAG derived from typical CFC values (baseline) and BAG derived from extreme CFC values across brain regions within each network, the latter speculatively reflecting the network's "capacity" for coordinated neural activity. We found that BAG based on these CFC capacity metrics showed robust associations with semantic memory performance. This suggests that measures capturing the peak or transient dynamics of neural coordination, rather than solely average background activity, are susceptible to an individual's current cognitive state. These findings strongly support the state hypothesis of brain ageing, indicating that brain age models emphasizing neural system capacity may offer a more sensitive approach for understanding brain–cognition relationships and identifying state-dependent alterations in brain health. This "capacity-focused" perspective on brain age holds promise for future brain-wide association studies seeking to capture dynamic aspects of cognitive function and decline.

Author Contributions: Conceptualization, V.A.V. and G.M.; Methodology, V.A.V. and S.M.; Formal analysis, T.L.; Investigation, V.A.V.; Resources, V.A.V., S.M.D., G.M. and S.M.; Data curation, H.L.; Writing—original draft, V.A.V.; Writing—review & editing, V.A.V., T.L., H.L., S.M.D., G.M. and S.M.; Visualization, T.L. and H.L.; Supervision, V.A.V., S.M.D. and S.M.; Project administration, V.A.V. and S.M.; Funding acquisition, S.M. All authors have read and agreed to the published version of the manuscript.

Funding: This research was funded by Age Well, grant number R679225.

Institutional Review Board Statement: This study was approved by the Simon Fraser University Research Ethics Board (Protocol #30001248).

Informed Consent Statement: Informed consent was not required for this study as it was based on the secondary use of publicly available and fully anonymized data.

Data Availability Statement: The data presented in this study are available from the Cam-CAN study at https://camcan-archive.mrc-cbu.cam.ac.uk/dataaccess/ (accessed on 11 June 2021).

Acknowledgments: This research was enabled in part by support provided by the Digital Research Alliance of Canada (alliancecan.ca).

Conflicts of Interest: The authors declare no conflict of interest.

Appendix A

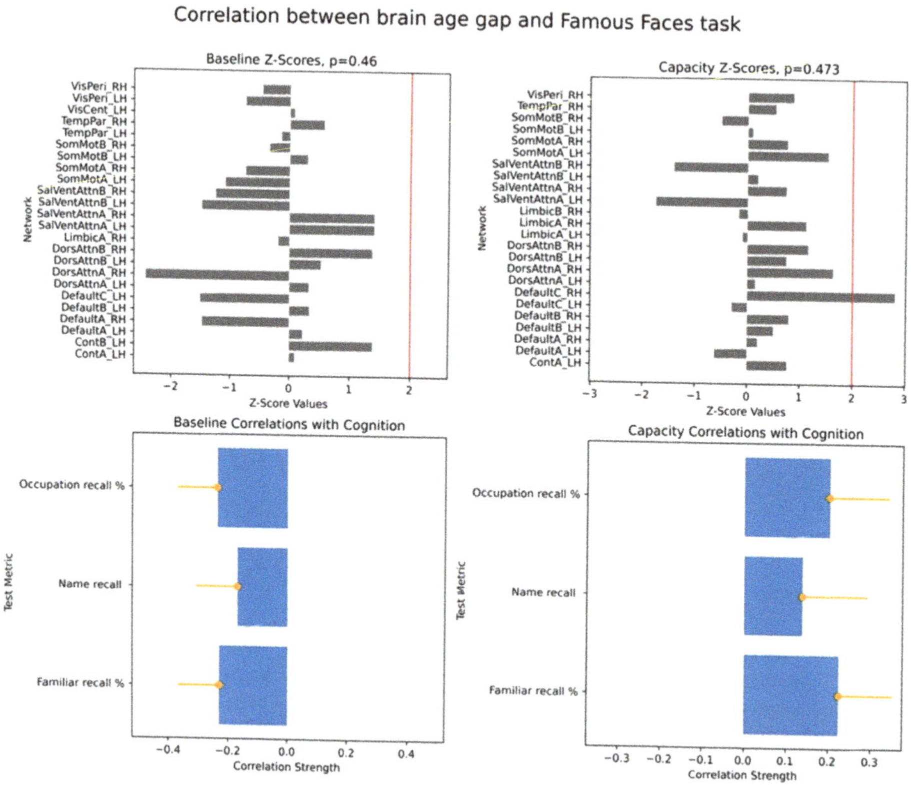

Figure A1. Relationships between Famous Faces task performance and brain age gap (BAG) from cross-frequency coupling metrics. This figure presents results from two separate partial least squares (PLS) analyses investigating the associations between cognitive performance on the Famous Faces Task and brain age gap (BAG). BAG was estimated using two distinct cross-frequency coupling (CFC) metrics: mean values (left column) and maximum values (right column). For each BAG type, the lower plot illustrates the overall correlation coefficients between individual cognitive scales of the Famous Faces task and the respective BAG measure. The upper plot displays the corresponding bootstrap ratios (Z-scores) for each contributing functional network, indicating the reliability of its contribution to the brain–behaviour correlation. The visual layout and interpretation of plot elements are consistent with Figure 1. Both PLS analyses yielded non-significant results ($p > 0.1$).

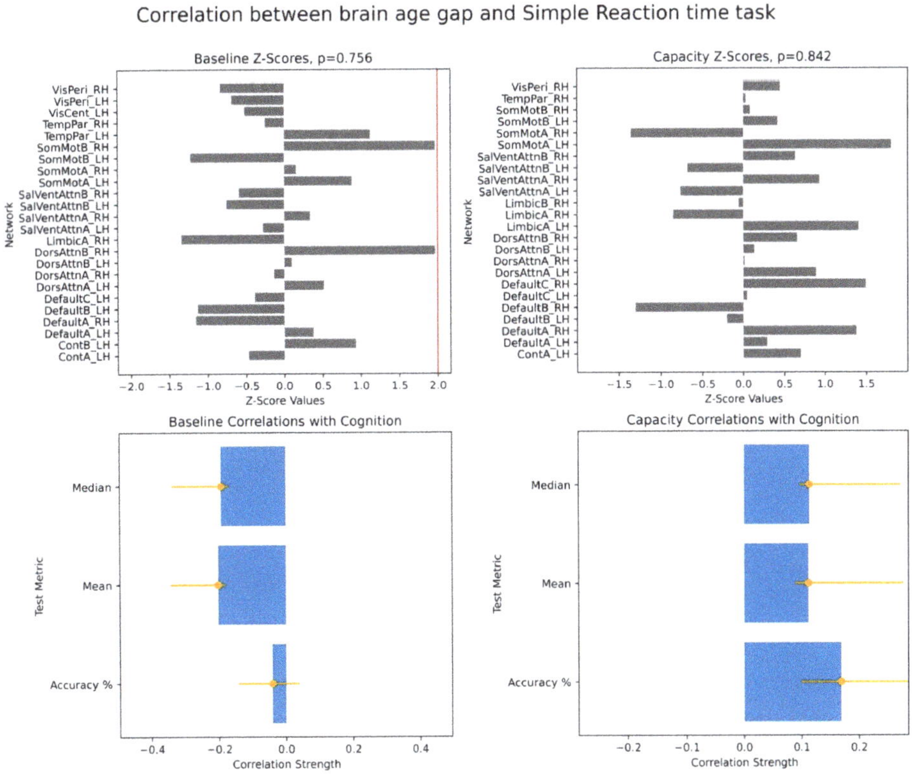

Figure A2. Relationships between Simple Reaction Time task performance and brain age gap (BAG) from cross-frequency coupling metrics. This figure presents results from two separate partial least squares (PLS) analyses investigating the associations between cognitive performance on the Simple Reaction Time (SRT) task and brain age gap (BAG). BAG was estimated using two distinct cross-frequency coupling (CFC) metrics: mean values or "baseline" (left column) and maximum values or "capacity" (right column). The lower plot illustrates the overall correlation coefficients between SRT performance metrics and the respective BAG measure for each BAG type. The upper plot displays the corresponding bootstrap ratios (Z-scores) for each contributing functional network, indicating the reliability of its contribution to the brain–behaviour correlation. The visual layout and interpretation of plot elements are consistent with Figure 1. Both PLS analyses yielded non-significant results ($p > 0.1$).

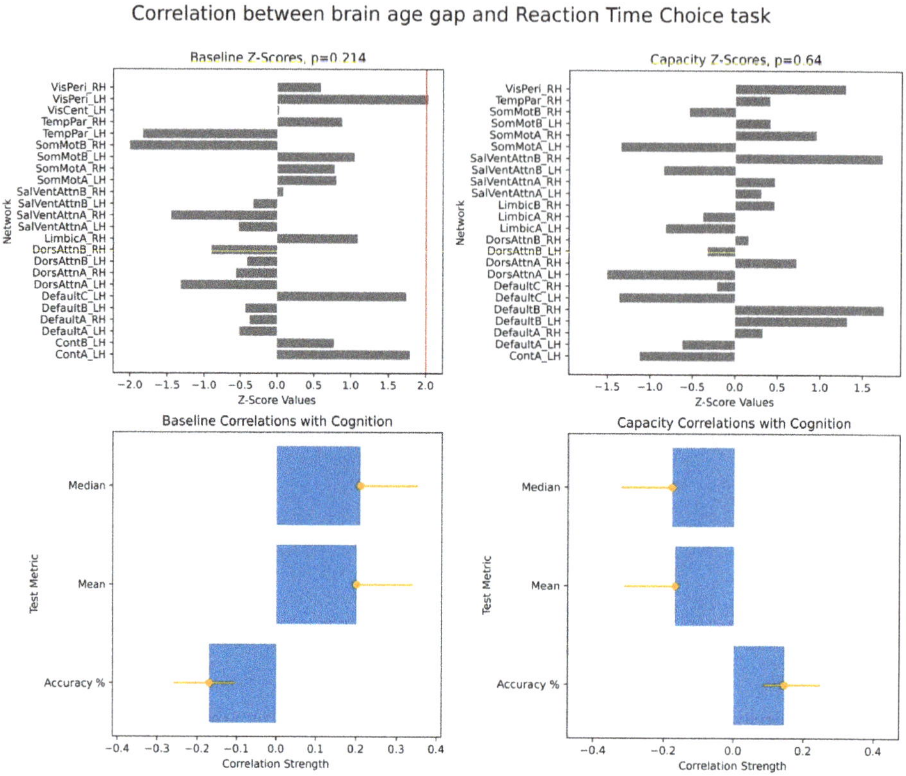

Figure A3. Correlations between cognitive performance on the Reaction Time Choice task and brain age gap (BAG) estimated from two cross-frequency coupling metrics. This figure presents results from two PLS analyses examining the relationship between cognitive performance on the Famous Faces Task and BAG estimated from two cross-frequency coupling metrics. The lower plots display overall correlations for each cognitive scale within the task, while the upper plots show the corresponding network-specific z-scores, following the same format as Figure 1. Both analyses yielded p-values greater than 0.1.

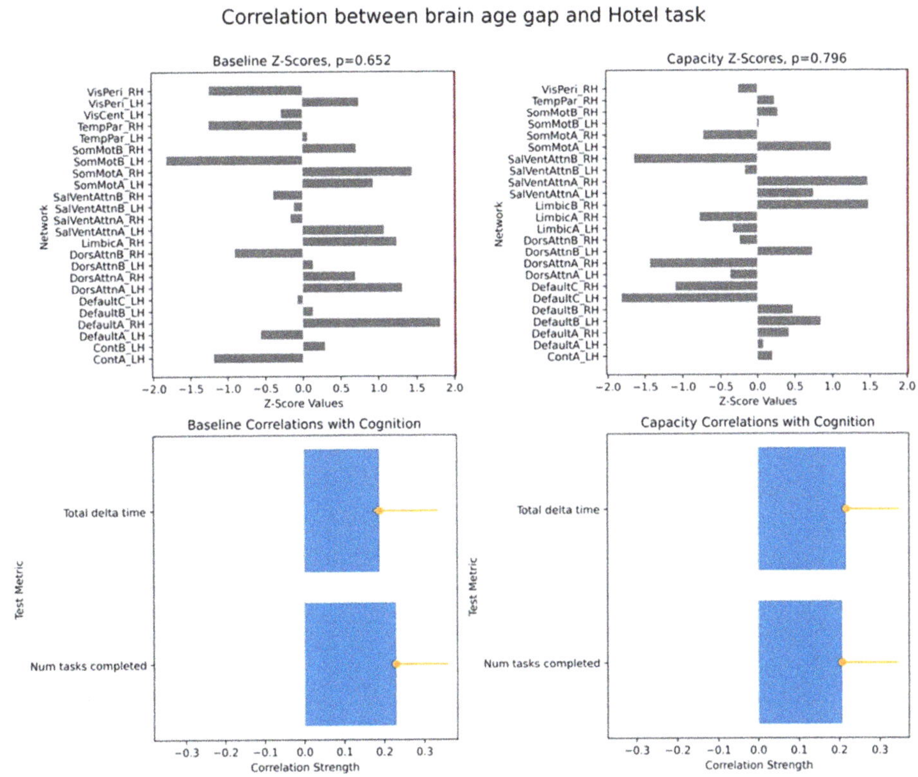

Figure A4. Correlations between cognitive performance on the Hotel task and brain age gap (BAG) estimated from two cross-frequency coupling metrics. This figure presents results from two PLS analyses examining the relationship between cognitive performance on the Famous Faces Task and BAG estimated from two cross-frequency coupling metrics. The lower plots display overall correlations for each cognitive scale within the task, while the upper plots show the corresponding network-specific z-scores, following the same format as Figure 1. Both analyses yielded *p*-values greater than 0.1.

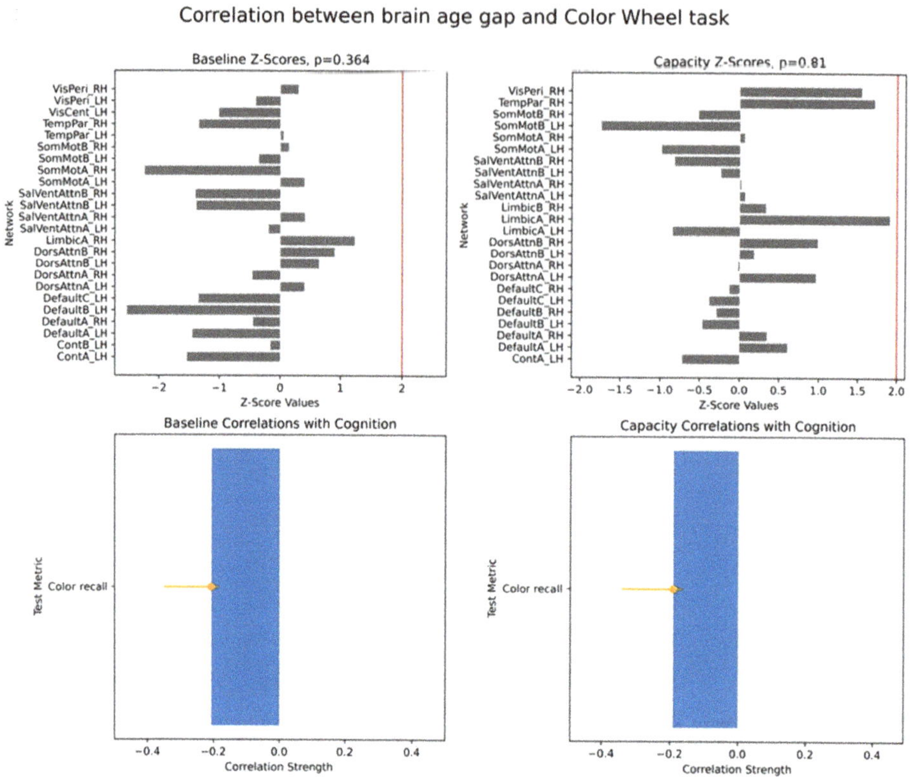

Figure A5. Correlations between cognitive performance on the Colour Wheel task and brain age gap (BAG) estimated from two cross-frequency coupling metrics. This figure presents results from two PLS analyses examining the relationship between cognitive performance on the Famous Faces Task and BAG estimated from two cross-frequency coupling metrics. The lower plots display overall correlations for each cognitive scale within the task, while the upper plots show the corresponding network-specific z-scores, following the same format as Figure 1. Both analyses yielded *p*-values greater than 0.1.

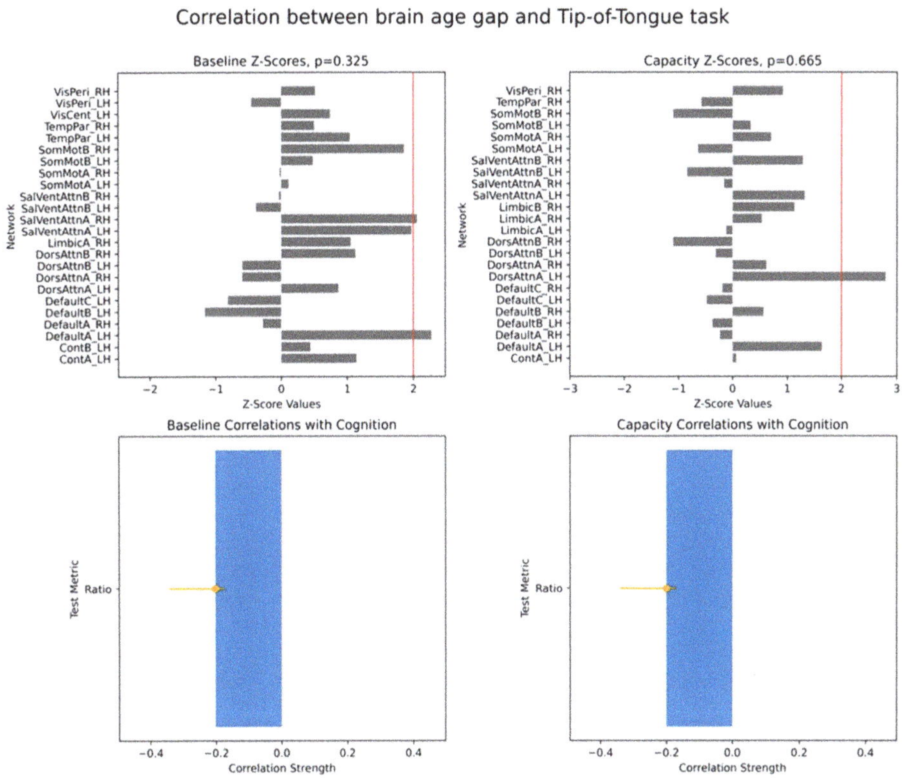

Figure A6. Correlations between cognitive performance on the Tip-of-Tongue task and brain age gap (BAG) estimated from two cross-frequency coupling metrics. This figure presents results from two PLS analyses examining the relationship between cognitive performance on the Famous Faces Task and BAG estimated from two cross-frequency coupling metrics. The lower plots display overall correlations for each cognitive scale within the task, while the upper plots show the corresponding network-specific z-scores, following the same format as Figure 1. Both analyses yielded *p*-values greater than 0.1.

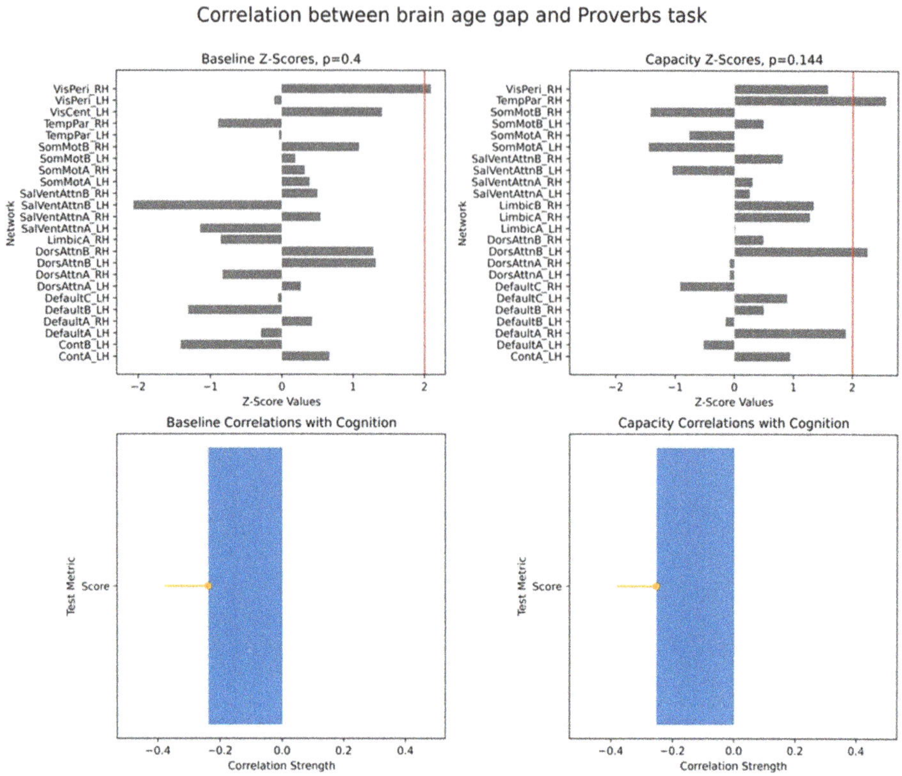

Figure A7. Correlations between cognitive performance on the Proverbs task and brain age gap (BAG) estimated from two cross-frequency coupling metrics. This figure presents results from two PLS analyses examining the relationship between cognitive performance on the Famous Faces Task and BAG estimated from two cross-frequency coupling metrics. The lower plots display overall correlations for each cognitive scale within the task, while the upper plots show the corresponding network-specific z-scores, following the same format as Figure 1. Both analyses yielded *p*-values greater than 0.1.

References

1. Gaser, C.; Kalc, P.; Cole, J.H. A Perspective on Brain-Age Estimation and Its Clinical Promise. *Nat. Comput. Sci.* **2024**, *4*, 744–751. [CrossRef] [PubMed]
2. Franke, K.; Gaser, C. Ten Years of BrainAGE as a Neuroimaging Biomarker of Brain Aging: What Insights Have We Gained? *Front. Neurol.* **2019**, *10*, 789. [CrossRef] [PubMed]
3. Vidal-Pineiro, D.; Wang, Y.; Krogsrud, S.K.; Amlien, I.K.; Baaré, W.F.; Bartres-Faz, D.; Bertram, L.; Brandmaier, A.M.; Drevon, C.A.; Düzel, S.; et al. Individual Variations in 'Brain Age' Relate to Early-Life Factors More than to Longitudinal Brain Change. *eLife* **2021**, *10*, e69995. [CrossRef] [PubMed]
4. Gemein, L.A.W.; Schirrmeister, R.T.; Boedecker, J.; Ball, T. Brain Age Revisited: Investigating the State vs. Trait Hypotheses of EEG-Derived Brain-Age Dynamics with Deep Learning. *Imaging Neurosci.* **2024**, *2*, 1–22. [CrossRef]
5. Hua, J.P.Y.; Abram, S.V.; Loewy, R.L.; Stuart, B.; Fryer, S.L.; Vinogradov, S.; Mathalon, D.H. Brain Age Gap in Early Illness Schizophrenia and the Clinical High-Risk Syndrome: Associations with Experiential Negative Symptoms and Conversion to Psychosis. *Schizophr. Bull.* **2024**, *50*, 1159–1170. [CrossRef]
6. Lee, J.; Burkett, B.J.; Min, H.-K.; Senjem, M.L.; Lundt, E.S.; Botha, H.; Graff-Radford, J.; Barnard, L.R.; Gunter, J.L.; Schwarz, C.G.; et al. Deep Learning-Based Brain Age Prediction in Normal Aging and Dementia. *Nat. Aging* **2022**, *2*, 412–424. [CrossRef]
7. Dular, L.; Špiclin, Ž.; for the Alzheimer's Disease Neuroimaging Initiative; the Australian Imaging Biomarkers and Lifestyle Flagship Study of Ageing. Analysis of Brain Age Gap across Subject Cohorts and Prediction Model Architectures. *Biomedicines* **2024**, *12*, 2139. [CrossRef]

8. Jawinski, P.; Markett, S.; Drewelies, J.; Düzel, S.; Demuth, I.; Steinhagen-Thiessen, E.; Wagner, G.G.; Gerstorf, D.; Lindenberger, U.; Gaser, C.; et al. Linking Brain Age Gap to Mental and Physical Health in the Berlin Aging Study II. *Front. Aging Neurosci.* **2022**, *14*, 791222. [CrossRef]
9. Kalapun, O.; Moiseev, A.; Medvedev, G.; Doesburg, S.M.; Xi, P.; Moreno, S.; Ribary, U.; Vakorin, V.A. Biological Age from Electroencephalographic Activity Could Be Decoupled from a Gradual Stochastic Accumulation of Pathologies over Adulthood. In Proceedings of the 2024 IEEE International Conference on Bioinformatics and Biomedicine (BIBM), Lisbon, Portugal, 3 December 2024; pp. 4172–4178.
10. Geerligs, L.; Rubinov, M.; Henson, R.N. State and Trait Components of Functional Connectivity: Individual Differences Vary with Mental State. *J. Neurosci.* **2015**, *35*, 13949–13961. [CrossRef]
11. Waschke, L.; Wöstmann, M.; Obleser, J. States and Traits of Neural Irregularity in the Age-Varying Human Brain. *Sci. Rep.* **2017**, *7*, 17381. [CrossRef]
12. Roberts, J.A.; Boonstra, T.W.; Breakspear, M. The Heavy Tail of the Human Brain. *Curr. Opin. Neurobiol.* **2015**, *31*, 164–172. [CrossRef] [PubMed]
13. Buzsáki, G.; Mizuseki, K. The Log-Dynamic Brain: How Skewed Distributions Affect Network Operations. *Nat. Rev. Neurosci.* **2014**, *15*, 264–278. [CrossRef] [PubMed]
14. Banerjee, S.; Scheirer, W.J.; Li, L. An Extreme Value Theory Model of Cross-Modal Sensory Information Integration in Modulation of Vertebrate Visual System Functions. *Front. Comput. Neurosci.* **2019**, *13*, 3. [CrossRef] [PubMed]
15. Halliday, D.W.R.; Mulligan, B.P.; Garrett, D.D.; Schmidt, S.; Hundza, S.R.; Garcia-Barrera, M.A.; Stawski, R.S.; MacDonald, S.W.S. Mean and Variability in Functional Brain Activations Differentially Predict Executive Function in Older Adults: An Investigation Employing Functional near-Infrared Spectroscopy. *Neurophotonics* **2017**, *5*, 1. [CrossRef]
16. Northoff, G.; Tumati, S. "Average Is Good, Extremes Are Bad"—Non-Linear Inverted U-Shaped Relationship between Neural Mechanisms and Functionality of Mental Features. *Neurosci. Biobehav. Rev.* **2019**, *104*, 11–25. [CrossRef]
17. Timme, N.M.; Marshall, N.J.; Bennett, N.; Ripp, M.; Lautzenhiser, E.; Beggs, J.M. Criticality Maximizes Complexity in Neural Tissue. *Front. Physiol.* **2016**, *7*, 425. [CrossRef]
18. Fiedler, A.; Careau, V. Individual (Co)Variation in Resting and Maximal Metabolic Rates in Wild Mice. *Physiol. Biochem. Zool.* **2021**, *94*, 338–352. [CrossRef]
19. Lemoine, É.; Neves Briard, J.; Rioux, B.; Gharbi, O.; Podbielski, R.; Nauche, B.; Toffa, D.; Keezer, M.; Lesage, F.; Nguyen, D.K.; et al. Computer-Assisted Analysis of Routine EEG to Identify Hidden Biomarkers of Epilepsy: A Systematic Review. *Comput. Struct. Biotechnol. J.* **2024**, *24*, 66–86. [CrossRef]
20. Shafto, M.A.; Tyler, L.K.; Dixon, M.; Taylor, J.R.; Rowe, J.B.; Cusack, R.; Calder, A.J.; Marslen-Wilson, W.D.; Duncan, J.; Dalgleish, T.; et al. The Cambridge Centre for Ageing and Neuroscience (Cam-CAN) Study Protocol: A Cross-Sectional, Lifespan, Multidisciplinary Examination of Healthy Cognitive Ageing. *BMC Neurol.* **2014**, *14*, 204. [CrossRef]
21. Taylor, J.R.; Williams, N.; Cusack, R.; Auer, T.; Shafto, M.A.; Dixon, M.; Tyler, L.K.; Cam-CAN; Henson, R.N. The Cambridge Centre for Ageing and Neuroscience (Cam-CAN) Data Repository: Structural and Functional MRI, MEG, and Cognitive Data from a Cross-Sectional Adult Lifespan Sample. *NeuroImage* **2017**, *144*, 262–269. [CrossRef]
22. Xifra-Porxas, A.; Ghosh, A.; Mitsis, G.D.; Boudrias, M.-H. Estimating Brain Age from Structural MRI and MEG Data: Insights from Dimensionality Reduction Techniques. *NeuroImage* **2021**, *231*, 117822. [CrossRef] [PubMed]
23. Gramfort, A.; Luessi, M.; Larson, E.; Engemann, D.A.; Strohmeier, D.; Brodbeck, C.; Parkkonen, L.; Hämäläinen, M.S. MNE Software for Processing MEG and EEG Data. *NeuroImage* **2014**, *86*, 446–460. [CrossRef] [PubMed]
24. Dammers, J.; Schiek, M.; Boers, F.; Silex, C.; Zvyagintsev, M.; Pietrzyk, U.; Mathiak, K. Integration of Amplitude and Phase Statistics for Complete Artifact Removal in Independent Components of Neuromagnetic Recordings. *IEEE Trans. Biomed. Eng.* **2008**, *55*, 2353–2362. [CrossRef] [PubMed]
25. Frost, O.L. An Algorithm for Linearly Constrained Adaptive Array Processing. *Proc. IEEE* **1972**, *60*, 926–935. [CrossRef]
26. Ilmoniemi, R.J.; Sarvas, J. *Brain Signals: Physics and Mathematics of MEG and EEG*; MIT Press: Cambridge, UK, 2019; ISBN 978-0-262-35281-9.
27. Dale, A.M.; Fischl, B.; Sereno, M.I. Cortical Surface-Based Analysis. *NeuroImage* **1999**, *9*, 179–194. [CrossRef]
28. Fischl, B. Automatically Parcellating the Human Cerebral Cortex. *Cereb. Cortex* **2004**, *14*, 11–22. [CrossRef]
29. Fischl, B. FreeSurfer. *NeuroImage* **2012**, *62*, 774–781. [CrossRef]
30. Schaefer, A.; Kong, R.; Gordon, E.M.; Laumann, T.O.; Zuo, X.-N.; Holmes, A.J.; Eickhoff, S.B.; Yeo, B.T.T. Local-Global Parcellation of the Human Cerebral Cortex from Intrinsic Functional Connectivity MRI. *Cereb. Cortex* **2018**, *28*, 3095–3114. [CrossRef]
31. Thomas Yeo, B.T.; Krienen, F.M.; Sepulcre, J.; Sabuncu, M.R.; Lashkari, D.; Hollinshead, M.; Roffman, J.L.; Smoller, J.W.; Zöllei, L.; Polimeni, J.R.; et al. The Organization of the Human Cerebral Cortex Estimated by Intrinsic Functional Connectivity. *J. Neurophysiol.* **2011**, *106*, 1125–1165. [CrossRef]
32. Jirsa, V.; Müller, V. Cross-Frequency Coupling in Real and Virtual Brain Networks. *Front. Comput. Neurosci.* **2013**, *7*, 78. [CrossRef]

33. Hyafil, A.; Giraud, A.-L.; Fontolan, L.; Gutkin, B. Neural Cross-Frequency Coupling: Connecting Architectures, Mechanisms, and Functions. *Trends Neurosci.* **2015**, *38*, 725–740. [CrossRef] [PubMed]
34. Canolty, R.T.; Knight, R.T. The Functional Role of Cross-Frequency Coupling. *Trends Cogn. Sci.* **2010**, *14*, 506–515. [CrossRef] [PubMed]
35. Özkurt, T.E.; Schnitzler, A. A Critical Note on the Definition of Phase–Amplitude Cross-Frequency Coupling. *J. Neurosci. Methods* **2011**, *201*, 438–443. [CrossRef] [PubMed]
36. Roehri, N.; Bréchet, L.; Seeber, M.; Pascual-Leone, A.; Michel, C.M. Phase-Amplitude Coupling and Phase Synchronization Between Medial Temporal, Frontal and Posterior Brain Regions Support Episodic Autobiographical Memory Recall. *Brain Topogr.* **2022**, *35*, 191–206. [CrossRef]
37. Lee, G.; Gommers, R.; Waselewski, F.; Wohlfahrt, K.; O'Leary, A. PyWavelets: A Python Package for Wavelet Analysis. *J. Open Source Softw.* **2019**, *4*, 1237. [CrossRef]
38. Tipping, M.E. Sparse Bayesian Learning and the Relevance Vector Machine. *J. Mach. Learn. Res.* **2001**, *1*, 211–244.
39. Székely, G.J.; Rizzo, M.L. The Distance Correlation t -Test of Independence in High Dimension. *J. Multivar. Anal.* **2013**, *117*, 193–213. [CrossRef]
40. McIntosh, A.R.; Lobaugh, N.J. Partial Least Squares Analysis of Neuroimaging Data: Applications and Advances. *NeuroImage* **2004**, *23*, S250–S263. [CrossRef]
41. Krishnan, A.; Williams, L.J.; McIntosh, A.R.; Abdi, H. Partial Least Squares (PLS) Methods for Neuroimaging: A Tutorial and Review. *NeuroImage* **2011**, *56*, 455–475. [CrossRef]
42. Vakorin, V.A.; McIntosh, A.R.; Mišić, B.; Krakovska, O.; Poulsen, C.; Martinu, K.; Paus, T. Exploring Age-Related Changes in Dynamical Non-Stationarity in Electroencephalographic Signals During Early Adolescence. *PLoS ONE* **2013**, *8*, e57217. [CrossRef]
43. Vakorin, V.A.; Doesburg, S.M.; Leung, R.C.; Vogan, V.M.; Anagnostou, E.; Taylor, M.J. Developmental Changes in Neuromagnetic Rhythms and Network Synchrony in Autism. *Ann. Neurol.* **2017**, *81*, 199–211. [CrossRef] [PubMed]
44. Broido, A.D.; Clauset, A. Scale-Free Networks Are Rare. *Nat. Commun.* **2019**, *10*, 1017. [CrossRef] [PubMed]
45. Joshua Davis, J.J.; Schubeler, F.; Ji, S.; Kozma, R. Discrimination Between Brain Cognitive States Using Shannon Entropy and Skewness Information Measure. In Proceedings of the 2020 IEEE International Conference on Systems, Man, and Cybernetics (SMC), Toronto, ON, Canada, 11 October 2020; pp. 4026–4031.
46. Quitadamo, L.R.; Mai, R.; Gozzo, F.; Pelliccia, V.; Cardinale, F.; Seri, S. Kurtosis-Based Detection of Intracranial High-Frequency Oscillations for the Identification of the Seizure Onset Zone. *Int. J. Neural Syst.* **2018**, *28*, 1850001. [CrossRef] [PubMed]
47. Wang, G.; Shepherd, S.J.; Beggs, C.B.; Rao, N.; Zhang, Y. The Use of Kurtosis De-Noising for EEG Analysis of Patients Suffering from Alzheimer's Disease. *Biomed. Mater. Eng.* **2015**, *26*, S1135–S1148. [CrossRef]
48. Wu, X.; Ye, Z. The Study of Classification of Motor Imaginaries Based on Kurtosis of EEG. In *Neural Information Processing*; King, I., Wang, J., Chan, L.-W., Wang, D., Eds.; Lecture Notes in Computer Science; Springer: Berlin/Heidelberg, Germany, 2006; Volume 4234, pp. 74–81. ISBN 978-3-540-46484-6.
49. Vakorin, V.A.; Liaqat, H.; Doesburg, S.M.; Moreno, S. Extreme Signal Amplitude Events in Neuromagnetic Oscillations Reveal Brain Aging Processing across Adulthood. *Front. Aging Neurosci.* **2025**, *17*, 1498400. [CrossRef]
50. Riahi, N.; Vakorin, V.A.; Menon, C. Estimating Fugl-Meyer Upper Extremity Motor Score from Functional-Connectivity Measures. *IEEE Trans. Neural Syst. Rehabil. Eng.* **2020**, *28*, 860–868. [CrossRef]
51. Weng, T.-W.; Zhang, H.; Chen, P.-Y.; Yi, J.; Su, D.; Gao, Y.; Hsieh, C.-J.; Daniel, L. Evaluating the Robustness of Neural Networks: An Extreme Value Theory Approach. *arXiv* **2018**, arXiv:1801.10578.
52. Chuderski, A. Fluid Intelligence and the Cross-Frequency Coupling of Neuronal Oscillations. *Span. J. Psychol.* **2016**, *19*, E91. [CrossRef]
53. Ceni, A.; Olmi, S.; Torcini, A.; Angulo-Garcia, D. Cross Frequency Coupling in next Generation Inhibitory Neural Mass Models. *Chaos Interdiscip. J. Nonlinear Sci.* **2020**, *30*, 053121. [CrossRef]
54. Qin, Y.; Sheremet, A. Mesoscopic Collective Activity in Excitatory Neural Fields: Cross-Frequency Coupling. *arXiv* **2022**, arXiv:2207.02833.
55. Yuste, R.; Cossart, R.; Yaksi, E. Neuronal Ensembles: Building Blocks of Neural Circuits. *Neuron* **2024**, *112*, 875–892. [CrossRef] [PubMed]
56. Bybee, C.; Belsten, A.; Sommer, F.T. Cross-Frequency Coupling Increases Memory Capacity in Oscillatory Neural Networks. *arXiv* **2022**, arXiv:2204.07163.
57. Kong, R.; Spreng, R.N.; Xue, A.; Betzel, R.F.; Cohen, J.R.; Damoiseaux, J.S.; De Brigard, F.; Eickhoff, S.B.; Fornito, A.; Gratton, C.; et al. A Network Correspondence Toolbox for Quantitative Evaluation of Novel Neuroimaging Results. *Nat. Commun.* **2025**, *16*, 2930. [CrossRef] [PubMed]
58. Gilbert, J.R.; Gotts, S.J.; Carver, F.W.; Martin, A. Object Repetition Leads to Local Increases in the Temporal Coordination of Neural Responses. *Front. Hum. Neurosci.* **2010**, *4*, 1252. [CrossRef]

59. Daume, J.; Gruber, T.; Engel, A.K.; Friese, U. Phase-Amplitude Coupling and Long-Range Phase Synchronization Reveal Frontotemporal Interactions During Visual Working Memory. *J. Neurosci.* **2017**, *37*, 313–322. [CrossRef]
60. Marconi, P.L.; Bandinelli, P.L.; Penna, M.P.; Pessa, E. Cross-Frequency Modulation, Network Information Integration and Cognitive Performance in Complex Systems. In *Towards a Post-Bertalanffy Systemics*; Minati, G., Abram, M.R., Pessa, E., Eds.; Contemporary Systems Thinking; Springer International Publishing: Cham, Switzerland, 2016; pp. 17–30. ISBN 978-3-319-24389-4.
61. Biel, A.L.; Minarik, T.; Sauseng, P. EEG Cross-Frequency Phase Synchronization as an Index of Memory Matching in Visual Search. *NeuroImage* **2021**, *235*, 117971. [CrossRef]

applied sciences

MDPI

Article

18F-Fluorodeoxyglucose Uptake in Cerebrospinal Fluid Reflects Both Brain Glucose Demand and Impaired Blood–Brain Barrier Transport in Alzheimer's Disease

Caterina Motta [1], Chiara Giuseppina Bonomi [1], Martina Poli [1], Nicola Biagio Mercuri [2], Alessandro Martorana [1] and Agostino Chiaravalloti [3,4,*]

1 Memory Clinic, Policlinico Tor Vergata, University of Rome Tor Vergata, 00133 Rome, Italy; caterinamotta86@hotmail.it (C.M.); chiarag.bonomi@gmail.com (C.G.B.); poli.martina95@gmail.com (M.P.); martorana@med.uniroma2.it (A.M.)
2 Neurology Unit, Policlinico Tor Vergata, University of Rome Tor Vergata, 00133 Rome, Italy; mercurin@med.uniroma2.it
3 Department of Biomedicine and Prevention, University of Rome Tor Vergata, 00133 Rome, Italy
4 IRCCS Neuromed, 86077 Pozzilli, Italy
* Correspondence: agostino.chiaravalloti@uniroma2.it

Abstract: Glucose delivery to the brain requires transporters at the blood–brain barrier (BBB), whose downregulation may be associated with neuronal deficits in Alzheimer's disease (AD). Whether this downregulation is due to reduced demand or primary BBB dysfunction remains unclear. We investigated novel 18F-Fluorodeoxyglucose Positron Emission Tomography (18F-FDG-PET) measures, namely ventricles ($FDG_{Ventricles}$) and cortical uptake (FDG_{Cortex}), and the $FDG_{Ventricles}/FDG_{Cortex}$ ratio in 224 patients with AD compared to those in 35 controls (CTRLs). AD patients showed lower FDG_{Cortex} and $FDG_{Ventricles}$ and higher cerebrospinal fluid (CSF) lactates than CTRLs. We found a positive correlation between FDG_{Cortex} and $FDG_{Ventricles}$ in both groups, although this was less strong in AD patients (AD: r = 0.358; $p < 0.001$; CTRL: r = 0.516; $p = 0.003$). Multivariate regression analyses showed that only older age was associated with reduced FDG_{Cortex} and $FDG_{Ventricles}$ in CTRLs. Conversely, lower FDG_{Cortex} was associated with higher Qalb and higher plasma glucose levels within the AD group. Moreover, lower $FDG_{Ventricles}$ and $FDG_{Ventricles}/FDG_{Cortex}$ ratios were associated with elevated CSF lactates in this group. Stratifying AD patients by Apolipoprotein E (APOE) genotype revealed distinct patterns. In APOE ε3 homozygotes, FDG_{Cortex} showed no associations, while $FDG_{Ventricles}$ and $FDG_{Ventricles}/FDG_{Cortex}$ were negatively associated with CSF lactate. In APOE ε4 carriers, lower FDG_{Cortex} was linked to higher plasma glucose and QAlb, whereas $FDG_{Ventricles}$ and $FDG_{Ventricles}/FDG_{Cortex}$ were positively associated with CSF p-tau/Aβ42. Our findings suggest that, in patients with AD, $FDG_{Ventricles}$ and the $FDG_{Ventricles}/FDG_{Cortex}$ ratio may reflect alterations in brain metabolism and glucose extraction capacity. These parameters are differently linked with age, BBB integrity, and metabolic dysfunction (CSF lactates), according to APOE genotype.

Keywords: Alzheimer's disease; glucose metabolism; FDG-PET; blood–brain barrier; APOE genotype; CSF

1. Introduction

Alzheimer's disease (AD) is characterized by the deposition of amyloid-β (Aβ) plaques and hyperphosphorylation of tau proteins, in parallel with significant impairments in neurovascular regulation [1], blood–brain barrier (BBB) integrity [2], and brain glucose metabolism [3]. Positron emission tomography with [18F]fluoro-2-deoxyglucose (18F-FDG-PET) has become a key tool in AD diagnosis, as it detects a decline in the cerebral metabolic rate of glucose—a phenomenon that can be observed even before the onset of clinical symptoms in individuals who will later develop dementia due to AD [4,5]. Importantly, the decline in glucose metabolism occurs earlier and is more pronounced than the appearance of brain atrophy [6,7]. One potential explanation for this early alteration in cerebral glucose uptake could be an abnormal delivery of glucose to the brain. Indeed, the loss of glucose transporters (GLUTs) from the BBB has been documented to occur before the onset of AD symptoms [8]. Moreover, postmortem studies on brains from patients with AD revealed that the decreased expression of GLUTs was significantly correlated with abnormal tau hyperphosphorylation, suggesting a link between impaired glucose transport across the BBB and the progression of AD pathology [9].

With its highly selective permeability, the BBB separates circulating blood from the brain's extracellular fluid. While not all patients with AD exhibit an overtly BBB disruption, it has been hypothesized that altered receptor-mediated transport could impair not only the clearance of toxic amyloid and tau proteins [10] but also the transport of glucose from plasma to the brain, contributing to the early metabolic changes observed in AD [11].

Studies examining brain glucose uptake predominantly leverage 18F-FDG-PET imaging. Indeed, 18F-FDG emits a radioactive signal upon entry into the central nervous system—which can be detected in both the parenchyma and cerebrospinal fluid (CSF). Once inside the brain cells, 18F-FDG is phosphorylated to FDG-6-phosphate which cannot be further metabolized [12]. Thus, this method enables the measurement of glucose uptake without fully reflecting subsequent glucose metabolism. Moreover, recent evidence supports the idea that the 18F-FDG signal in CSF spaces—particularly the ventricles—may convey meaningful information about altered glucose transport and metabolism in neurodegenerative diseases. Specifically, Zhou et al. [13] demonstrated that dynamic 18F-FDG PET can quantify ventricular CSF clearance, serving as a surrogate marker for glymphatic function, particularly relevant to amyloid and tau clearance in AD.

The exact mechanisms governing neuronal glucose uptake remain poorly understood, raising critical questions about the interpretation of 18F-FDG-PET data. In the context of AD, is the cortical hypometabolism due to the loss of neuronal and glial cells, or is it linked to reduced 18F-FDG transport caused by BBB dysfunction? Moreover, the extent to which 18F-FDG uptake is influenced by alterations in glucose metabolism—potentially contributing to the pathogenesis of late-onset AD—has yet to be fully elucidated [14]. These questions should be examined in the context of Apolipoprotein E (APOE) genotype, the strongest genetic risk factor for late-onset AD. The presence of the APOE ε4 allele not only increases the risk of developing AD but is also linked to a more aggressive disease progression, with more pronounced amyloid and tau pathology [15]. Recent studies have further demonstrated that APOE ε4 affects brain cell metabolism, leading to reduced glycolytic activity and impaired mitochondrial respiration [16,17].

To attempt to address some of these issues, the present study investigated the role of novel 18F-FDG-PET parameters in late-onset AD. We evaluated both the 18F-FDG signal in the cerebrospinal fluid of ventricles ($FDG_{Ventricles}$), not directly related to cellular uptake, and the same parameter with respect to the total amount of cortical glucose uptake ($FDG_{Ventricles}/FDG_{Cortex}$). Moreover, we considered the relationship between these parameters, BBB permeability, and glucose metabolism, as well as the possible modulating role played by APOE genotype.

2. Materials and Methods

2.1. Patients Enrolment and Study Design

Between January 2021 and June 2024, we enrolled 362 consecutive outpatients in active follow-up at the Memory Clinic of the University Hospital "Tor Vergata" in Rome. We considered eligible for the study all patients aged 60 to 85 that had obtained a biomarker-based diagnosis of AD—defined as the presence of decreased CSF Aβ42 with or without increase CSF p-tau181.

The criteria for retrospective inclusion were as follows: (1) a complete diagnostic workup, including standardized neurological examination, laboratory testing, magnetic resonance imaging, 18F-FDG-PET scanning, neuropsychological assessment, APOE genotyping, and CSF analysis, and (2) the fulfillment of the diagnostic criteria for dementia [18] or mild cognitive impairment due to AD [19]. Specifically, the neuropsychological battery consists of the following: Mini Mental State Examination (MMSE) to assess global cognitive functions; the Rey Auditory Verbal Learning Test—immediate and delayed recall for verbal episodic memory; the Rey–Osterrieth Complex Figure Test—copy and recall for visuospatial memory; Raven Colored Progressive Matrices for abstract reasoning; and the Stroop test and verbal fluency test for attention and executive functions. The exclusion criteria were as follows: (1) a Hachinski scale score > 4 at baseline MRI, considered suggestive of vascular co-pathology; (2) a history of traumatic brain injury within 6 months before lumbar puncture; (3) the use of antipsychotics or antidepressants. These criteria were selected to minimize potential confounding factors that could affect brain glucose metabolism or CSF biomarker levels. Eventually, 224 patients were enrolled in this retrospective study.

Furthermore, 35 age-matched controls (CTRLs) were also enrolled in this study. Specifically, CTRLs were recruited among patients admitted to the Neurology Department of the University Hospital "Tor Vergata" in Rome between January 2021 and June 2024, whose CSF samples were collected in accordance with standard hospital practice. Upon discharge, all 35 subjects had received a diagnosis of either functional neurological disorder (n = 33) or tensive-type headache (n = 2). Residual CSF samples obtained during routine diagnostic evaluations were used for the analyses. Active infections, incidental presence of cognitive impairment, and other primary neurological conditions had been ruled out, including non-specific CSF changes such as an increased CSF cell count (>4 cells/mmc) or altered AD biomarker profile. Subjects underwent also MRI and 18F-FDG-PET scanning for diagnostic purposes. Eventually, only subjects with normal CSF and 18F-FDG-PET findings and without structural or functional brain abnormalities were selected for the CTRL group.

Written informed consent was acquired from all participants or legally authorized representatives. All procedures were performed according to the Declaration of Helsinki. The local ethical committee considered the study protocol an observational retrospective design (57.25CET2PTV).

2.2. CSF Collection and Biomarker Analysis

All lumbar punctures were performed between 8 and 10 a.m. An 8 mL CSF sample was collected for each patient in polypropylene tubes. A total of 2 mL of CSF was used for routine biochemical analysis including the calculation of the Albumin Quotient (Qalb) as the BBB permeability index—considering CSF/serum albumin—and levels of lactates. A second aliquot of 2 mL was used for CSF AD biomarkers. We used commercially available kits for biochemical analysis. CSF amyloid-β 1–42 (Aβ42), phosphorylated-tau (p-tau), and total tau (t-tau) concentrations were determined using a sandwich enzyme-linked immunosorbent assay (EUROIMMUN ELISA©, Waltham, MA, USA). Amyloid groups were defined according to EUROIMMUN guidelines: CSF Aβ42 was Aβ-positive if Aβ < 600 pg/mL or Aβ-negative if Aβ42 $\geq$ 600 pg/mL.

Blood samples were also drawn in EDTA tubes. The DNA was extracted automatically and APOE genotyping was conducted by allelic discrimination technology with real-time PCR, according to the manufacturer's instructions (TaqMan; Applied Biosystems, Foster City, CA, USA).

2.3. F-FDG-PET Data

All PET scans were conducted at the Nuclear Medicine Unit of the University Hospital "Policlinico Tor Vergata" in Rome using a General Electric VCT PET/CT scanner (GE Medical Systems, Powell, TN, USA). Participants fasted for at least 5 h before the intravenous administration of 18F-FDG, and serum glucose concentrations were confirmed to be within acceptable ranges, as recommended by the European Association of Nuclear Medicine guidelines [20]. Patients received an intravenous injection of 18F-FDG (dose ranging from 185 to 295 MBq), followed by hydration with 500 mL of saline solution (0.9% sodium chloride). Imaging began 30 min post-injection and lasted for ten minutes. Detailed acquisition and reconstruction parameters were consistent with those described in the cited guidelines [20].

Brain uptake of 18F-FDG was assessed using Statistical Parametric Mapping (SPM) 12 software (Wellcome Department of Cognitive Neurology, London, UK; available at https://www.fil.ion.ucl.ac.uk/spm/software/spm12/, accessed on 1 January 2025) running on MATLAB 2022b (Mathworks, Natick, MA, USA). PET data underwent conversion from DICOM to Nifti format utilizing MRIcron software (accessible at https://www.nitrc.org/projects/mricron/, accessed on 1 January 2025), followed by normalization procedures. To minimize image intensity distortions caused by spatially varying artifacts and enhance the accuracy of automated processing, a bias regularization factor of 0.0001 was applied. Additionally, the Gaussian smoothing kernel's full width at half maximum (FWHM) was restricted to 60 mm to prevent the algorithm from modeling intensity variations attributed to distinct tissue types. For image processing, a bias regularization factor and a smoothing kernel of an 8 mm full width at half maximum (FWHM) were selected to balance sensitivity and specificity, minimizing noise while preserving regional anatomical accuracy, consistently with standard PET imaging recommendations. The tissue probability map embedded in SPM12 was internally developed from brain 18F-FDG PET scans, including data from 285 Alzheimer's disease patients and 121 healthy control subjects.

An affine registration with mutual information with the tissue probability maps [21] was used to achieve approximate alignment with the ICBM spatial template European brains [22]. Warping regularization was set with 1×5 arrays (0, 0.001, 0.5, 0.05, 0.2); smoothing (to cope with functional anatomical variability not compensated by spatial normalization and to improve signal-to-noise ratio) was set to 5 mm; and the sampling distance (encoding the approximate distance between sampled points in estimating model parameters) was set to 3.

The clusters containing the cortex, ventricles (CSF), and pons (see below) were exported by the means of the WFU PickAtlas tool implemented in SPM 12 (WFU PickAtlas (RRID:SCR_007378); available online: https://www.nitrc.org/projects/wfu_pickatlas/ (accessed on 23 April 2025)) [23] (Figure 1). Specifically, the mean signal intensities calculated from the cortex and ventricles within each subject were normalized to the average intensities of the pons volume of interest. The use of normalization based on activity in the pons, rather than the cerebellum, brainstem, or primary sensorimotor cortex, has been reported to result in greater accuracy in discriminating patients from controls in neurodegenerative diseases [20,24]. Normalization to the pons was chosen due to its demonstrated stability and robustness in AD, as metabolism in this region is relatively preserved even in early stages of neurodegeneration. Previous studies have confirmed that the pons provides reliable reference values for normalization, helping to reduce inter-subject variability and enhance sensitivity to pathological changes in cortical metabolism, which is critical for detecting subtle disease-related effects [20,25].

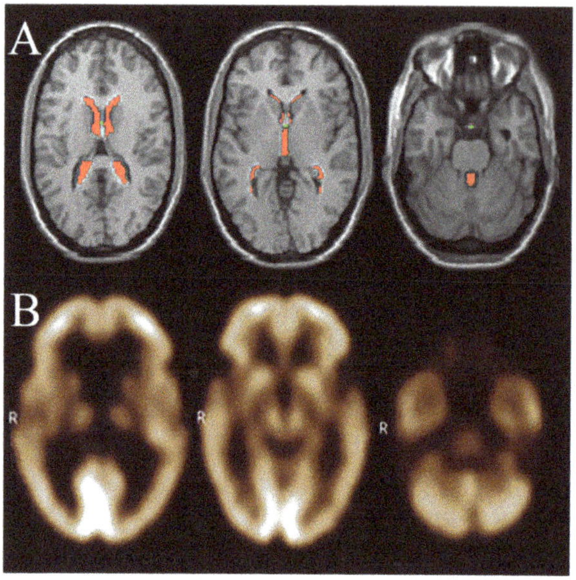

Figure 1. An exemplificative representation of the methodological approach used in this study for defining regions of interest (ROIs) using the WFU PickAtlas tool implemented in SPM12. Upper row (**A**): anatomical MRI images displaying ventricular ROIs (highlighted in red). These ventricular ROIs, derived through the WFU PickAtlas, were subsequently applied to PET scans of the patients included in the analysis. Lower row (**B**): corresponding 18F-FDG PET images, onto which the ventricular ROIs were mapped to measure 18F-FDG in CSF. The 18F-FDG values were normalized to pons activity to correct for individual metabolic variability, thus enabling the assessment of cerebrospinal fluid glucose dynamics in relation to Alzheimer's disease pathophysiology. The ventricles in PET images may appear larger than those in MRI due to the lower 18F-FDG uptake in surrounding white matter. This image is intended purely to clarify the methodological approach and does not represent the exact methodological procedure used in this study (see text for detailed information). R: right side.

ROIs for ventricles specifically included lateral ventricles as predefined by the WFU PickAtlas. Cortical ROIs were defined as a composite cortical mask encompassing frontal, parietal, temporal, and occipital lobes using the anatomical label atlas in WFU PickAtlas [23]. A dataset of normalized 18F-FDG values relevant to the cluster under study was exported. To determine whether the normalized 18F-FDG values for the studied cluster were Gaussian distributed, the D'Agostino K-squared normality test was applied (with the null hypothesis being normal distribution).

2.4. Statistical Analysis

Continuous variables are presented as the mean ± standard deviation (SD) if normally distributed and as the median (Interquartile range_IQR) if not normally distributed. Categorical variables are expressed as percentages (%). Patients with AD were assigned to the APOE ε4 subgroup when carrying at least one ε4 allele (ε4/ε4 or ε3/ε4), while all the remaining patients exclusively carried APOE ε3 alleles (ε3/ε3).

Statistical differences in continuous variables between groups were tested using *t* tests or the Mann–Whitney U test in case of a non-normal distribution. Pearson's X2 squared was used for categorical variables. Spearman correlation tests were performed to analyze the relationships between non-parametric variables. To compare the strength of independent correlation coefficients between groups, we used Fisher's r-to-z transformation (http://vassarstats.net/rdiff.html, accessed on 23 April 2025). To test if factors including Qalb, p-tau/Ab42, and CSF lactates were linked to changes in the 18F-FDG-PET parameters, we performed different multivariate regression analyses. The analyses were performed in both AD and CTRL groups. We then used separate models, considering only AD patients stratified according to APOE genotype.

Statistical analysis was performed via JASP© (Version 0.18.3-Computer Software-JASP TEAM 2020, https://jasp-stats.org/, accessed on 30 April 2025) and GraphPad Prism© (Version 9.5.0, GraphPad Software, San Diego, CA, USA, www.graphpad.com, accessed on 8 April 2025).

All results were computed with two-tailed tests of significance; *p*-values < 0.05 were considered statistically significant.

3. Results

3.1. Participant Characteristics

Statistically significant differences in age and the prevalence of the APOE ε4 genotype were observed between the CTRL and the all AD group (Table 1). As expected, the cognitive status (MMSE) as well as all CSF biomarker levels were also significantly different between groups. Both FDG_{Cortex} and $FDG_{Ventricles}$ were significantly lower in AD patients with respect to CTRLs, with a larger effect size for $FDG_{Ventricles}$ (FDG_{Cortex} $r = -0.362$; $p < 0.001$; $FDG_{Ventricles}$ $r = -0.774$; $p < 0.001$) (Figure 2A and 2B). Accordingly, $FDG_{Ventricles}/FDG_{Cortex}$ was also significantly lower in AD patients ($FDG_{Ventricles}/FDG_{Cortex}$ $r = -0.679$; $p > 0.001$) (Figure 2C). No difference was found in Qalb, but higher CSF lactate levels were observed in AD patients with respect to CTRLs (r= 0.282; $p = 0.007$).

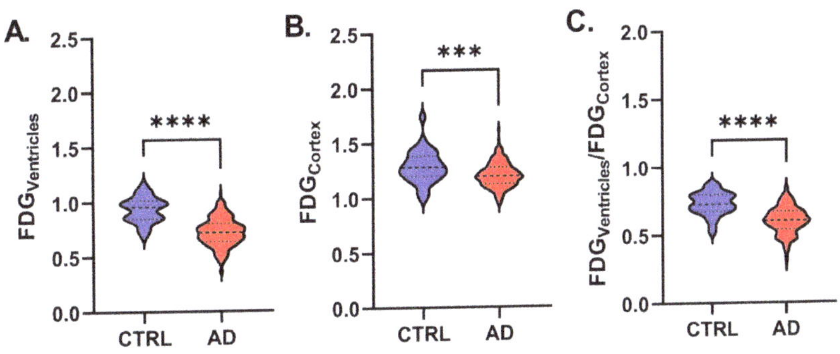

Figure 2. Violin plots depicting differences in FDG_{Cortex} (**A**), $FDG_{Ventricles}$ (**B**), and $FDG_{Ventricles}/FDG_{Cortex}$ (**C**) in the CTRL (blue) and AD (red) groups. *** means a *p*-value < 0.01, **** means a *p*-value < 0.001. The thick dotted line within each violin represents the median, while the thinner dotted lines indicate the first and third quartiles.

Table 1. Comparisons of demographic features and CSF dosages between CTRLs and patients with AD.

	CTRL (n = 35)		All AD (n = 224)		*p*
	Mean ± SD	**Median (IQR)**	**Mean ± SD**	**Median (IQR)**	
MMSE	28.60 ± 1.22	29.00 (2.22)	22.73 ± 2.79	23.50 (3.00)	**<0.001**
CSF Aβ42 (pg/mL)	1028.22 ± 290.34	1005.00 (334.00)	391.57 ± 140.25	390.50 (182.20)	**<0.001**
CSF p-tau (pg/mL)	35.75 ± 15.19	32.35 (23.00)	84.57 ± 44.37	73.50 (53.00)	**<0.001**
CSF t-tau (pg/mL)	212.05 ± 133.80	169.00 (158.95)	616.98 ± 351.76	555.81 (457.65)	**<0.001**
CSF p-tau/Aβ42	0.036 ± 0.018	0.034 (0.020)	0.252 ± 0.201	0.200 (0.147)	**<0.001**
Qalb	7.21 ± 2.15	6.98 (2.42)	6.72 ± 2.87	6.05 (3.54)	**<0.001**
CSF lactates (pg/mL)	1.63 ± 0.40	1.50 (0.69)	1.79 ± 0.35	1.80 (0.40)	**<0.001**
Age (y)	67.09 ± 7.33	67.00 (11.50)	70.25 ± 6.83	71.00 (9.00)	0.085
	%		%		
Sex (F)	40.0		53.5		**0.007**
APOE (ε4)	14.3		44.3		**0.014**
	mean ± SD		mean ± SD		
$FDG_{Ventricles}$	0.94 ± 0.22		0.72 ± 0.13		**<0.001**
FDG_{Cortex}	1.28 ± 0.15		1.20 ± 0.11		**<0.001**
$FDG_{Ventricles}/FDG_{Cortex}$	0.71 ± 0.11		0.60 ± 0.10		**<0.001**

CTRL: control; AD: Alzheimer's disease; n: number; *p*: *p*-values; SD: standard deviation; IQR: interquartile range; y: years; F: female; CSF: cerebrospinal fluid; APOE: Apolipoprotein E; Qalb: Albumin Quotient; FDG: fluorodeoxyglucose. Continuous variables are presented as mean ± SD if normally distributed and as median (IQR) if not normally distributed. Categorical variables are expressed as percentages (%). Bold *p*-values denote statistical significance.

Stratifying AD patients according to APOE genotype, we did not retrieve any significant difference between APOE ε3 (n = 122) and APOE ε4 (n = 97) in clinical–demographical characteristics and CSF biomarker levels, nor in 18F-FDG parameters ($p > 0.05$ for all comparisons) (Table S1).

3.2. Correlation Analysis Between 18F-FDG Parameters

A significant strong correlation was found between FDG_{Cortex} and $FDG_{Ventricles}$ in the CTRL group (r = 0.516; $p = 0.003$). When considering AD patients, we retrieved a less strong but still significant correlation between the two parameters (r = 0.358; $p < 0.001$) (Figure 3). No significant differences were observed between the two correlations (z-score 1.038; $p = 0.299$).

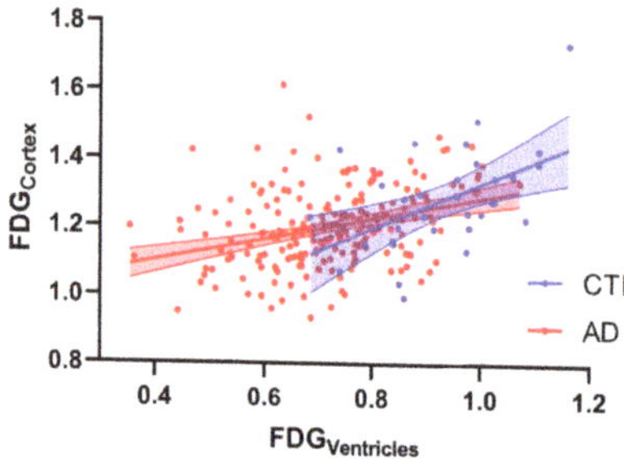

Figure 3. Scatter plots depicting the correlation between FDG_{Cortex} and $FDG_{Ventricles}$ in the CTRL (blue dots) and AD (red dots) groups. Colored areas represent 95% confidence intervals.

3.3. Regression Analyses

We performed multivariate regression analysis to evaluate which factors among BBB integrity (Qalb), the burden of AD pathology (CSF p-tau/Aβ42), and mitochondrial dysfunction (CSF lactates) were associated with variation in FDG_{Cortex}, $FDG_{Ventricles}$, and $FDG_{Ventricles}/FDG_{Cortex}$. We added in the model possible confounding factors, namely age, sex, and plasma glucose levels (Table 2).

Table 2. Multivariate regression analyses in CTRLs and AD patients.

	CTRL		All AD	
	β	*p*	β	*p*
FDG_{Cortex}				
Age	−0.455	**0.044**	−0.013	0.854
Sex	n.a.	0.532	n.a.	0.480
Plasma glucose (mg/dL)	−0.080	0.689	−0.147	**0.036**
Qalb	0.098	0.588	−0.158	**0.033**
CSF p-tau/$A\beta_{42}$	0.037	0.869	−0.071	0.295
CSF lactates	0.082	0.698	0.011	0.876
Adjusted R^2	0.019		0.036	
$FDG_{Ventricles}$				
Age	−0.608	**0.004**	−0.255	**<0.001**
Sex	n.a.	0.188	n.a.	0.053
Plasma glucose (mg/dL)	0.013	0.940	−0.002	0.978
Qalb	0.226	0.173	−0.067	0.333
CSF p-tau/$A\beta_{42}$	0.254	0.207	0.075	0.238
CSF lactates	0.026	0.893	−0.207	**0.002**
Adjusted R^2	0.212		0.148	
$FDG_{Ventricles}/FDG_{Cortex}$				
Age	−0.200	0.397	−0.264	**<0.001**
Sex	n.a.	0.481	n.a.	0.100
Plasma glucose (mg/dL)	0.081	0.707	0.082	0.213
Qalb	0.152	0.439	0.010	0.885
CSF p-tau/$A\beta_{42}$	0.213	0.375	0.136	**0.034**
CSF lactates	−0.047	0.838	−0.214	**0.002**
Adjusted R^2	−0.139		0.144	

CTRL: control; *p*: *p*-values; FDG: fluorodeoxyglucose; CSF: cerebrospinal fluid; Qalb: Albumin Quotient. Bold values denote statistical significance.

In the CTRL group, both lower FDG_{Cortex} (β = −0.455; p = 0.044) and lower $FDG_{Ventricles}$ (β = −0.608; p = 0.004) were associated with older age.

In the AD group, FDG_{Cortex} was negatively associated with glucose plasma levels (β = −0.147; p = 0.036) and Qalb (β = −0.158; p = 0.033). Lower $FDG_{Ventricles}$ was associated with older age (β = −0.255; p < 0.001) and more CSF lactates (β = −0.207; p = 0.002). A lower $FDG_{Ventricles}/FDG_{Cortex}$ ratio was again associated with older age (β = −0.264; p < 0.001) and more CSF lactates (β = −0.214; p = 0.002) but also with lower CSF p-tau/Aβ42 (β = 0.136; p = 0.034).

Finally, we stratified AD patients according to APOE genotype and performed separate multivariate regression analyses considering, all factors which were significantly associated with FDG_{Cortex}, $FDG_{Ventricles}$, and $FDG_{Ventricles}/FDG_{Cortex}$ in the all AD group analysis (Table 3).

Table 3. Multivariate regression analyses in APOE ε3 and APO ε4 patients.

	APOE ε3		**APOE ε4**	
	β	*p*	β	*p*
FDG_{Cortex}				
Age	−0.126	0.900	−0.032	0.747
Plasma glucose (mg/dL)	−0.059	0.544	−0.268	**0.010**
Qalb	−0.138	0.166	−0.240	**0.019**
CSF p-tau/$Aβ_{42}$	−0.096	0.311	−0.071	0.468
CSF lactates	0.141	0.151	−0.085	0.431
Adjusted R^2	−0.001		0.126	
$FDG_{Ventricles}$				
Age	−0.247	**0.008**	−0.289	**0.002**
Plasma glucose (mg/dL)	0.061	0.511	−0.080	0.395
Qalb	−0.032	0.742	−0.218	**0.020**
CSF p-tau/$Aβ_{42}$	−0.057	0.531	0.253	**0.006**
CSF lactates	−0.201	**0.035**	−0.142	0.155
Adjusted R^2	0.069		0.261	
$FDG_{Ventricles}/FDG_{Cortex}$				
Age	−0.238	**0.010**	−0.303	**0.002**
Plasma glucose (mg/dL)	0.089	0.341	0.071	0.457
Qalb	0.041	0.669	−0.120	0.202
CSF p-tau/$Aβ_{42}$	0.015	0.866	0.334	**<0.001**
CSF lactates	−0.260	**0.006**	−0.118	0.243
Adjusted R^2	0.086		0.248	

APOE: Apolipoprotein E; *p*: *p*-values; FDG: fluorodeoxyglucose; CSF: cerebrospinal fluid; Qalb: Albumin Quotient. Bold values denote statistical significance.

Considering the APOE ε3 group, we did not retrieve any significant association with FDG_{Cortex}, but we confirmed negative associations between $FDG_{Ventricles}$ and both age ($β = −0.247$; $p = 0.008$) and CSF lactates ($β = −0.201$; $p = 0.035$). Similarly, we confirmed the negative associations of both age ($β = −0.238$; $p = 0.010$) and CSF lactates ($β = −0.260$; $p = 0.006$) with $FDG_{Ventricles}/FDG_{Cortex}$.

In the APOE ε4 group, we retrieved negative associations between both plasma glucose ($β = −0.268$; $p = 0.010$) and Qalb ($β = −0.240$; $p = 0.019$) and FDG_{Cortex}. As for $FDG_{Ventricles}$, we confirmed a significant association with age ($β = −0.289$; $p = 0.002$), but we also found a significant negative association with Qalb ($β = −0.218$; $p = 0.020$) and a positive association with CSF p-tau/Aβ42 ($β = 0.253$; $p = 0.006$). No association was found between CSF lactates and $FDG_{Ventricles}$ in this subgroup. Finally, the regression analysis for $FDG_{Ventricles}/FDG_{Cortex}$ confirmed the association with age ($β = −0.303$; $p = 0.002$) and CSF p-tau/Aβ42 ($β = 0.334$; $p < 0.001$) but, again, not with CSF lactates.

4. Discussion

Our study highlights the potential utility of two newly introduced 18F-FDG-PET parameters—$FDG_{Ventricles}$ and the $FDG_{Ventricles}/FDG_{Cortex}$ ratio—in exploring metabolic dysfunctions in late-onset AD.

In our cohort, as expected, no difference was found in BBB permeability, evaluated as Qalb, between the AD and CTRL groups. While the literature does not support the notion of a fully disrupted BBB in AD, it has been shown that early BBB disfunction occurs and exacerbates protein misfolding and neurodegeneration [5]. On the other hand, our AD patients showed higher levels of CSF lactates, supporting the hypothesis that changes in energy metabolism—a hallmark of normal aging processes—may also be involved in late-onset AD [26]. Indeed, when neurons experience an increased demand for oxidative substrates, particularly lactate, astrocytes may supply this demand through their high glycolytic capacity leveraging the Astrocyte–Neuron Lactate Shuttle [27].

Considering the 18F-FDG-PET parameters, both cortical and cerebrospinal fluid 18F-FDG uptake (FDG_{Cortex} and $FDG_{Ventricles}$) were significantly lower in patients with AD compared to CTRLs. Reduced 18F-FDG uptake in cortical gray matter is widely recognized as a key feature of AD and a valuable marker for predicting cognitive decline [3]. The underlying concept is that neurodegeneration leads to the loss of neurons and glial cells responsible for glucose uptake or to a reduction in their metabolic efficiency. However, $FDG_{Ventricles}$ does not represent cellular uptake itself. Its reduction in patients with AD may potentially correspond to a reduced glucose demand from the brain. Indeed, we found a significant correlation between the $FDG_{Ventricles}$ and the FDG_{Cortex} levels in both CTRLs and patients, supporting the hypothesis of a physiological "brain pull effect" in which the brain's glucose demand influences its CSF levels [28]. In this context, it has been shown that GLUT expressions at the BBB and blood–CSF barrier in the choroid plexus are strictly regulated by brain metabolic needs and can be differently modulated in several pathological conditions [29,30]. Interestingly, we noted a weaker correlation between $FDG_{Ventricles}$ and FDG_{Cortex} in patients with AD and a more pronounced decrease in $FDG_{Ventricles}$ compared to that in FDG_{Cortex}. As a result, the $FDG_{Ventricles}/FDG_{Cortex}$ ratio was also significantly lower in patients with AD. This finding may suggest that in such a neurodegenerative condition, while there is a global reduction in cortical glucose uptake, there is an extra rate of glucose reduction in the CSF. This glucose could be extracted and directed to astrocytes, in an attempt to compensate for impaired neuronal energy metabolism, to enhance their glycolytic activity and boost lactate production [31]. Thus, the lactate increase that we observed in the CSF could be linked to the neuronal metabolic impairment reflected by 18F-FDG-PET hypometabolism, as has already been observed in AD pathology [32]. Alternatively, the downregulation of GLUT1 occurring in AD could be different at the cortex and the choroid plexus levels [33], explaining the discrepancy between $FDG_{Ventricles}$ and FDG_{Cortex}. Indeed, while bioenergetic failure in aging and neurodegeneration results from increased mitochondrial dysfunction, deficiencies in glucose transport may also be involved. Recent studies in mice suggest that GLUT1 deficiency in the endothelial cells triggers BBB disruption and accelerates the progression of AD neuropathology [9], and reduced GLUT expression has also been detected at the BBB and in the cerebral cortex of patients with AD [11]. Overall, the weaker correlation between 18F-FDG uptake in the cortex and in CSF compared to that in controls could indicate that in AD, the mechanisms regulating glucose transport across the blood–brain and blood–CSF barriers are compromised, leading to a greater dysregulation of glucose homeostasis in the CSF relative to the cortex. These disruptions may reflect underlying pathological changes in the brain's ability to meet its metabolic demands, potentially contributing to the neurodegenerative process.

To investigate which factors were associated with our 18F-FDG-PET parameters, we conducted a multivariate regression analysis considering BBB permeability (Qalb), AD-related pathology (p-tau/Aβ42), and brain metabolic state (CSF lactates) and including potential confounders like age, sex, and basal plasma glucose levels. In CTRLs, none of these factors significantly influenced the 18F-FDG-PET parameters, except for age, with older individuals showing reduced glucose uptake in both FDG_{Cortex} and $FDG_{Ventricles}$. In contrast, within the AD group, lower FDG_{Cortex} was not linked to age but rather to higher BBB permeability. In a previous study, we already observed a similar inverse relation between BBB permeability and glucose consumption in the temporal lobes of patients with AD [34]. It is possible that progressive BBB dysfunction may be associated with reduced GLUT expression, causing an impairment in glucose transport that could exacerbate cerebral metabolic dysfunction. Further, lower $FDG_{Ventricles}$ was associated with increased

BBB permeability, as well as with older age and elevated CSF lactates. The relationship with advanced age may be attributed to neuronal mitochondrial failure, which is common with aging and induces astrocytes to extract glucose from both blood and CSF to produce lactate. Additionally, reduced 18F-FDG uptake in the CSF could be linked to a weakened "pull effect" from the brain due to mitochondrial energy dysfunction, which is further associated with elevated CSF lactate levels. Interestingly, a lower $FDG_{Ventricles}/FDG_{Cortex}$ ratio was also associated with older age, more CSF lactates, and also with lower AD-related pathology (p-tau/Aβ42). We hypothesize that this parameter may reflect the excess of "glucose extraction" from CSF in an attempt to support brain metabolism, an ability that could be progressively lost as the burden of AD pathology increases. These multivariate analyses may partly explain the weaker correlation between FDG_{Cortex} and $FDG_{Ventricles}$ we found in the AD group. Indeed, while only age was associated with these parameters in the CTRLs, in patients with AD, the relationship was influenced by several additional factors, including BBB permeability, the burden of amyloid and tau pathologies, and bioenergetic dysfunctions, which may disrupt the normal metabolic coupling between the cortex and the CSF. Given the cross-sectional design of our study, no causal inferences can be drawn regarding the directionality of the observed associations—i.e., whether BBB dysfunction precedes or follows cerebral metabolic changes. However, our findings raise the hypothesis that alterations in the $FDG_{Ventricles}/FDG_{Cortex}$ ratio may reflect early pathophysiological processes in AD. Future longitudinal studies are warranted to explore whether dynamic changes in this ratio over time are predictive of cognitive decline or disease progression.

To further explore these associations, we stratified AD patients according to APOE genotype, which highlighted the presence of distinct patterns of association. In the APOE ε3 group, no significant link was found between any of the variables and FDG_{Cortex}. This lack of association was reflected by the very low adjusted R^2 observed for this model. However, negative associations with both age and CSF lactates were confirmed for $FDG_{Ventricles}$ and for the $FDG_{Ventricles}/FDG_{Cortex}$ ratio. In contrast, in the APOE ε4 group, lower FDG_{Cortex} was significantly associated with higher plasma glucose levels and BBB permeability. These findings are consistent with the literature, which has demonstrated a strong link between cortical hypometabolism and insulin resistance [35], as well as a higher degree of vascular abnormalities which can impact AD pathology and cognitive decline in APOE ε4 carriers [16,36,37]. In this subgroup of patients, lower $FDG_{Ventricles}$ was associated with older age, higher BBB permeability, and higher degree of AD-related pathology, while no association was observed with CSF lactates. Finally, regression analysis of the $FDG_{Ventricles}/FDG_{Cortex}$ ratio in the APOE ε4 group confirmed significant associations with age and AD-related pathology, but again, no relationship was found with CSF lactates. Collectively, these findings suggest that APOE genotype may modulate the association between amyloid pathology, vascular changes, and brain metabolic dysfunction. Specifically, while brain glucose metabolism in APOE ε3 patients appears to be primarily associated with bioenergetic failure (e.g., elevated CSF lactates), in APOE ε4 carriers, it seems to be influenced by the combined effects of BBB dysfunction and AD-related pathology. Metabolic failure appears to be a common feature among all AD patients, but it is likely more pronounced in APOE ε4 carriers [16], in whom additional factors—such as a well-documented higher burden of amyloid and tau pathology [15], as well as greater microvascular damage accompanied by reduced GLUT expression [9]—may further exacerbate metabolic dysfunction.

A key strength of this study lies in the introduction and evaluation of novel 18F-FDG-PET parameters related to CSF glucose dynamics, which, to the best of our knowledge, have not been previously described in the literature. However, there are notable limitations to consider. CSF regions were defined using the WFU PickAtlas toolbox, which allows the identification of intraventricular spaces (lateral, third, and fourth ventricles), but does not include subarachnoid spaces. As such, our findings specifically reflect intraventricular FDG dynamics and cannot be extrapolated to extracerebral CSF compartments. Future studies employing dedicated segmentation pipelines or dynamic imaging may provide a more comprehensive assessment of global CSF glucose metabolism. Additionally, the selection of the control group, consisting of individuals with functional neurological disorder or tension-type headache, represent a potential limitation. Nonetheless, these subjects were carefully screened to exclude any abnormal CSF findings or structural/metabolic brain alterations. Additionally, the cross-sectional nature of this study limits our ability to assess causal relationships or temporal dynamics. In particular, the lack of longitudinal follow-up prevented us from determining whether the 18F-FDG-PET alterations observed were associated with a more severe trajectory of disease progression. Future longitudinal studies involving larger and more diverse populations, such as those with preclinical AD, as well as consideration of other physiological variables (e.g., diet, physical activity) that may influence brain bioenergetics and glucose transport, will be crucial to validate and expand upon these findings.

5. Conclusions

Overall, our findings suggest that $FDG_{Ventricles}$ and the $FDG_{Ventricles}/FDG_{Cortex}$ ratio may reflect both metabolic dysfunction and altered glucose dynamics in patients with AD. Furthermore, these measures are associated with key factors such as age, BBB integrity, and mitochondrial dysfunction, with distinct patterns observed across different APOE genotypes.

While our cross-sectional design does not allow for causal inference, these associations are consistent with the hypothesis that bioenergetic imbalance plays a role in the pathophysiology of AD. If validated in longitudinal studies, biomarkers sensitive to early metabolic shifts could contribute to identifying individuals at increased risk of AD. Moreover, given that several therapeutic strategies for AD aim to target metabolic dysfunction, such parameters might prove useful in identifying treatment-responsive individuals and in monitoring therapeutic efficacy.

Supplementary Materials: The following supporting information can be downloaded at https://www.mdpi.com/article/10.3390/app15105677/s1. Table S1: Comparisons of demographic features and CSF dosages between APOE ε3 and APOE ε4 groups.

Author Contributions: Conceptualization, C.M. and A.C.; methodology, C.M. and C.G.B.; software, A.C.; validation, A.M. and N.B.M.; formal analysis, C.M. and A.C.; investigation, C.G.B. and M.P.; resources, N.B.M.; data curation, M.P.; writing—original draft preparation, C.M.; writing—review and editing, C.G.B. and A.C.; visualization, C.M.; supervision, A.M. All authors have read and agreed to the published version of the manuscript.

Funding: This research received no external funding.

Institutional Review Board Statement: This study was conducted in accordance with the Declaration of Helsinki and approved by the Ethics Committee of Policlinico di Roma Tor Vergata (57.25CET2PTV) on 27 February 2025.

Informed Consent Statement: Informed consent was obtained from all subjects involved in this study.

Data Availability Statement: Data are available upon reasonable request from the corresponding author.

Acknowledgments: We thank Carlangelo Carrese for his kind support in the refinement of the figures included in this article.

Conflicts of Interest: The authors declare no conflicts of interest.

Abbreviations

The following abbreviations are used in this manuscript:

AD	Alzheimer's disease
APOE	Apolipoprotein E
Aβ42	amyloid-β 1–42
CSF	cerebrospinal fluid
F	female
FDG	fluorodeoxyglucose
GLUT	glucose transporter
CTRL	control
n	number
p	p-value
p-tau	phosphorylated-tau
Qalb	Albumin Quotient
t-tau	total tau
18F-FDG-PET	positron emission tomography with [18F]fluoro-2-deoxyglucose

References

1. Sweeney, M.D.; Montagne, A.; Sagare, A.P.; Nation, D.A.; Schneider, L.S.; Chui, H.C.; Harrington, M.G.; Pa, J.; Law, M.; Wang, D.J.J.; et al. Vascular Dysfunction-The Disregarded Partner of Alzheimer's Disease. *Alzheimers Dement.* **2019**, *15*, 158–167. [CrossRef]
2. Zlokovic, B.V. Neurovascular Pathways to Neurodegeneration in Alzheimer's Disease and Other Disorders. *Nat. Rev. Neurosci.* **2011**, *12*, 723–738. [CrossRef]
3. Salmon, E.; Collette, F.; Bastin, C. Cerebral Glucose Metabolism in Alzheimer's Disease. *Cortex* **2024**, *179*, 50–61. [CrossRef]
4. Mosconi, L.; Berti, V.; Glodzik, L.; Pupi, A.; De Santi, S.; De Leon, M.J. Pre-Clinical Detection of Alzheimer's Disease Using FDG-PET, with or without Amyloid Imaging. *J. Alzheimers Dis.* **2010**, *20*, 843–854. [CrossRef]
5. Nordberg, A.; Rinne, J.O.; Kadir, A.; Lngström, B. The Use of PET in Alzheimer Disease. *Nat. Rev. Neurol.* **2010**, *6*, 78–87. [CrossRef]
6. Hunt, A.; Schönknecht, P.; Henze, M.; Seidl, U.; Haberkorn, U.; Schröder, J. Reduced Cerebral Glucose Metabolism in Patients at Risk for Alzheimer's Disease. *Psychiatry Res.* **2007**, *155*, 147–154. [CrossRef]
7. Ibáñez, V.; Pietrini, P.; Alexander, G.E.; Furey, M.L.; Teichberg, D.; Rajapakse, J.C.; Rapoport, S.I.; Schapiro, M.B.; Horwitz, B. Regional Glucose Metabolic Abnormalities Are Not the Result of Atrophy in Alzheimer's Disease. *Neurology* **1998**, *50*, 1585–1593. [CrossRef] [PubMed]
8. Patching, S.G. Glucose Transporters at the Blood-Brain Barrier: Function, Regulation and Gateways for Drug Delivery. *Mol. Neurobiol.* **2017**, *54*, 1046–1077. [CrossRef]
9. Winkler, E.A.; Nishida, Y.; Sagare, A.P.; Rege, S.V.; Bell, R.D.; Perlmutter, D.; Sengillo, J.D.; Hillman, S.; Kong, P.; Nelson, A.R.; et al. GLUT1 Reductions Exacerbate Alzheimer's Disease Vasculo-Neuronal Dysfunction and Degeneration. *Nat. Neurosci.* **2015**, *18*, 521–530. [CrossRef]
10. Ma, Q.; Zhao, Z.; Sagare, A.P.; Wu, Y.; Wang, M.; Owens, N.C.; Verghese, P.B.; Herz, J.; Holtzman, D.M.; Zlokovic, B.V. Blood-Brain Barrier-Associated Pericytes Internalize and Clear Aggregated Amyloid-B42 by LRP1-Dependent Apolipoprotein E Isoform-Specific Mechanism. *Mol. Neurodegener.* **2018**, *13*, 57. [CrossRef]
11. Kalaria, R.H.N.; Harik, S.I. Reduced Glucose Transporter at the Blood-Brain Barrier and in Cerebral Cortex in Alzheimer Disease. *J. Neurochem.* **1989**, *53*, 1083–1088. [CrossRef] [PubMed]
12. Patching, S. Roles of Facilitative Glucose Transporter GLUT1 in [18F]FDG Positron Emission Tomography (PET) Imaging of Human Diseases. *J. Diagn. Imaging Ther.* **2015**, *2*, 30–102. [CrossRef]
13. Zhou, Z.; Zhao, T.; Gardus, J.; Wen, Q.; Feng, Y.; DeLorenzo, C.; Parsey, R.; Huang, C. Quantifying Ventricular CSF Clearance in the Human Brain Using Dynamic 18F-FDG PET: Insights into Age-Related Glymphatic Impairment. *medRxiv* **2025**. [CrossRef]
14. Yao, J.; Rettberg, J.R.; Klosinski, L.P.; Cadenas, E.; Brinton, R.D. Shift in Brain Metabolism in Late Onset Alzheimer's Disease: Implications for Biomarkers and Therapeutic Interventions. *Mol. Asp. Med.* **2011**, *32*, 247–257. [CrossRef] [PubMed]

15. Serrano-Pozo, A.; Das, S.; Hyman, B.T. APOE and Alzheimer's Disease: Advances in Genetics, Pathophysiology, and Therapeutic Approaches. *Lancet Neurol.* **2021**, *20*, 68–80. [CrossRef]
16. Johnson, L.A. APOE and Metabolic Dysfunction in Alzheimer's Disease. *Int. Rev. Neurobiol.* **2020**, *154*, 131–151. [CrossRef]
17. Wu, L.; Zhang, X.; Zhao, L. Human ApoE Isoforms Differentially Modulate Brain Glucose and Ketone Body Metabolism: Implications for Alzheimer's Disease Risk Reduction and Early Intervention. *J. Neurosci.* **2018**, *38*, 6665–6681. [CrossRef]
18. McKhann, G.M.; Knopman, D.S.; Chertkow, H.; Hyman, B.T.; Jack, C.R.; Kawas, C.H.; Klunk, W.E.; Koroshetz, W.J.; Manly, J.J.; Mayeux, R.; et al. The Diagnosis of Dementia Due to Alzheimer's Disease: Recommendations from the National Institute on Aging-Alzheimer's Association Workgroups on Diagnostic Guidelines for Alzheimer's Disease. *Alzheimers Dement.* **2011**, *7*, 263–269. [CrossRef]
19. Albert, M.S.; DeKosky, S.T.; Dickson, D.; Dubois, B.; Feldman, H.H.; Fox, N.C.; Gamst, A.; Holtzman, D.M.; Jagust, W.J.; Petersen, R.C.; et al. The Diagnosis of Mild Cognitive Impairment Due to Alzheimer's Disease: Recommendations from the National Institute on Aging-Alzheimer's Association Workgroups on Diagnostic Guidelines for Alzheimer's Disease. *Alzheimers Dement.* **2011**, *7*, 270–279. [CrossRef]
20. Guedj, E.; Varrone, A.; Boellaard, R.; Albert, N.L.; Barthel, H.; van Berckel, B.; Brendel, M.; Cecchin, D.; Ekmekcioglu, O.; Garibotto, V.; et al. EANM Procedure Guidelines for Brain PET Imaging Using [18F]FDG, Version 3. *Eur. J. Nucl. Med. Mol. Imaging* **2022**, *49*, 632–651. [CrossRef]
21. D'Agostino, E.; Maes, F.; Vandermeulen, D.; Suetens, P. Atlas-to-Image Non-Rigid Registration by Minimization of Conditional Local Entropy. *Inf. Process. Med. Imaging* **2007**, *20*, 320–332. [CrossRef] [PubMed]
22. Mazziotta, J.C.; Toga, A.W.; Evans, A.; Fox, P.; Lancaster, J. A Probabilistic Atlas of the Human Brain: Theory and Rationale for Its Development. *Neuroimage* **1995**, *2*, 89–101. [CrossRef] [PubMed]
23. Maldjian, J.A.; Laurienti, P.J.; Kraft, R.A.; Burdette, J.H. An Automated Method for Neuroanatomic and Cytoarchitectonic Atlas-Based Interrogation of FMRI Data Sets. *Neuroimage* **2003**, *19*, 1233–1239. [CrossRef]
24. Verger, A.; Doyen, M.; Campion, J.Y.; Guedj, E. The Pons as Reference Region for Intensity Normalization in Semi-Quantitative Analysis of Brain 18FDG PET: Application to Metabolic Changes Related to Ageing in Conventional and Digital Control Databases. *EJNMMI Res.* **2021**, *11*, 31. [CrossRef]
25. Minoshima, S.; Mosci, K.; Cross, D.; Thientunyakit, T. Brain [F-18]FDG PET for Clinical Dementia Workup: Differential Diagnosis of Alzheimer's Disease and Other Types of Dementing Disorders. *Semin. Nucl. Med.* **2021**, *51*, 230–240. [CrossRef]
26. Demetrius, L.A.; Magistretti, P.J.; Pellerin, L. Alzheimer's Disease: The Amyloid Hypothesis and the Inverse Warburg Effect. *Front. Physiol.* **2015**, *5*, 522. [CrossRef]
27. Pellerin, L.; Pellegri, G.; Bittar, P.G.; Charnay, Y.; Bouras, C.; Martin, J.L.; Stella, N.; Magistretti, P.J. Evidence Supporting the Existence of an Activity-Dependent Astrocyte-Neuron Lactate Shuttle. *Dev. Neurosci.* **1998**, *20*, 291–299. [CrossRef]
28. Peters, A.; Sprengell, M.; Kubera, B. The Principle of "brain Energy on Demand" and Its Predictive Power for Stress, Sleep, Stroke, Obesity and Diabetes. *Neurosci. Biobehav. Rev.* **2022**, *141*, 104847. [CrossRef]
29. Cornford, E.M.; Hyman, S.; Swartz, B.E. The Human Brain GLUT1 Glucose Transporter: Ultrastructural Localization to the Blood-Brain Barrier Endothelia. *J. Cereb. Blood Flow Metab.* **1994**, *14*, 106–112. [CrossRef]
30. Lee, W.H.; Bondy, C.A. Ischemic Injury Induces Brain Glucose Transporter Gene Expression. *Endocrinology* **1993**, *133*, 2540–2544. [CrossRef]
31. Magistretti, P.J.; Pellerin, L. Cellular Mechanisms of Brain Energy Metabolism and Their Relevance to Functional Brain Imaging. *Philos. Trans. R. Soc. B Biol. Sci.* **1999**, *354*, 1155–1163. [CrossRef] [PubMed]
32. Liguori, C.; Chiaravalloti, A.; Sancesario, G.; Stefani, A.; Sancesario, G.M.; Mercuri, N.B.; Schillaci, O.; Pierantozzi, M. Cerebrospinal Fluid Lactate Levels and Brain [18F]FDG PET Hypometabolism within the Default Mode Network in Alzheimer's Disease. *Eur. J. Nucl. Med. Mol. Imaging* **2016**, *43*, 2040–2049. [CrossRef] [PubMed]
33. Redzic, Z. Molecular Biology of the Blood-Brain and the Blood-Cerebrospinal Fluid Barriers: Similarities and Differences. *Fluids Barriers CNS* **2011**, *8*, 3. [CrossRef]
34. Chiaravalloti, A.; Fiorentini, A.; Francesco, U.; Martorana, A.; Koch, G.; Belli, L.; Torniolo, S.; Di Pietro, B.; Motta, C.; Schillaci, O. Is Cerebral Glucose Metabolism Related to Blood-Brain Barrier Dysfunction and Intrathecal IgG Synthesis in Alzheimer Disease? A 18 F-FDG PET/CT Study. *Medicine* **2016**, *95*, e4206. [CrossRef]
35. Rhea, E.M.; Leclerc, M.; Yassine, H.N.; Capuano, A.W.; Tong, H.; Petyuk, V.A.; Macauley, S.L.; Fioramonti, X.; Carmichael, O.; Calon, F.; et al. State of the Science on Brain Insulin Resistance and Cognitive Decline Due to Alzheimer's Disease. *Aging Dis.* **2024**, *15*, 1688–1725. [CrossRef]

36. Onos, K.; Lin, P.B.; Pandey, R.S.; Persohn, S.A.; Burton, C.P.; Miner, E.W.; Eldridge, K.; Kanyinda, J.N.; Foley, K.E.; Carter, G.W.; et al. Assessment of Neurovascular Uncoupling: APOE Status Is a Key Driver of Early Metabolic and Vascular Dysfunction. *Alzheimer's Dement.* **2024**, *20*, 4951–4969. [CrossRef]
37. Williams, H.C.; Farmer, B.C.; Piron, M.A.; Walsh, A.E.; Bruntz, R.C.; Gentry, M.S.; Sun, R.C.; Johnson, L.A. APOE Alters Glucose Flux through Central Carbon Pathways in Astrocytes. *Neurobiol. Dis.* **2020**, *136*, 104742. [CrossRef]

applied sciences

MDPI

Article

Multi-Graph Assessment of Temporal and Extratemporal Lobe Epilepsy in Resting-State fMRI

Dimitra Amoiridou [1], Kostakis Gkiatis [1,2], Ioannis Kakkos [1], Kyriakos Garganis [2] and George K. Matsopoulos [1,*]

1 Biomedical Engineering Laboratory, National Technical University of Athens, 15773 Athens, Greece; amidimi@biomed.ntua.gr (D.A.); gkiatisk@biomed.ntua.gr (K.G.); ikakkos@biomed.ntua.gr (I.K.)
2 Epilepsy Monitoring Department, St. Luke's Hospital, 55236 Thessaloniki, Greece; kyriakosgarganis@gmail.com
* Correspondence: gmatsopoulos@biomed.ntua.gr; Tel.: +30-210-772-2288

Abstract: Epilepsy is a common neurological disorder that affects millions of people worldwide, disrupting brain networks and causing recurrent seizures. In this regard, investigating the distinctive characteristics of brain connectivity is crucial to understanding the underlying neural processes of epilepsy. However, the various graph-theory frameworks and different estimation measures may yield significant variability among the results of different studies. On this premise, this study investigates the brain network topological variations between patients with temporal lobe epilepsy (TLE) and extratemporal lobe epilepsy (ETLE) using both directed and undirected network connectivity methods as well as different graph-theory metrics. Our results reveal distinct topological differences in connectivity graphs between the two epilepsy groups, with TLE patients displaying more disassortative graphs at lower density levels compared to ETLE patients. Moreover, we highlight the variations in the hub regions across different network metrics, underscoring the importance of considering various centrality measures for a comprehensive understanding of brain network dynamics in epilepsy. Our findings suggest that the differences in brain network organization between TLE and ETLE patients could be attributed to the unique characteristics of each epilepsy type, offering insights into potential biomarkers for type-specific epilepsy diagnosis and treatment.

Keywords: epilepsy; network topology; graph theory; fMRI; centrality measures

1. Introduction

Epilepsy is one of the most common neurological disorders, affecting around 50 million people worldwide [1] and characterized by recurrent, unprovoked seizures (elicited by sudden, abnormal bursts of neural activity in the brain) that can result in alterations in behavior, motor functions, sensory perceptions, or levels of consciousness [2]. Despite multiple antiepileptic drugs, over 30% of epilepsy patients continue to experience seizures and report lower quality of life (QoL) [3–5]. In children, epilepsy surgery has shown a higher rate of seizure freedom compared to pharmacological treatment [6]. On the other hand, although surgical resection can improve seizure freedom, it carries cognitive impairment risks [6,7]. In this regard, the identification of brain areas involved in epilepsy is particularly important due to their unique characteristics and challenges in each individual patient [8]. Accurate non-invasive localization can improve surgical outcomes while preserving cognitive functions, allowing more patients to successfully undergo epilepsy surgery [9,10].

Although several modalities have been implemented in epilepsy research and clinical management, functional magnetic resonance imaging (fMRI) can optimize spatial resolution, providing insights into brain activity and seizure propagation in the brain [11]. As such, by utilizing the blood oxygenation level-dependent (BOLD) signal, fMRI can track how the brain reorganizes itself in response to seizures or treatments.

The recent review by Feng et al. [12] presents a comprehensive overview of biomarkers related to pediatric epilepsy, including 62 studies that exclusively utilized fMRI data. According to the review, integrating fMRI analysis with graph theory provides a robust framework for enhancing predictions related to cognitive performance, assessing surgical outcomes, and evaluating symptom severity in focal epilepsy. In fact, neural plasticity can elucidate the brain's ability to adapt and reorganize itself, both functionally and structurally [13]. Furthermore, chronic seizures can lead to changes in functional neural networks, making certain regions more prone to generating and spreading seizures.

On this premise, functional connectivity (FC) analysis allows the brain to be represented as a graph, where edges can represent the temporal dependencies of signals from the nodes, which can be the anatomical brain regions [14]. In this network, different brain regions have specialized functions and work together to perform higher cognitive functions [15]. Epilepsy has been characterized as a brain network disorder since the seizures not only affect the epileptogenic regions but can also spread to anatomically distant brain regions [16]. From this standpoint, multiple studies underline the importance of the resting-state fMRI (rs-fMRI) FC analysis in decision-making about epilepsy surgery [17,18]. Foit et al. [18] suggested that connectivity-based biomarkers could enhance preoperative evaluation, emphasizing the role of FC in predicting cognitive outcomes post-epilepsy surgery and in localizing the epileptogenic zone. Boerwinkle et al. [17] demonstrated that incorporating rs-fMRI connectivity into the presurgical evaluation resulted in a 50% increase in positive decisions for surgery. In this regard, the incorporation of graph-theory metrics can increase our comprehension of how distant brain regions contribute to information processing, introducing new biomarkers associated with brain network disorders such as epilepsy [19].

This aspect has sprung an increase in research attention towards the topological features that distinguish the epileptic brain network from the healthy one. Specifically, since epilepsy often involves abnormal connectivity patterns in the brain, network metrics, such as global efficiency, clustering coefficients, and path lengths, can highlight changes in brain connectivity over time, providing insights into the neural disruptions, aiding in diagnosis, treatment or surgical planning, and understanding the underlying mechanisms of epilepsy [19,20]. The use of graph-theory fMRI biomarkers is especially important in clinical applications [21]. For instance, the study of Doucet et al. [22] displays the value of graph theoretical features in the prediction of neurocognitive outcomes through the presurgical evaluation of TLE patients, proposing a model of three topological features (distance, local efficiency, and participation) capable of explaining more than 68% of the variance observed in the neurocognitive functions. Moreover, longer path lengths (indicating a lower global efficiency) in the graphs of patients with temporal lobe epilepsy (TLE) compared to controls are frequently observed [23–28].

Most notably, graph-theory metrics help in characterizing the overall architecture of brain networks, identifying hubs (the regions that have a key role in the information flow since they facilitate communication between other regions), and understanding how different regions interact. In fact, hub-related analysis can reflect adaptive changes in network connectivity, demonstrating distinct functional or structural characteristics in individuals with neurological disorders [29]. In this context, numerous studies emphasize the significance of hub reorganization in epilepsy, highlighting its role as a critical biomarker associated with patients' cognitive impairments. A meta-analysis of Crossley et al. [30] concludes that the hub regions are prone to turning into lesions in a variety of brain disorders, and their increased biological cost is combined with higher vulnerability compared to the non-hub regions. Interestingly, they report significantly increased values of degree centrality in the lesion voxels compared to the non-lesion voxels in 9 of the 26 disorders analyzed, where 3 of them include different types of epilepsy (right and left TLE, juvenile myoclonic epilepsy). Regarding TLE patients, the study of Bernhardt et al. [23] reports different hub organization in patients compared to controls, with most of the TLE hubs localized in the paralimbic and the temporal lobe. Another relevant study also presented a shift in the hub localization of TLE patients, with the 98% of the hub regions of the right TLE group located in the contralateral hemisphere to the epilepsy onset [31]. Moreover, the study by Royer et al. [32] summarizes the utility of hub-mapping analysis in the diagnosis, the formation of a treatment plan, and the presurgical evaluation of patients with epilepsy, as well as highlighting the necessity for further investigation of the hub-mapping techniques.

There is a need for extensive validation and refinement of our understanding of brain reorganization in epilepsy, and for accurately identifying hub nodes and predicting the spread of epileptic discharge in the individual patient's brain. However, most studies employ only degree centrality to identify hub nodes, neglecting other measures that consider the distance between regions or their ability to facilitate connections. Furthermore, most relevant research analyzes average graphs constructed from mean connectivity values within each group, thus disregarding individual-specific brain networks.

In this study, we aim to address these constraints by evaluating the importance of the regions of interest (ROIs), employing four centrality measures for the hub definition in three different connectivity techniques. Moreover, instead of analyzing the mean brain network, we analyze the hubs of each subject separately and then create the hub probability based on their presence within each group, minimizing the averaging bias. Specifically, we introduce a methodological framework to do the following:

- Explore the brain reorganization characteristics and their differences between epileptic patients and healthy controls.
- Reveal the hub regions, utilizing different brain connectivity methods and assessing the centrality measures to quantify the importance of a node.
- Uncover the regional alterations in hub probability, identifying the ROIs that play a critical role in both healthy individuals and TLE and/or ETLE patients.

For this purpose, we employ three different connectivity techniques (Pearson correlation, mutual information (MI), and Granger causality) to construct graphs for each subject among TLE and ETLE patients as well as healthy controls. Subsequently, we compare the global centralization features between all subjects and groups at different density level thresholds. Further, we calculate the hub probability of each ROI for each group and assess their differences to offer a comprehensive understanding of the areas controlling the information flow in each type of epilepsy and give prominence to the hub reorganization as a possible biomarker to predict network restructuring in epilepsy.

2. Materials and Methods

2.1. Subjects

Data were gathered retrospectively and include resting-state fMRI recordings from 65 individuals, 44 of which were epilepsy patients (21 females), and 21 healthy controls (12 females). Subsequently, the patient group dataset was divided into temporal lobe epilepsy (TLE—28 subjects, 12 females, mean age 28.4 $\pm$ 8.9) and extratemporal lobe epilepsy (ETLE—16 subjects, 9 females, mean age 21.3 $\pm$ 6.4) (Tables 1 and 2). The control group (mean age 31.6 years ($\pm$7.4)) was screened prior to data acquisition to ensure no history of neurological diseases or drug abuse, and each reported normal or corrected-to-normal vision and right hand as dominant [33]. Patient group inclusion criteria were as follows: (1) epilepsy patient, either temporal epilepsy or extratemporal epilepsy, (2) ability to complete all necessary recordings, (3) MRI negative or small lesion that does not interfere with the brain architecture/structures, and (4) no anesthesia during the fMRI acquisition. Assignment to either the temporal (TLE) or the extratemporal (ETLE) group was determined following a comprehensive presurgical evaluation protocol in the "St. Luke's" Epilepsy Center on the basis of localizing information obtained through prolonged Video-EEG monitoring, structural MRI, and in selected cases, additional studies such as EEG-FMRI and PET. Data recording was performed in accordance with the Declaration of Helsinki, while the study was approved by the Institution's Review Board (3 April 2024). Prior to all processing and analysis procedures, data anonymization was performed.

Table 1. Demographic and clinical characteristics of healthy controls.

Characteristics	
Number of Individuals	21
Female, N (%)	12 (57.1)
Age, mean in years (std, range)	31.6 (7.4, 18–44)
Handedness: left, N (%)	0 (0)
Handedness: right, N (%)	20 (100)
Language Lateralization: left, N (%)	19 (95)
Language Lateralization: right, N (%)	1 (5)

Table 2. Demographic and clinical characteristics of patient groups.

Characteristics	TLE	ETLE
Number of Individuals	28	16
Age, mean in years (range)	28.4 ± 8.9 (14–41) *, **	21.3 ± 6.4 (13–40) **
Age of seizure onset,	17.1 ± 12.1 (0.2–39) **	2.5–39
Affected Hemisphere		
Left, N (%)	19 (67.9)	11 (68.75)
Right, N (%)	7 (25)	5 (31.25)
Bilateral, N (%)	2 (7.1)	0 (0)
Pathology		
Grade I astrocytoma	3 (10.7)	2 (12.5)
Ganglioglioma	2 (7.1)	0 (0)
Gliosis	0 (0)	1 (6.25)
MTS	5 (17.9)	0 (0)
Meningioma	0 (0)	1 (6.25)
DNET	3 (10.7)	1 (6.25)
FCD	3 (10.7)	3 (18.75)
unknown	12 (42.9)	5 (31.25)

Note: Asterisk (*) denotes a statistical difference with the control group ($p < 0.05$); Double asterisk (**) denotes a statistical difference between the patient groups ($p < 0.05$); MTS: Medial Temporal Sclerosis; DNET: Dysembryoblastic Neuroepithelial Tumor; FCD: Focal Cortical Dysplasia.

2.2. Data Acquisition

All fMRI recordings were performed at "St. Luke's" Hospital, Thessaloniki, Greece. Since bias can be introduced if different instrumentation is used [34], all data acquisition utilized the same scanner, i.e., a 1.5T AVANTO FIT MRI scanner with a standard Siemens 20-channel head coil. The MRI protocol included(a) resting-state fMRI, $2 \times 2 \times 2$ mm^3 voxel size, TR 1700 ms, TE 50 ms, 530 volumes, 15 min, (b) gradient field map, $2 \times 2 \times 2$ mm^3 voxel size, $228 \times 228 \times 170$ mm^3 field of view, 4.76 ms and 9.52 ms the two echo times, (c) T1 MPRAGE, $1 \times 1 \times 1$ mm^3 voxel size, $250 \times 250 \times 192$ mm^3 field of view, (d) T2 FLAIR, $1 \times 1 \times 1$ mm^3 voxel size, $260 \times 252 \times 176$ mm^3 field of view. All resting-state fMRI were acquired at the beginning of each session. Subjects were instructed to keep their eyes closed throughout the acquisition while trying not to fall asleep. Pneumatic headphones were used to insulate from the scanner noises, but no music or other audio was playing during fMRI acquisition. More details of the data acquisition procedures can be found in [33].

2.3. Preprocessing

The FMRIB's Software Library (FSL; v 6.0.1; https://www.fMRIb.ox.ac.uk/fsl; accessed on 15 January 2024) was utilized for the preprocessing pipeline of fMRI data [35]. Gradient field maps were utilized to estimate the scanner inhomogeneities for data correction and better registration procedures. The effective echo spacing of the fMRI data was at 0.78 ms. The preprocessing pipeline included brain extraction, motion correction via linear registration with 6 DOF to the middle fMRI volume of the acquisition, spatial smoothing with a kernel of 4 mm FWHM, grand mean intensity normalization, and a high pass filter at 100 s. Registration was performed from the fMRI template image to the T1 image with the BBR algorithm and inverted to apply the transformation to the label image and capture the time series of each region in the fMRI space to avoid unnecessary interpolations [36,37]. Afterwards, independent component analysis was performed for each session and the components related to any artifact were manually recognized and regressed out of the signal for the subsequent analysis [38,39]. Registration to a template was not performed, as the time series were extracted to the fMRI space for each subject.

2.4. Segmentation and Time Series Extraction

The atlas utilized in the current study is the Destrieux atlas [40], as they were parcellated from FreeSurfer [41]. For the segmentation, both T1 isotropic and T2 FLAIR images were utilized, with a total of 160 regions being parcellated into cortical and subcortical regions (the names and abbreviations of the brain regions can be found in Appendix A). The mean value of each region in each timepoint was calculated to create the time series of the regions for each subject from the preprocessed signal.

2.5. Brain Connectivity Estimation

For each subject, the temporal dependencies between the BOLD time series extracted from the ROIs of the Destrieux atlas were estimated using three connectivity methods: the Pearson correlation, the mutual information (MI), and the Granger causality. The mathematical formulas for each method can be found in Appendix B. Each method output was a connectivity matrix with size 160 × 160. The corresponding graphs were constructed at a subject level with 160 nodes denoting the ROIs used for the analysis, and the edges between those ROIs represented the strength of their dependence, expressed by the values of the connectivity matrices. The correlation matrices corresponding to the Pearson correlation [42] and the mutual information (MI) [43] are undirected (symmetric), while the causation matrix corresponding to the Granger causality [44] is directed (asymmetric).

To ensure stability of the graph-theory metrics, the same number of edges were taken into account, applying a proportional threshold on the edges of the graph to preserve a specific density level [45]. Specifically, we define as density level the percentage of the edges/connections preserved in the final graph, while all other edges were set to 0. In this study, ten (10) density levels were utilized, from 5% to 50% with a step of 5%, resulting in the construction of 30 graphs in total for each subject (10 of each density level × 3 connectivity metrics × 65 subjects = 1950 graphs). A schematic of the applied framework is presented in Figure 1.

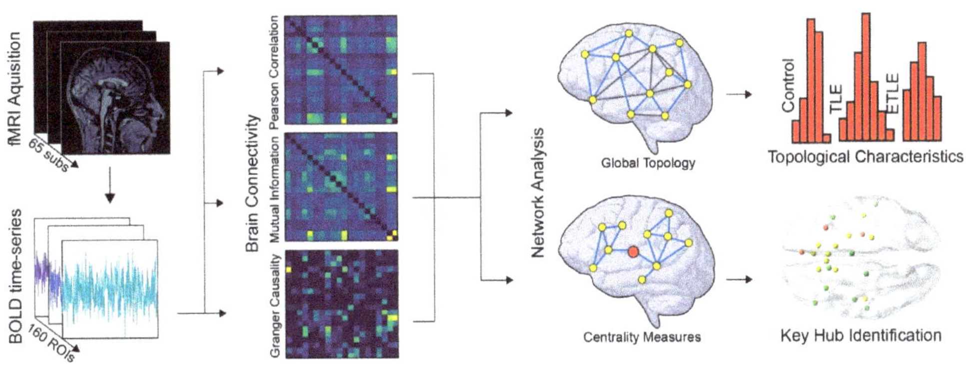

Figure 1. The steps followed for the analysis pipeline.

2.6. Global Network Analysis

Accordingly, each graph was investigated for its topological characteristic via the graph metrics of (i) global assortativity, (ii) global efficiency, (iii) global (mean) clustering coefficient, and (iv) number of components. The definitions of the aforementioned global topological features are included in Table 3. Those global features are calculated for each of the $3 \times 10 = 30$ graphs corresponding to each subject (3 types of graphs—Pearson, MI and Granger, and 10 density levels) resulting in 120 total features (30 graphs and four global topological metrics) for each subject. For each of the global topological features, an ANOVA test among the groups was implemented ($p < 0.05$ FDR corrected). Tukey–Kramer post hoc test was implemented to infer the statistically significant changes between each pair of groups ($p < 0.05$).

Table 3. Definitions of the global topological features.

Features	Description
Global assortativity	Assesses the similarity of the nodes connected in the graph, concerning their degree centrality. High positive values of assortativity indicate that the nodes tend to connect with nodes with similar centrality degrees. Negative values indicate that the hubs of the network tend to connect with low-degree nodes [46].
Global efficiency	Measures how easily information can travel between any pair of nodes. High values indicate a more integrated and well-connected network. Lower global efficiency suggests a network where information transfer is less efficient, indicating potential disruptions or fragmentation in network connectivity [47]. Mathematically, it can be formed as follows: $$E_{Global} = \frac{1}{N(N-1)} \sum_{i \neq j \in G} \frac{1}{d_{ij}}$$ where d_{ij} is the shortest path length that connects the nodes i and j in the graph.
Mean clustering coefficient	Quantifies the degree to which nodes tend to cluster together. High mean clustering coefficient values suggest that there are many localized clusters of interconnected brain regions, which might be indicative of specialized functional modules within the brain [48]. Low values of the mean clustering coefficient indicate that the graph cannot be divided into clusters, and most of its nodes participate in closed triangles [49]. The mean clustering coefficient can be calculated as the average of the nodal clustering coefficient from all the nodes in the graph: $$CC_{Global} = \frac{1}{N} \sum_{i \in G} c_i$$ c_i is the clustering coefficient of the node i, which is calculated according to the following formula: $$c_i = \frac{2T_i}{d_i(d_i - 1)}$$ T_i denotes the number of triangles that include the node i, and d_i is the degree of this node.
Number of components	Evaluates the maximal connected subgraphs within the entire network. A graph with only one component is fully connected, while a graph with multiple components is disconnected or has distinct clusters.

2.7. Hub-Related Analysis

2.7.1. Probability Distribution in Regions of Interest

Following the global topological characteristics, we aimed to identify the important brain areas (hubs) associated with TLE and ETLE. To do so, we focused on various centrality measures which quantify the importance of each node in the graph (Table 4). We calculated the (i) degree centrality, (ii) betweenness centrality, (iii) local efficiency, and iv) eigenvector centrality for every ROI of all three graph methods, with the exception being the eigenvector centrality, which is defined only for undirected graphs (i.e., Pearson correlation and MI). If the centrality value of the ROI differs from the mean centrality values (i.e., the average value of the 160 nodes included in the graph) for more than one standard deviation, then this ROI was considered a hub region for the graph. This was implemented for all ROIs for each subject, for each group (TLE, ETLE, and control), and for each centrality measure. To ensure reliable node identification and to preserve all the ROIs as possible hubs of the graphs, centrality measures were handled at the density level of 50%. The hub probability within each group was defined as the number of subjects within the group that have this ROI as a hub divided by the total number of subjects in this group, as shown in Equation (1).

$$p_{R,G} = \frac{\#S_G \text{ with R as a hub}}{\#S_G} \tag{1}$$

where $p_{R,G}$ denotes the hub probability of the ROI R ∈ {1, . . ., 160} in the specified group G ∈ {TLE, ETLE, Controls}, and $\#S_G$ denotes the number of subjects within group G.

Table 4. Definitions of the nodal topological features.

Measures	Description
Degree centrality	Measures the number of edges starting from the region. For the directed graph of Granger causality, we used the sum of the in-degree (number of incoming edges) and out-degree (number of outgoing edges) of each node. Nodes with higher degree centrality have a higher number of connections within the network and are considered to be more influential [50].
Betweenness centrality	Quantifies the ability of a node to act as a bridge along the shortest paths between pairs of other nodes. Nodes with high betweenness centrality act as intermediaries that facilitate the flow of information or resources between other nodes [51].
Local efficiency	Evaluates the network's ability to maintain communication despite potential node failures. Nodes with high local efficiency have neighbors that can communicate with each other via multiple paths, ensuring robust communication pathways within local clusters or neighborhoods. [47].
Eigenvector centrality	Assesses the significance of a node within a network based on its connections to other highly central nodes. Nodes with high eigenvector centrality are deemed influential due to their direct connections with high centrality score nodes [52].

The result is a map of 3 × 160 hub probabilities for each of the three groups and each ROI. This procedure was also implemented separately for each of the four centrality measures, resulting in 5 hub probabilities corresponding to a ROI within a group: hub probability (regardless of the centrality used), degree hub probability, betweenness hub probability, local efficiency hub probability, and eigenvector hub probability. Finally, we defined the hubs of the group as the ROIs with the relative higher values of hub probability. Since we have different sample sizes between the groups, we do not apply a common threshold for the hub probabilities. Instead, the important hub regions are defined as the ROIs which have a hub probability higher than the mean hub probability within the specified group by more than one standard deviation:

$$p_{R,G} > \text{mean}_G\left(p_{i,G}\right) + \text{std}_G\left(p_{i,G}\right) \tag{2}$$

where $p_{R,G}$ denotes the hub probability of the ROI R ∈ {1, . . ., 160} in the specified group G ∈ {TLE, ETLE, Controls}, $\text{mean}_G\left(p_{i,G}\right)$ denotes the mean hub probability across the ROIs within group G, and $\text{std}_G\left(p_{i,G}\right)$ is the standard deviation of the hub probability values of the ROIs within group G.

2.7.2. Probability Distribution in Brain Sectors

To further elucidate the centrality-specific hub probabilities of the different graphs, we employed the same analysis on distinct anatomical subdivisions within the brain, encompassing various regions or structures that share functional and/or structural characteristics

(Brain Sectors). The exact categorization can be found in Appendix A. As such, the 160 ROIs were divided into eight brain sectors to properly assess the results. The brain sectors include the following: CNG (cingulate cortex), FR (frontal lobe), INS (insula), MOT (motor area), OCC (occipital lobe), PAR (parietal lobe), SUB (subcortical area), and TMP (temporal lobe). For every brain sector, we compute the mean degree hub probability, the mean betweenness hub probability, the mean local efficiency hub probability, and the mean eigenvector hub probability of the ROIs of this brain sector. Similar to global network analysis, we compare those probabilities within every brain sector between the three groups using the ANOVA test, followed by the Tukey–Kramer post hoc test.

3. Results

3.1. Topological Characteristics

The comparison between the global topological features, including global assortativity, efficiency, the mean clustering coefficient, and the number of components, was implemented for the three types of graphs (Pearson, MI, Granger causality) for all density levels. The results of all the global topological features comparison are presented in Table 5.

Table 5. Global topological features comparison.

Global Topological Features	HC vs. TLE	HC vs. ETLE	TLE vs. ETLE	Metric	Density Level
Global assortativity	↑			Pearson correlation	30–50%
			↓	Mutual information	5–10%
			↓	Granger causality	5–45%
Global efficiency	↑			Pearson correlation	45–50%
			↓	Pearson correlation	30–50%
			↑	Granger causality	5%
Global clustering coefficient		↓		Pearson correlation	5%
	↑			Pearson correlation	25%
			↓	Pearson correlation	30–35%
		↓		Mutual information	40–45%
			↑	Granger causality	15–40%
		↓		Pearson correlation	5%
	↑			Pearson correlation	25%
Number of components	↓			Pearson correlation	40–50%
			↑	Pearson correlation	30–50%
		↑		Granger causality	5%

Note: The upwards/downwards arrow (↑/↓) indicates that the group has statistically significant higher/lower values, respectively. When there is no arrow, there is no statistically significant result. The metric of the test is shown in the fourth column. The fifth column presents the graph's density level when there was a statistically significant result. The dash denotes that the difference was observed in all the density levels from the starting percentage to the final with a step of 5%.

From the topological characteristics analysis, the features that displayed significant differences among the three groups in all the connectivity methodologies were the global assortativity and the mean clustering coefficient (Figure 2). Specifically, in the Pearson correlation analysis, the healthy control displayed positive values of assortativity in all density levels. On the contrary, TLE patients exhibited disassortative functional networks for density levels higher than 30%. In higher density levels (above 45%), the TLE group showed lower values of global efficiency when compared to the other two groups. Interestingly, ETLE patients presented an increased mean clustering coefficient compared to the other two groups. Further, when the density level is more than 40%, TLE patients presented more components compared to the other two groups.

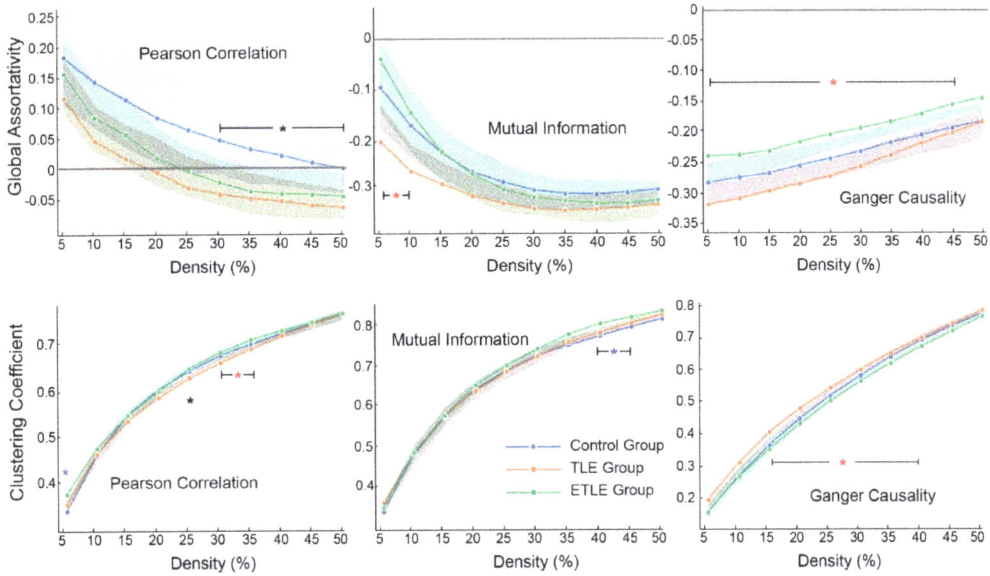

Figure 2. The global topological features presented significant differences between the three groups for all the types of graphs: Pearson correlation, MI, and Granger causality, for the various density levels. Blue lines correspond to the control group, orange to the TLE group, and green to the ETLE group. Statistically significant differences ($p < 0.05$) are presented with the asterisk (*) symbol. Red asterisk: difference between TLE and ETLE; blue asterisk: difference between ETLE and control; black asterisk: difference between TLE and control.

The graphs that evaluate the functional connectivity with the non-linear metric of MI present comparable topological features between the control and the TLE group. When 5% or 10% of the highest total edges are preserved, ETLE patients have higher assortativity compared to the TLE group. All the groups have disassortative graphs in all the density levels. In higher levels of density (40–45%) of the MI graphs, ETLE patients presented higher levels of clustering coefficient on average, compared to the graphs of the control group.

Concerning the Granger causality network, controls present similar topological features with each of the two patient groups, with significant differences being observed between TLE and ETLE groups. Specifically, TLE patients present graphs with higher clustering coefficient, and lower values of assortativity compared to ETLE group in most density levels. When we keep only the highest 5% of the connectivity values in the graph, the TLE patients show a more integrated graph compared to the ETLE patients (higher global efficiency), and the number of components is significantly lower in the ETLE group compared to controls.

3.2. Hub Probabilities in Regions of Interest

The threshold on the hub probability of the ROIs is defined within each of the three groups and it is different for the three types of connectivity metrics, resulting in 9 different thresholds. In the Pearson correlation graph network analysis, the thresholds that were used to identify the hubs of each group were: $p_{R,ETLE} > 0.55$, $p_{R,TLE} > 0.48$, and $p_{R,C} > 0.56$. According to those thresholds, the hub regions presented in all groups in both hemispheres were as follows: the superior frontal gyrus (SFG), the superior parietal gyrus (SPG), the postcentral (PostCG) and precentral gyrus (PreCG), the precuneus (PCUN), the middle temporal gyrus (MTG), and the lateral aspect of the superior temporal gyrus (STGlat). In the left hemisphere, all groups have a hub with the supramarginal gyrus (SMG), and in the right hemisphere, the middle occipital gyrus (MOG) and the superior temporal sulcus (STS) (Figure 3). The highest probabilities in all groups are reported in Appendix C.

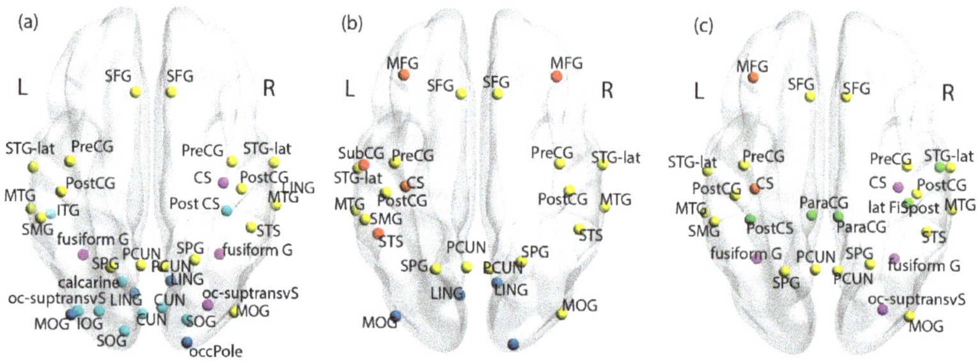

Figure 3. The hub regions with the highest probabilities in the Peason correlation graphs for: (**a**) the control group, (**b**) the TLE group, and (**c**) the ETLE group. Node color denotes the inclusion of each node to control, TLE and ETLE groups. i.e., Light blue: node included only in the control group, Red: node included only in the TLE group, Green: node included only in the ETLE group, Purple: node included in the control and ETLE groups, Blue: node included in control and TLE groups, Orange: node included in the TLE and ETLE groups, and Yellow: node included in all groups.

We observed a hub reorganization between the control group and the patients, explained through the differences in hub probabilities. The probability distribution indicated that the superior occipital gyrus (SOG) was identified in the control group with a hub probability of more than 61% in both hemispheres (61.9% the left SOG, 71.4% the right SOG), while the same probability was less than 36% for the TLE group (35.7% the left SOG, 32.1% the right SOG) and less than 50% for the ETLE group (37.5% the left SOG, 50% the right SOG). Furthermore, the bilateral Cuneus (CUN) demonstrated decreased probability in both patient groups compared to controls. Specifically, the left CUN hub probability was 66.7% in controls, 46.4% in TLE patients, and 31% in ETLE patients, while the right CUN hub probability was 71% for the controls, 42.9% in the TLE, and 50% in the ETLE groups. Another deviation that was observed in the patients' hubs was in the left inferior occipital gyrus (IOG), where the control group showed a hub probability of 76.2% while the patient groups were significantly lower (39% for the TLE patients, 43.8% for the ETLE patients). Finally, the paracentral gyrus (ParaCG) was deemed as a region with a key role in the ETLE graphs, with 68.75% hub probability bilateral, while the bilateral hub probability for the control group was 38% and for the TLE was lower than 50% (46.43% in the left hemisphere and 42.86% in the right).

Regarding the MI graph networks, most of the hub regions did not display significant differences among the three groups (Figure 4). In the MI graph analysis, the thresholds that corresponded to the hub probabilities for each of the three groups were: $p_{R,ETLE} > 0.56$, $p_{R,TLE} > 0.48$, and $p_{R,C} > 0.49$. The highest probabilities in all groups are reported in Appendix C. The brain regions found as hubs in both hemispheres in all groups were more posterior regions, mainly in the occipital lobe as well as the superior parietal gyrus and the posterior central gyrus. The hubs presented unilaterally in all groups were the left CUN and the right ParaCG. Interestingly, the right CUN was presented in most patients' MI graphs with higher hub probability values compared to the control group (hub probability was 60.7% in the TLE and 81% in the ETLE group, but only 47.6% in the controls). Similarly, the right angular gyrus (ANG) was identified as an important brain region in all ETLE patients and in 64.3% of TLE patients but displayed lower hub probability (47.6%) in the controls. On the contrary, the regions of left PCUN, left subparietal sulcus (subparS) and right posterior-ventral part of the cingulate gyrus (vPCC) demonstrated over 50% hub probability in the control group but less than 20% hub probability in the ETLE group. Those regions are not identified as hubs in the TLE group either, since they present hub probabilities lower than 48% (46.43% for the left PCUN and the right vPCC, 32.15% for the left subparS).

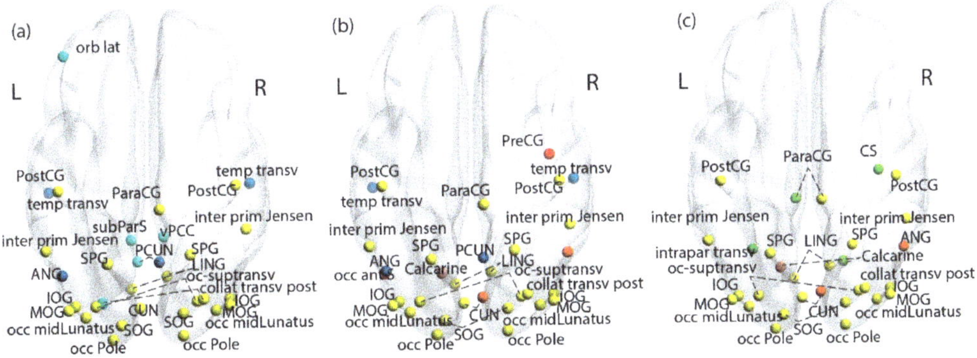

Figure 4. The hub regions with the highest probabilities in the MI graphs for: (**a**) the control group, (**b**) the TLE group, and (**c**) the ETLE group. Node color denotes the inclusion of each node to control, TLE and ETLE groups. i.e., Light blue: node included only in the control group, Red: node included only in the TLE group, Green: node included only in the ETLE group, Purple: node included in the control and ETLE groups, Blue: node included in control and TLE groups, Orange: node included in the TLE and ETLE groups, and Yellow: node included in all groups.

Concerning the Granger causality graph analysis, we employed three of the four centrality measures since eigenvector centrality can only be applied to undirected graphs. The thresholds applied to reveal the hubs in the directed graphs for the three groups were: $p_{R,ETLE} > 0.57$, $p_{R,TLE} > 0.53$, and $p_{R,C} > 0.52$. Notably, the hubs identified from the Granger causality directional graphs differ significantly from those reported in the Pearson correlation and MI graphs (Figure 5). In detail, the hub ROIs observed in both hemispheres in all groups include the subcallosal gyrus, the anterior transverse collateral sulcus of the occipital lobe, the medial olfactory orbital sulcus, and the pallidum. Additionally, hub nodes common to all groups in the right hemisphere include the gyrus rectus of the frontal lobe, the right amygdala, and the right thalamus. Notably, the regions with the highest probabilities in all groups are the right olfactory sulcus of the frontal lobe and the right pallidum, with the latter consistently identified as a hub ROI in all individuals of the TLE and ELTE groups, and in 90.48% of the controls. However, there are regions presented as hubs only in the control group, such as the left parahippocampal gyrus (Parahip G) and the left planum polare of the superior temporal gyrus (STG polar). ETLE patients present low hub probability (31.25%) in the left parahippocampal gyrus, while it is identified as a hub for 50% of the TLE patients and 57.14% of the controls. The pole of STG presents 61.9% hub probability in the control group but less than 50% hub probability in both patient groups (46.4% in the TLE group, 43.75% in the ETLE group). On the other hand, the right parahippocampal gyrus presents high hub probabilities in the patients (64.3% in TLE group, 68.75% in ETLE group), but it is not a hub for the controls (38.1% hub probability). In the subcortical areas, the right putamen and the left thalamus are identified as hubs in the TLE and ETLE groups with hub probabilities higher than 60%, but not in the controls where the hub probability is 47.6%.

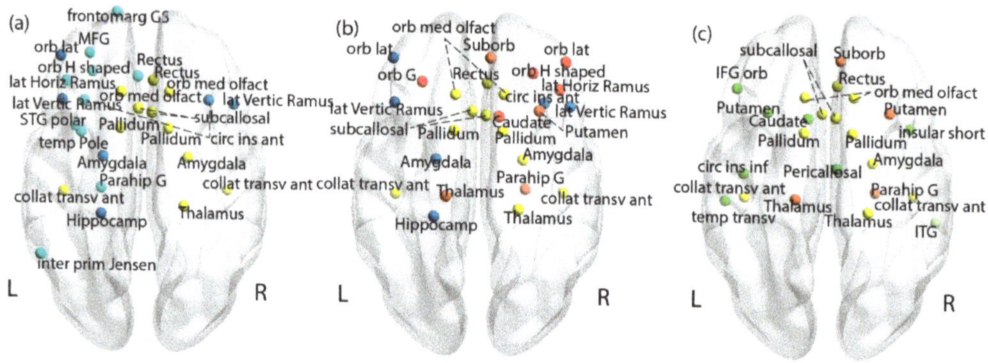

Figure 5. The hub regions with the highest probabilities in the Granger causality graphs for: (**a**) the control group, (**b**) the TLE group, and (**c**) the ETLE group. Node color denotes the inclusion of each node to control, TLE, and ETLE groups. i.e., Light blue: node included only in the control group, Red: node included only in the TLE group, Green: node included only in the ETLE group, Purple: node included in the control and ETLE groups, Blue: node included in control and TLE groups, Orange: node included in the TLE and ETLE groups, and Yellow: node included in all groups.

3.3. Hub Probabilities in Brain Sectors

We employed a statistical evaluation of anatomical subdivisions within the brain (brain sectors), each incorporating several ROIs (extended results can be found in Supplementary Materials). This procedure was applied separately for each centrality measure and each graph, while additionally implemented for the overall hub probability, defined as the mean hub probability of the ROIs included in each brain sector, regardless of the centrality measure. Figure 6 presents the mean centrality-specific hub probabilities within the eight brain sectors, for each of the three groups (C, TLE, ETLE). Similarly, Figure 7 displays the overall mean hub probability within the brain sectors.

In the Pearson graphs, the hubs based on the degree centrality hub probability are mainly located in the motor area for both patient groups. On the other hand, the control group presents most of the degree centrality hubs in the temporal lobe. The highest betweenness centrality hub probability is observed in the subcortical areas in all three groups. In the local efficiency hub probability, significant differences were indicated between ETLE and TLE patients in the frontal lobe and the cingulate cortex. Similarly, the control group displayed statistically significantly higher values compared to the TLE group in the motor areas. In the MI graphs, the three groups displayed similar hub probabilities, regardless of the centrality measure. Most of the hubs based on the degree centrality hub probability were observed in the parietal lobe. Interestingly, although all centrality measures indicated very low values and therefore close to zero hubs in the subcortical area, the local efficiency hub probability had its highest values there.

In the Granger causality graphs, the subcortical regions presented higher values in the degree centrality hub probability. This was also observed in the local efficiency hub probability values. The betweenness centrality hub probability displayed extremely low (close to zero) hub probability values. Despite this, there were a few hubs presented in the motor and parietal areas that were able to influence the mean values and thus display significant differences among the groups.

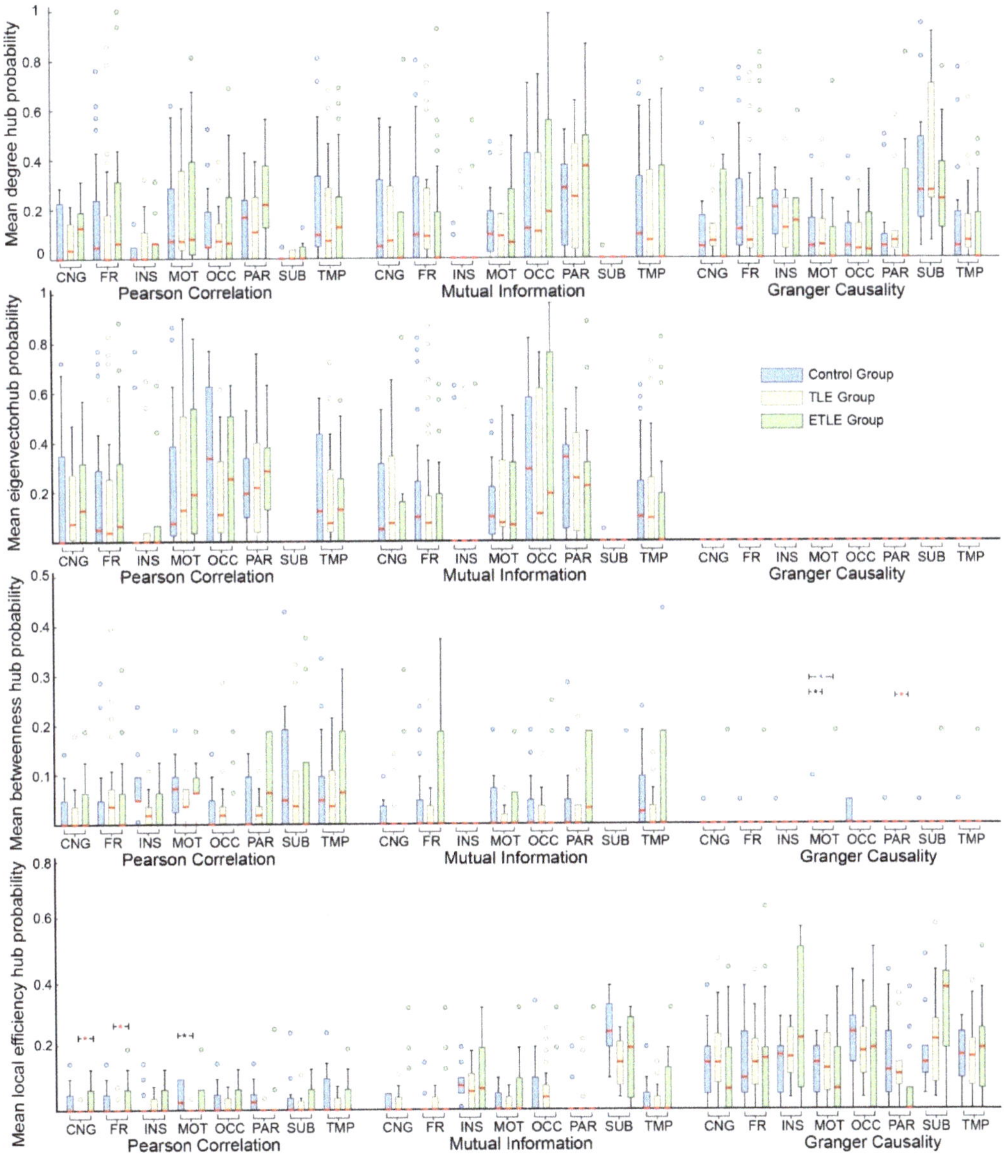

Figure 6. Mean centrality-specific hub probabilities within the eight brain sectors. The bottom and top edges of each box indicate the 25th and 75th percentiles, while the red line indicates the median value. The whiskers extend to the most extreme values, whereas outliers are presented with a circle. Significant differences ($p < 0.05$) are presented with the asterisk (*) symbol. Red: difference between TLE and ETLE; blue: difference between ETLE and control; black: difference between TLE and control.

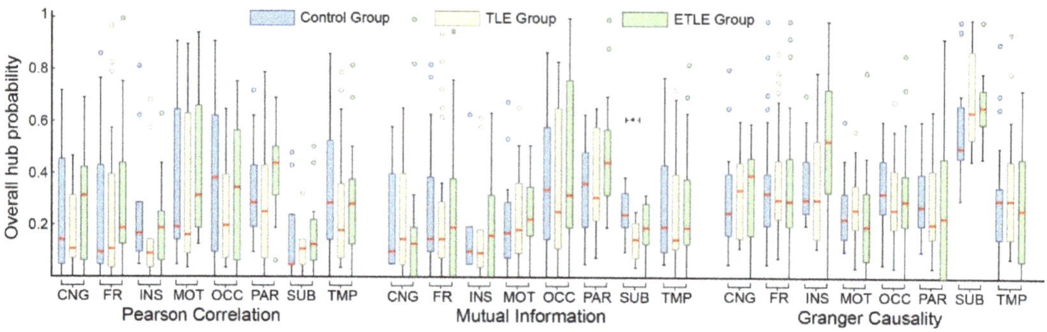

Figure 7. Overall hub probability within the brain sectors. The bottom and top edges of each box indicate the 25th and 75th percentiles, while the red line indicates the median value. The whiskers extend to the most extreme values, whereas outliers are presented with a circle. Significant differences ($p < 0.05$) are presented with the asterisk (*) symbol.

Of note is that the eigenvector centrality hub probability had comparable results with the degree centrality in both Pearson correlation and MI. As we mentioned in the methodology section, the eigenvector centrality can be calculated only for nodes of undirected graphs and hence, it is not defined for the graphs of Granger causality.

In the overall hub probability, the highest values in the Pearson correlation graphs were indicated in the motor area and the parietal lobe for TLE and ETLE patients, while in the control group, the occipital lobe presented the highest values, compared to the other brain sectors. The occipital lobe predominance was also observed in the MI graphs, although a significantly higher overall hub probability was found in the subcortical regions of the control group compared to the TLE patients. Meanwhile, Granger causality indicated the highest hub overall probability values for all groups in the subcortical regions.

4. Discussion

This study introduces a methodological approach to overcome existing limitations in the analysis of hub regions in epilepsy research. Unlike traditional approaches that solely rely on degree centrality, we employ four different centrality measures to evaluate the importance of regions of interest (ROIs) in connectivity networks. This inclusion allows the emergence of hubs based not only on the number of their connections but on a variety of nodal features that reveal their significance in the graph, such as their ability to bridge regions through a short path length. Additionally, instead of averaging connectivity values across groups, we analyze hubs at the individual level to maintain each subject's connectivity patterns and minimize averaging bias. The proposed scheme offers promising findings that could potentially pave the way for future research in epilepsy biomarker discovery.

4.1. Brain Reorganization

In the comparison of global topological features, the healthy control group presents statistical differences with the TLE group only in the Pearson correlation graphs. Our analysis revealed lower values of global assortativity, global efficiency, and mean clustering coefficient values in TLE patients compared to healthy controls, particularly at density levels exceeding 25%. Although the reduction of global efficiency observed in TLE patients is supported by similar studies [23–28], it is worth mentioning that other researchers report contradictory findings [31,53]. In a similar manner, TLE patients display lower mean clustering coefficient values, with some studies reporting comparable results [24,26–28,31,53], while others suggest a higher global clustering coefficient for patients with focal epilepsy compared to controls [25].

This could be related to the fact that there are methodological differences between those studies, such as the definition of the ROIs with different atlases, and the threshold methods applied [54]. For instance, Mazrooyisebdani et al. [31] applied automated anatomical labeling (AAL) and temporal correlations and asserted that individuals with temporal lobe epilepsy (TLE) exhibit functional networks characterized by higher global efficiency and lower clustering coefficient compared to healthy controls. However, most of the studies report similar results to our analysis, where TLE patients present lower global integration (reduced efficiency) and segregation (reduced mean clustering coefficient) in their graphs compared to the controls. Interestingly, those topological features have been related to cognitive deficits observed in chronic epilepsy [28]. The ETLE group presents increased mean clustering coefficient compared to the control group in the undirected graphs of Pearson correlation and MI, at different density levels (40–45% in the MI and 5% in the Pearson correlation graph). Previous studies report similar changes in children with frontal lobe epilepsy, where patients presented efficient communication within specific clusters of ROIs but limited connection between the clusters [55]. Since half of our cohort in the ETLE group are patients with frontal lobe epilepsy, no generalization can be made in all extratemporal lobe epilepsies, but further investigation should be performed with bigger cohorts. We find no difference in the global efficiency of the ETLE groups compared to controls. In line with our results, Pedersen et al. [56] employed a brain mask consisting of 278 nodes and a Pearson correlation method, reporting no difference in the path length but increased segregation in patients with ETLE compared to controls. We find statistical differences in the brain topological features of the TLE and ETLE group that may include useful information regarding the organization of the epileptic network in the two types of epilepsy. TLE patients exhibit more disassortative graphs compared to ETLE in lower density levels of the MI graphs (5–10%) and in almost all the density levels of the Granger causality graphs. Further, the TLE group exhibits a greater number of components in relation to the other two groups at density levels exceeding 40%. This could be due to network edges present in the TLE graphs with weak connectivity strength although acting as bridges. In the undirected graphs of Pearson correlation, at 30–35% density level, ETLE graphs show topological characteristics more aligned with small-world architecture (higher clustering coefficients and global efficiency) in comparison to the TLE group. As such, ETLE patients can form clusters of regions and ensure communication between distant parts of the brain by short path lengths concurrently. On the opposite, the directed graphs of Granger causality indicate that ETLE patients present lower mean clustering coefficient in most of the density levels (from 15% to 40%) and lower global efficiency at the lowest density level (5%) compared to TLE group. The meta-analysis of Diessen et al. [25] suggests that patients with focal epilepsy have increased clustering coefficient and decreased efficiency compared to controls, tending towards a regular network topology, but contrary to our results, they find no difference between the subgroups of TLE and ETLE patients. In-vivo studies reveal that the segregated topology observed in drug-resistant TLE patients, with increased local connectivity and disrupted long-distance connectivity could serve as a mechanism that regulates seizure propagation [57]. Our study suggests that the mean clustering coefficient is a measure presenting differences between the two epilepsy groups (temporal and extratemporal), but the outcome is conflicting between the directed and undirected graphs. This observation is in line with the study of Prando et al. [58] where they compare the topological features between the effective connectivity graphs (directed connectivity), estimated using dynamic causal modeling and the functional connectivity graphs (undirected connectivity) of healthy subjects, and report differences between the two connectivity methods especially in the nodal clustering coefficient. In the future, more studies should investigate the impact of directionality in functional connectivity networks, and on the tendency of brain regions to form clusters.

Although overall differences among the various techniques are to be expected due to the dissimilar algorithmic designs, it should be noted that the patients displayed a disassortative graph in thresholds higher than 25% in all cases. In addition to this, TLE patients have more disassortative graphs compared to the other groups, with significant differences reported in all connectivity methods. Since disassortative graphs are more vulnerable to targeted attacks [59], the removal or the dysfunction of some brain regions would create more isolated regions in the TLE group compared to the ETLE group, with consequences on the size of the largest connected component of the graph. However, it is not clear if this disruption in connectivity, expressed by the vulnerability of the graph, would be beneficial for the epilepsy group [60]. As such, the vulnerability may assist in the control of the seizure propagation reducing the possible pathways of the epileptic discharge but could also imply the dysfunction of brain regions and disorganization of functional networks (such as the default mode network), with an impact on the pathophysiology of patients. The importance of global assortativity as a clinical biomarker for epilepsy has been highlighted in previous studies, reporting that patients with drug-resistance focal epilepsy had more disassortative graphs compared to patients who had positive responses in anti-epileptic drugs [61]. In our study, all the patients (TLE and ETLE) present drug resistance, but only TLE patients present lower assortativity compared to controls.

4.2. Key Region Mapping

Regarding specific brain areas from the analysis of the hub regions SFG, SPG, PCUN, MOG, SOG, ParaCG, Calcarine, and ITG were identified in Pearson correlation and MI graphs. The significance of these regions is corroborated by various studies [62–65]. The review of Heuvel and Sporns designates the PCUN, SFG, and SPG regions as both structural and functional hubs [64]. In line with this review, Lin et al. [65] suggested that bilateral PCUN, SFG and SPG are part of a set of regions that are strongly connected to each other and they observed that TLE patients show decreased functional connectivity especially between ROIs that belong to rich-club. On the other hand, bilateral CUN presented low probability values of being a hub in TLE and ETLE patients, while it is identified as a hub in most control subjects. This is supported by other studies indicating that surgery in children with epilepsy improved significantly the nodal efficiency of cuneus [23,66]. Also, the study of Galovic et al. [67] reports a progressive cortical thinning in the left CUN, observed in patients with left temporal lobe and right frontal lobe epilepsies.Furthermore, in the Pearson graphs an alteration in the hubs of the occipital lobe was observed in the bilateral SOG and IOG, demonstrating low hub probability in both TLE and ETLE groups while in healthy controls this probability was high. Previous studies have also found decreased regional homogeneity in TLE patients compared to controls in these regions (SOG, IOG, and CUN), suggesting that abnormalities in brain regions which act as central nodes could deter efficient communication and integration of information across different parts of the brain [68]. On the contrary, in the MI graphs, ANG was identified as a hub region for all ETLE patients and 64% of the TLE patients but displayed less than 50% probability of being a hub region in the control group. Similar to this outcome, previous studies report increased nodal parameters, such as betweenness and degree centrality, in the ANG of patients with temporal lobe and juvenile myoclonic epilepsies compared to controls [69,70]. Since the angular gyrus is part of the inferior parietal cluster of the default mode network (DMN), the increased connectivity of this region may lead to the spread of the seizure within the DMN and can be the reason for cognitive difficulties observed in patients with TLE [71–73].

Regarding the causal relationship between ROIs, the Granger causality graphs in the control group identified the left parahippocampal gyrus and the left STG polar as hubs. However, those regions show lower hub probability in both TLE and ETLE epilepsy groups. In the study by Galovic et al. [67] the left STG presents progressive atrophy in patients with left TLE and they provide evidence of increased progressive atrophy of the parahippocampal gyrus in the TLE compared to patients with frontal lobe epilepsy. In our study, the significance of parahippocampal gyrus is also highlighted in the patient groups but only in the right hemisphere. Since the majority of the patients have left hemisphere lateralized epilepsy (68.2% of total patients), this could imply a disturbance of this region ipsilateral to the seizure onset. Moreover, in the directed Granger causality graphs, the subcortical regions were significant across all groups, while the amygdala and the pallidum were the most prominent. This is corroborated by recent studies indicating the subcortical areas of pallidum, putamen, and thalamus as brain connector hubs, suggesting their multiple roles in the efficient information flow of the healthy brain [63,74]. The subcortical regions play a key role in seizure propagation, especially the thalamus [75,76]. There is evidence supporting that reduction in the grey matter of thalamus and hippocampus may be related to the severity of epilepsy [77]. A study that investigated the presurgical resting-state fMRI networks of TLE patients proposed the nodal topological features of thalamus as biomarkers for the prediction of surgery outcomes, as patients who did not achieve seizure freedom after surgery presented higher nodal degree and eigenvector centrality values in the thalamus compared to patients who achieved seizure freedom [78]. The study of Park et al. [79] underlines the significance of this area as a possible biomarker for focal epilepsy since it presents increased nodal topological features in patients compared to the healthy controls. Further, they suggest that the increased connectivity of the pallidum can be responsible for the generation of epileptic seizures. Contrary to this assumption, the pallido-cortical pathway has been assumed to be the mechanism that regulates the absence seizures [80]. In our study, we found that Granger causality can depict the importance of subcortical areas both in patients and controls, with most prominent regions the right amygdala and thalamus, as well as the bilateral pallidum. Those regions are identified as hubs in most of the subjects due to their high values of degree centrality and hence, they regulate the communication within the graph through out-flow and in-flow connections

4.3. Network Analysis Comparisons

Further investigation regarding the eight brain sectors revealed that hub regions are dependent on the type of graph. For instance, the subcortical regions had almost zero mean degree hub probability in the undirected graphs of Pearson and MI, while in the directed graph of Granger causality, their role was crucial across all three groups. The causal influence exerted from deep brain structures to the cortical areas has been highlighted in various studies [81,82]. Although, this causality has not been found before in graph characteristics, or it was not as prominent as in our study. On the other hand, the subcortical regions presented non-zero probability values both in the mean betweenness hub probability of the Pearson connectivity graphs and in the mean local efficiency hub probability of the MI graphs, in all groups. In the directed graphs of Granger causality, all the groups present high hub probability levels in the subcortical area, based on the degree centrality and the local efficiency. This suggests that the subcortical regions are significant in every type of graph, but their role differs based on the metric. In the Pearson graph, the subcortical regions may display a higher role as network mediators, minimizing the distances in the interaction between other regions. The review of Herbet and Duffau proposes the meta-networks of the brain, based both on anatomic and functional connectivity, and suggested the mediating role of subcortical regions (especially thalamus, amygdala, and hippocampus) in pathways supporting higher cognitive functions [83].

In the MI and Granger graphs, the mean local efficiency hub probability denotes that the removal of those nodes would affect the efficient information flow in the whole brain network. Further, in the Granger graphs, this brain sector includes most of the hubs based on the degree centrality, suggesting that the regions with most connections (inflow and outflow) are included in the deep brain structures. On this premise, the study of van den Heuvel and Sporns [84] analyzed the anatomical connectivity of the brain, showing that the subcortical regions of the thalamus, putamen, and hippocampus are recognized as hubs using all types of centralities mentioned in our study. They concluded by classifying those regions as 'provincial hubs,' meaning that their value is profound mainly at a local level. In our study, we confirm this regional importance of the subcortical regions, both in the MI and Granger causality graphs, where the values of the mean local efficiency hub probability are high compared to the other brain sectors.

Concerning the mean local efficiency hub probability of the Brain Sectors in the Pearson graphs, the TLE patients show near-zero values, with statistical differences compared to the ETLE group in the cingulate cortex and the frontal lobe, as well as in the motor area compared to the controls. This outcome, combined with the fact that we found the lowest global efficiency in the TLE group compared to the other two groups, suggests that the brain regions of TLE patients do not communicate locally in such an efficient way, and this nodal observation also affects the overall efficiency in the communication of the brain network. The study of Liu et al. [85] verifies this, suggesting that TLE patients present lower nodal efficiency in many regions, resulting in significant reduction of global efficiency metrics compared to healthy controls. On the other hand, the mean local efficiency hub probability of the ETLE patients presents similar values with the control group. The divergence observed between the two types of epilepsy may include information regarding the complexity of the epileptic network, where the different patterns of hub reorganization might correspond to the attempt of the epileptic brain to mitigate neurocognitive deficits. Although many studies underlie the altered hub distribution as a biomarker of epilepsy, there is limited knowledge regarding the connection of this mechanism to the clinical features of the disease [32].

Regarding the overall hub probability, the insula and the subcortical regions presented conflicting results, being the ones with the higher values in the Granger causality and the lowest in the undirected graphs. We can hence conclude that the ROIs included in the insula and the subcortical structures have a causal influence on (or from) other regions of the graph. In a similar manner, the parietal lobe displayed higher values in the Pearson and MI graphs and the lowest overall hub probability in the Granger causality. This further supports the premise that the topology of the brain graph can vary depending on whether the graph is based on functional or effective connectivity measures [86]. This variability underscores the need for using multiple connectivity measures to obtain a comprehensive understanding of brain network dynamics and hub regions in patients

4.4. Limitations and Future Work

When interpreting the results of this study, certain limitations need to be acknowledged. Initially, it is important to note that the size of the cohort, particularly in the ETLE group (16 patients), reduces the statistical power of our results. Further studies with larger cohorts would narrow the confidence intervals in the evaluation of hub probabilities. Secondly, our study does not differentiate between left and right TLE and ETLE patients. This decision was made because of the small sample size and after ensuring there is no statistical significance in the metrics within the TLE and ETLE groups with left/right seizure onset hemispheres. However, graph-theory metrics may vary, and future investigation is encouraged to explore whether hub localization depends on the hemisphere of the epilepsy onset. Lastly, our evaluation focuses solely on static connectivity using linear and non-linear metrics to quantify correlation and causality between ROIs. Future research incorporating dynamic analysis of hubs could enhance our understanding of the stability of nodes, which play crucial roles in information flow within different types of graphs.

5. Conclusions

This study investigates the brain network topological variations between patients with TLE and ETLE using both directed and undirected network connectivity methods alongside various graph-theory metrics. Given that different graph-theory frameworks and estimation measures can lead to significant variability among study results, we address these constraints by evaluating the importance of regions of interest (ROIs) using four centrality measures across three different connectivity techniques (Pearson correlation, Mutual Information, and Granger causality). Instead of averaging brain network data, we analyze the hubs of each subject individually and then create a hub probability based on their presence within each group, thereby minimizing averaging bias. Our findings reveal distinct topological differences between the two epilepsy groups, with TLE patients displaying more disassortative graphs at lower density levels compared to ETLE patients. We also uncover variations in hub regions across different network metrics, highlighting the need to consider various centrality measures for a comprehensive understanding of brain network dynamics in epilepsy. Furthermore, we explore the global centralization features and hub probabilities of each ROI across all subjects and groups at different density thresholds, revealing regional alterations that underscore the unique characteristics of each epilepsy type. This comprehensive approach offers insights into potential biomarkers for type-specific epilepsy diagnosis and treatment, emphasizing the importance of hub reorganization as a possible predictor of network restructuring in epilepsy.

Supplementary Materials: The following are available online at https://www.mdpi.com/article/10.3390/app14188336/s1, Table S1: Pearson's correlation, Table S2: Mutual Information. Table S3: Granger Causality.

Author Contributions: Conceptualization, K.G. (Kyriakos Garganis) methodology, D.A. and K.G. (Kostakis Gkiatis); software, D.A. and K.G. (Kostakis Gkiatis); validation, K.G. (Kostakis Gkiatis), I.K., and G.K.M.; formal analysis, K.G. (Kostakis Gkiatis) and K.G. (Kyriakos Garganis); investigation, D.A.; resources, G.K.M. and K.G. (Kyriakos Garganis); data curation, K.G. (Kostakis Gkiatis) and I.K.; writing—original draft preparation, D.A and I.K.; writing—review and editing, D.A., I.K., K.G. (Kostakis Gkiatis), and K.G. (Kyriakos Garganis); supervision, G.K.M. and K.G. (Kyriakos Garganis); project administration, G.K.M. All authors have read and agreed to the published version of the manuscript.

Funding: This research received no external funding.

Institutional Review Board Statement: This study was conducted in accordance with the Declaration of Helsinki and approved by the Ethics Committee of St. Luke's Hospital (3 April 2024).

Informed Consent Statement: Patient consent was waived as all patient data were analyzed retrospectively after being anonymized.

Data Availability Statement: The data presented in this study are available on reasonable request from the corresponding author.

Acknowledgments: This research work was supported by the Hellenic Foundation for Research and Innovation (HFRI) under the 4th Call for HFRI PhD Fellowships (Fellowship Number: 9375). The authors are grateful to St. Luke's Hospital, which provided the data for this study.

Conflicts of Interest: The authors declare no conflict of interest.

Appendix A

Below is the list with the ROIs used in the Destrieux atlas. It includes the ROI's long name, the abbreviation, and the brain sector where it belongs.

Table A1. List with the Destrieux atlas ROIs.

Index	Brainsector	ROI	Abbreviation
1	CNG	Anterior part of the cingulate gyrus and sulcus	ACC
2	CNG	Middle anterior part of the cingulate gyrus and sulcus	aMCC
3	CNG	Middle posterior part of the cingulate gyrus and sulcus	pMCC
4	FR	Fronto-marginal gyrus of Wernicke and sulcus	frontomarg GS
5	OCC	Inferior occipital gyrus and sulcus	IOG
6	MOT	Paracentral lobule and sulcus	ParaCG
7	MOT	Subcentral gyrus (central operculum) and sulci	SubCG
8	FR	Transverse frontopolar gyri and sulci	frontopoltransv GS
9	CNG	Posterior-dorsal part of the cingulate gyrus	dPCC
10	CNG	Posterior-ventral part of the cingulate gyrus (isthmus of the cingulate gyrus)	vPCC

Table A1. Cont.

Index	Brainsector	ROI	Abbreviation
11	OCC	Cuneus	CUN
12	FR	Opercular part of the inferior frontal gyrus	IFG Operc
13	FR	Orbital part of the inferior frontal gyrus	IFG orb
14	FR	Triangular part of the inferior frontal gyrus	IFG Triang
15	FR	Middle frontal gyrus	MFG
16	FR	Superior frontal gyrus	SFG
17	INS	Long insular gyrus and central sulcus of the insula	long insul G
18	INS	Short insular gyri	insular short
19	OCC	Middle occipital gyrus (lateral occipital gyrus)	MOG
20	OCC	Superior occipital gyrus	SOG
21	OCC	Lateral occipito-temporal gyrus (fusiform gyrus)	Fusiform G
22	OCC	Lingual gyrus, lingual part of the medial occipito-temporal gyrus	LING
23	TMP	Parahippocampal gyrus, parahippocampal part of the medial occipito-temporal gyrus	Parahip G
24	FR	Orbital gyri	orb G
25	PAR	Superior parietal lobule	SPG
26	PAR	Angular gyrus	ANG
27	PAR	Supramarginal gyrus	IPG supramar
28	MOT	Postcentral gyrus	PostCG
29	MOT	Precentral gyrus	PreCG
30	PAR	Precuneus	PCUN
31	FR	Straight gyrus, Gyrus rectus	Rectus
32	FR	Subcallosal area, subcallosal gyrus	subcallosal
33	TMP	Inferior temporal gyrus	ITG
34	TMP	Middle temporal gyrus	MTG
35	TMP	Anterior transverse temporal gyrus of Heschl	STG ant transv
36	TMP	Lateral aspect of the superior temporal gyrus	STG lat
37	TMP	Planum polare of the superior temporal gyrus	STG plan polar
38	TMP	Temporal plane of the superior temporal gyrus	STG plan tempo
39	FR	Horizontal ramus of the anterior segment of the lateral sulcus	latFis ant Hor
40	FR	Vertical ramus of the anterior segment of the lateral sulcus	latFis ant Vert
41	TMP	Posterior ramus of the lateral sulcus	latFis post
42	OCC	Occipital pole	occ Pole
43	TMP	Temporal pole	temp Pole
44	OCC	Calcarine sulcus	Calcarine
45	MOT	Central sulcus	CS
46	CNG	Marginal branch of the cingulate sulcus	cingulmarginalis
47	INS	Anterior segment of the circular sulcus of the insula	circ ins ant
48	INS	Inferior segment of the circular sulcus of the insula	circ ins inf
49	INS	Superior segment of the circular sulcus of the insula	circ ins sup
50	TMP	Anterior transverse collateral sulcus	collattransv ant
51	TMP	Posterior transverse collateral sulcus	collattransv post
52	FR	Inferior frontal sulcus	IFS
53	FR	Middle frontal sulcus	MFS
54	FR	Superior frontal sulcus	SFS
55	PAR	Sulcus intermedius primus of Jensen	inter prim Jensen
56	PAR	Intraparietal sulcus and transverse parietal sulci	
57	OCC	Anterior occipital sulcus and preoccipital notch	occ ant S
58	OCC	Middle occipital sulcus and lunatus sulcus	occ midLunatus
59	OCC	Superior occipital sulcus and transverse occipital sulcus	occ suptransv
60	TMP	Lateral occipito-temporal sulcus	oct lat S
61	OCC	Medial occipito-temporal sulcus (collateral sulcus) and lingual sulcus	oct medLING
62	FR	Orbital sulci (H-shaped sulci)	orb H shaped
63	FR	Lateral orbital sulcus	orb lat
64	FR	Medial orbital sulcus (olfactory sulcus)	orb med olfact
65	OCC	Parieto-occipital sulcus	par occ S
66	CNG	Pericallosal sulcus	Pericallosal

Table A1. *Cont.*

Index	Brainsector	ROI	Abbreviation
67	MOT	Postcentral sulcus	PostCS
68	MOT	Inferior part of the precentral sulcus	PreCS inf part
69	MOT	Superior part of the precentral sulcus	PreCS sup part
70	FR	Suborbital sulcus	Suborb
71	PAR	Subparietal sulcus	Subpar
72	TMP	Inferior temporal sulcus	ITS
73	TMP	Superior temporal sulcus (parallel sulcus)	STS
74	TMP	Transverse temporal sulcus	temp trasnv
75	SUB	Amygdala	Amygdala
76	SUB	Caudate	Caudate
77	SUB	Hippocampus	Hippocampus
78	SUB	Pallidum	Pallidum
79	SUB	Putamen	Putamen
80	SUB	Thalamus	Thalamus

Appendix B

The mathematical formulas for the 3 graph-theory methods are presented below.

Pearson's Correlation

The Pearson's correlation coefficient estimates the linear relationship between two variables. In the fMRI connectivity, Pearson's correlation quantifies the linear relationship between the signals of the ROIs i and j:

$$\rho_{ij} = \frac{cov_{ij}(t)}{\sqrt{var_i(t)var_j(t)}} \tag{A1}$$

The value of ρ_ij will vary from −1 to 1. When the absolute value of the correlation is equal to 1, there is a perfect correlation between the two ROIs, while 0 indicates that there is no relationship between the signals.

Mutual Information (MI)

Mutual information is a statistical measure that quantifies the amount of shared information between two signals. It is a non-linear measure to evaluate the functional connectivity between the signals of two ROIs. If X_i (t) is the signal of ROI i, then the information of the time series is defined from the Shannon entropy as follows:

$$I(X_i) = -\sum_{k=1}^{K} p_k \ln p_k \tag{A2}$$

where the X_i signal has been discretized into K bins, and p_k is the probability of the k = 1, 2 … K bin. The number of bins was defined according to the Freedman—Diaconis rule [87]. According to this definition, we can define the joint entropy of two signals X and Y:

$$I(X, Y) = -\sum_{i,j} p_{ij} \ln p_{ij} \tag{A3}$$

where p_ij is the joint probability of the two signals. The mutual information between the two signals can be calculated as:

$$MI(X, Y) = I(X) + I(Y) - I(X, Y) \tag{A4}$$

Granger Causality (GC)

Granger causality (GC) estimates the effective connectivity between two ROIs by quantifying the improvement in predictability of one random variable X, if we use not only its past values but also the information of another random variable Y. The GC is estimated using linear autoregressive models. If we have the time series of two ROIs (X_t, Y_t), then the estimated model of X_t using the information of Y_t is defined as follows:

$$X_t = a_1 X_{t-1} + \cdots + a_p X_{t-p} + b_1 Y_{t-1} + \cdots + b_p Y_{t-p} + e_t \quad \text{(A5)}$$

While the autoregressive model for X_t is:

$$X_t = a_1' X_{t-1} + \ldots + a_p' X_{t-p} + e_t' \quad \text{(A6)}$$

In these models, p is the order of the autoregressive models and the residuals e_t, e_t' must be independent and identically distributed. If Σ, Σ' are the estimators of the residuals' covariance matrices, then the GC measure can be expressed as:

$$GC_{Y \to X} = \ln\left(\frac{\det(\Sigma)}{\det(\Sigma')}\right) \quad \text{(A7)}$$

where $\det(\Sigma), \det(\Sigma')$ are the generalized variances that quantify the prediction errors. The appropriate order is defined after comparisons between the models, according to the Chi-square test [88].

Appendix C

The hubs with the highest overall hub probability within each group for each graph are presented below. These hubs are the ROIs presenting overall hub probability more than 2 standard deviations higher from the mean overall hub probability of the group. In the table we include the degree hub probability ($p^D{}_{R,G}$), the betweenness hub probability ($p^B{}_{R,G}$), the local efficiency hub probability ($p^{EF}{}_{R,G}$), the eigenvector hub probability ($p^{EV}{}_{R,G}$), and the overall hub probability ($p_{R,G}$) of each hub and the brain sector where it belongs.

Table A2. The hubs with the highest overall hub probability in the Pearson graph.

ROI	$p^D{}_{R,G}$	$p^B{}_{R,G}$	$p^{EF}{}_{R,G}$	$p^{EV}{}_{R,G}$	$p_{R,G}$	Brain Sector
			Healthy Controls			
R SFG	76.2%	23.8%	0	66.7%	85.7%	FR
R LING	52.4%	14.3%	0	76.2%	90.5%	OCT
R PreCG	61.9%	4.8%	0	85.7%	90.5%	S/M
R MTG	81%	4.8%	0	52.4%	85.7%	TEMP
			TLE group			
L SFG	78.6%	21.4%	0	57.1%	82.1%	FR
R SFG	85.7%	39.3%	0	71.4%	96.4%	FR
L SPG	35.7%	0	0	82.1%	82.1%	PAR
R SPG	25%	4.6%	0	75%	78.6%	PAR
L PreCG	60.7%	3.6%	0	89.3%	89.3%	S/M
R PreCG	46.4%	0	0	82.1%	85.7%	S/M
L IPG—Supramar	39.3%	10.7%	0	60.7%	75%	PAR
R PostCG	25%	0	3.6%	75%	78.6%	S/M
R PCUN	46.4%	7.1%	0	64.3%	75%	PAR
R MTG	67.9%	10.7%	0	42.9%	71.4%	TEMP
R STS	60.7%	14.3%	0	71.4%	78.6%	TEMP
			ETLE group			
L SFG	93.75%	50%	0	62.5%	100%	FR
R SFG	100%	31.25%	0	87.5%	100%	FR
L PreCG	81.25%	6.25%	0	75%	93.75%	S/M
R PreCG	56.25%	6.25%	0	81.25%	81.25%	S/M
R PCUN	81.25%	6.25%	0	75%	87.5%	PAR
R STG-lateral	62.5%	25%	0	31.25%	81.25%	TEMP
R STS	68.75%	18.75%	0	56.25%	81.25%	TEMP

Table A3. The hubs with the highest overall hub probability in the MI graph.

ROI	$p^{D}_{R,G}$	$p^{B}_{R,G}$	$p^{EF}_{R,G}$	$p^{EV}_{R,G}$	$p_{R,G}$	Brain Sector
		Healthy Controls				
L SOG	61.9%	14.3%	4.8%	66.7%	71.4%	OCC
R SOG	71.4%	9.5%	0	81%	85.7%	OCC
L SPG	57.1%	9.5%	0	76.2%	76.2%	PAR
R SPG	66.7%	9.5%	0	81%	81%	PAR
L IOG	66.7%	19%	0	71.4%	76.2%	OCC
L Jensen	81%	19%	0	38.1%	81%	PAR
L occ suptransv	57.1%	4.8%	0	71.4%	71.4%	OCC
L temp transverse	71.4%	23.8%	0	28.6%	76.2%	TEMP
R col transv post	66.7%	0	4.8%	61.9%	71.4%	TEMP
R occ midLunatus	61.9%	19%	0	66.7%	71.4%	OCC
		TLE group				
L SOG	71.4%	7.1%	3.6%	75%	82.1%	OCC
R SOG	75%	21.4%	3.6%	75%	78.6%	OCC
L SPG	71.4%	3.6%	0	85.7%	92.9%	PAR
R SPG	78.6%	7.1%	0	78.6%	82.1%	PAR
L LING	50%	3.6%	0	71.4%	71.4%	OCT
L Jensen	71.4%	25%	0	46.4%	71.4%	PAR
R IOG	60.7%	10.7%	0	64.3%	71.4%	OCC
R occ Pole	64.3%	14.3%	0	71.4%	71.4%	OCC
		ETLE group				
L SOG	56.25%	0	0	93.75%	93.75%	OCC
R SOG	100%	25%	0	100%	100%	OCC
L CUN	81.25%	18.75%	0	100%	100%	OCC
L LING	81.25%	0	0	100%	100%	OCT
L Jensen	93.75%	56.25%	0	25%	93.75%	PAR
R IPG—Ang	62.5%	56.25%	0	6.25%	100%	PAR
R PostCG	87.5%	6.25%	0	87.5%	87.5%	S/M

Table A4. The hubs with the highest overall hub probability in the Granger causality graph.

ROI	$p^{D}_{R,G}$	$p^{B}_{R,G}$	$p^{EV}_{R,G}$	$p_{R,G}$	Brain Sector
		Healthy Controls			
L Orb med Olf	81%	0	9.5%	95.2%	ORB
R Orb med Olf	76.2%	0	4.8%	81%	ORB
L Pallidum	100%	0	0	100%	SUB
R Pallidum	85.7%	4.8%	4.8%	90.5%	SUB
L subcallosal	76.2%	0	4.8%	81%	FR
L circ insula-ant	71.4%	0	4.8%	76.2%	INS
L Amygdala	81%	0	4.8%	85.7%	FR
		TLE group			
L Amygdala	82.1%	0	7.1%	89.3%	SUB
R Amygdala	82.1%	0	7.1%	89.3%	SUB
L Pallidum	96.4%	0	3.6%	100%	SUB
R Pallidum	96.4%	0	3.6%	100%	SUB
L col transv-ant	50%	0	25%	75%	TEMP
R subcallosal	78.6%	0	3.6%	82.1%	FR
R Lat-FiSant-Hor	67.9%	0	7.1%	75%	FR
R Lat-FiSant-Vert	75%	0	0	75%	PAR
R Orb med Olf	64.3%	0	10.7%	75%	ORB
		ETLE group			
L subcallosal	62.5%	0	18.75%	81.25%	FR
R subcallosal	75%	0	18.75%	93.75%	FR
L Orb med Olf	81.25%	18.75%	0	81.25%	ORB
R Orb med Olf	87.5%	0	0	87.5%	ORB
R rectus	87.5%	0	0	87.5%	FR
R col transv-ant	62.5%	0	31.25%	93.75%	TEMP
R pericallosal	0	0	81.25%	81.25%	CING
R Pallidum	81.25%	0	18.75%	100%	SUB

References

1. Epilepsy. Available online: https://www.who.int/news-room/fact-sheets/detail/epilepsy (accessed on 26 July 2024).
2. Fisher, R.S.; Acevedo, C.; Arzimanoglou, A.; Bogacz, A.; Cross, J.H.; Elger, C.E.; Engel Jr, J.; Forsgren, L.; French, J.A.; Glynn, M.; et al. ILAE Official Report: A Practical Clinical Definition of Epilepsy. *Epilepsia* **2014**, *55*, 475–482. [CrossRef] [PubMed]
3. Brodie, M.J.; Barry, S.J.E.; Bamagous, G.A.; Norrie, J.D.; Kwan, P. Patterns of Treatment Response in Newly Diagnosed Epilepsy. *Neurology* **2012**, *78*, 1548–1554. [CrossRef] [PubMed]
4. Strzelczyk, A.; Aledo-Serrano, A.; Coppola, A.; Didelot, A.; Bates, E.; Sainz-Fuertes, R.; Lawthom, C. The Impact of Epilepsy on Quality of Life: Findings from a European Survey. *Epilepsy Behav.* **2023**, *142*, 109179. [CrossRef] [PubMed]
5. Tani, A.; Adali, N. Cognitive Disorders, Depression and Anxiety in Temporal Lobe Epilepsy: An Overview. *J. Biosci. Med.* **2024**, *12*, 77–93. [CrossRef]
6. Jain, P.; Smith, M.L.; Speechley, K.; Ferro, M.; Connolly, M.; Ramachandrannair, R.; Almubarak, S.; Andrade, A.; Widjaja, E.; Team, P.S. Seizure Freedom Improves Health-Related Quality of Life after Epilepsy Surgery in Children. *Dev. Med. Child Neurol.* **2020**, *62*, 600–608. [CrossRef]
7. Swarup, O.; Waxmann, A.; Chu, J.; Vogrin, S.; Lai, A.; Laing, J.; Barker, J.; Seiderer, L.; Ignatiadis, S.; Plummer, C.; et al. Long-Term Mood, Quality of Life, and Seizure Freedom in Intracranial EEG Epilepsy Surgery. *Epilepsy Behav.* **2021**, *123*, 108241. [CrossRef]
8. Lehnertz, K.; Bröhl, T.; Wrede, R. von Epileptic-Network-Based Prediction and Control of Seizures in Humans. *Neurobiol. Dis.* **2023**, *181*, 106098. [CrossRef]
9. Gleichgerrcht, E.; Keller, S.S.; Drane, D.L.; Munsell, B.C.; Davis, K.A.; Kaestner, E.; Weber, B.; Krantz, S.; Vandergrift, W.A.; Edwards, J.C.; et al. Temporal Lobe Epilepsy Surgical Outcomes Can Be Inferred Based on Structural Connectome Hubs: A Machine Learning Study. *Ann. Neurol.* **2020**, *88*, 970–983. [CrossRef]
10. De Palma, L.; Benedictis, A.D.; Specchio, N.; Marras, C.E. Epileptogenic Network Formation. *Neurosurg. Clin.* **2020**, *31*, 335–344. [CrossRef]
11. Cohen, M.S.; Bookheimer, S.Y. Localization of Brain Function Using Magnetic Resonance Imaging. *Trends Neurosci.* **1994**, *17*, 268–277. [CrossRef]
12. Feng, X.; Piper, R.J.; Prentice, F.; Clayden, J.D.; Baldeweg, T. Functional Brain Connectivity in Children with Focal Epilepsy: A Systematic Review of Functional MRI Studies. *Seizure Eur. J. Epilepsy* **2024**, *117*, 164–173. [CrossRef] [PubMed]
13. Park, H.-J.; Friston, K. Structural and Functional Brain Networks: From Connections to Cognition. *Science* **2013**, *342*, 1238411. [CrossRef] [PubMed]
14. Friston, K.J.; Frith, C.D.; Liddle, P.F.; Frackowiak, R.S.J. Functional Connectivity: The Principal-Component Analysis of Large (PET) Data Sets. *J. Cereb. Blood Flow Metab.* **1993**, *13*, 5–14. [CrossRef] [PubMed]
15. van den Heuvel, M.P.; Hulshoff Pol, H.E. Exploring the Brain Network: A Review on Resting-State fMRI Functional Connectivity. *Eur. Neuropsychopharmacol.* **2010**, *20*, 519–534. [CrossRef] [PubMed]
16. Spencer, S.S. Neural Networks in Human Epilepsy: Evidence of and Implications for Treatment. *Epilepsia* **2002**, *43*, 219–227. [CrossRef]
17. Boerwinkle, V.L.; Mirea, L.; Gaillard, W.D.; Sussman, B.L.; Larocque, D.; Bonnell, A.; Ronecker, J.S.; Troester, M.M.; Kerrigan, J.F.; Foldes, S.T.; et al. Resting-State Functional MRI Connectivity Impact on Epilepsy Surgery Plan and Surgical Candidacy: Prospective Clinical Work. *J. Neurosurg. Pediatr.* **2020**, *25*, 574–581. [CrossRef]
18. Foit, N.A.; Bernasconi, A.; Bernasconi, N. Functional Networks in Epilepsy Presurgical Evaluation. *Neurosurg. Clin.* **2020**, *31*, 395–405. [CrossRef]
19. Bernhardt, B.C.; Bonilha, L.; Gross, D.W. Network Analysis for a Network Disorder: The Emerging Role of Graph Theory in the Study of Epilepsy. *Epilepsy Behav.* **2015**, *50*, 162–170. [CrossRef]
20. Farahani, F.V.; Karwowski, W.; Lighthall, N.R. Application of Graph Theory for Identifying Connectivity Patterns in Human Brain Networks: A Systematic Review. *Front. Neurosci.* **2019**, *13*, 585. [CrossRef]
21. Larivière, S.; Bernasconi, A.; Bernasconi, N.; Bernhardt, B.C. Connectome Biomarkers of Drug-Resistant Epilepsy. *Epilepsia* **2021**, *62*, 6–24. [CrossRef]
22. Doucet, G.E.; Rider, R.; Taylor, N.; Skidmore, C.; Sharan, A.; Sperling, M.; Tracy, J.I. Presurgery Resting-State Local Graph-Theory Measures Predict Neurocognitive Outcomes after Brain Surgery in Temporal Lobe Epilepsy. *Epilepsia* **2015**, *56*, 517–526. [CrossRef] [PubMed]
23. Bernhardt, B.C.; Chen, Z.; He, Y.; Evans, A.C.; Bernasconi, N. Graph-Theoretical Analysis Reveals Disrupted Small-World Organization of Cortical Thickness Correlation Networks in Temporal Lobe Epilepsy. *Cereb. Cortex* **2011**, *21*, 2147–2157. [CrossRef] [PubMed]
24. Ai, H.; Yang, C.; Lu, M.; Ren, J.; Li, Z.; Zhang, Y. Abnormal White Matter Structural Network Topological Property in Patients with Temporal Lobe Epilepsy. *CNS Neurosci. Ther.* **2024**, *30*, e14414. [CrossRef] [PubMed]
25. Van Diessen, E.; Zweiphenning, W.J.E.M.; Jansen, F.E.; Stam, C.J.; Braun, K.P.J.; Otte, W.M. Brain Network Organization in Focal Epilepsy: A Systematic Review and Meta-Analysis. *PLoS ONE* **2014**, *9*, e114606. [CrossRef]
26. Ma, K.; Zhang, X.; Song, C.; Han, S.; Li, W.; Wang, K.; Mao, X.; Zhang, Y.; Cheng, J. Altered Topological Properties and Their Relationship to Cognitive Functions in Unilateral Temporal Lobe Epilepsy. *Epilepsy Behav.* **2023**, *144*, 109247. [CrossRef]
27. Haneef, Z.; Chiang, S. Clinical Correlates of Graph Theory Findings in Temporal Lobe Epilepsy. *Seizure* **2014**, *23*, 809–818. [CrossRef]

28. Vlooswijk, M.C.G.; Vaessen, M.J.; Jansen, J.F.A.; de Krom, M.C.F.T.M.; Majoie, H.J.M.; Hofman, P.A.M.; Aldenkamp, A.P.; Backes, W.H. Loss of Network Efficiency Associated with Cognitive Decline in Chronic Epilepsy. *Neurology* **2011**, *77*, 938–944. [CrossRef]
29. Stam, C.J. Hub Overload and Failure as a Final Common Pathway in Neurological Brain Network Disorders. *Netw. Neurosci.* **2024**, *8*, 1–23. [CrossRef]
30. Crossley, N.A.; Mechelli, A.; Scott, J.; Carletti, F.; Fox, P.T.; McGuire, P.; Bullmore, E.T. The Hubs of the Human Connectome Are Generally Implicated in the Anatomy of Brain Disorders. *Brain* **2014**, *137*, 2382–2395. [CrossRef]
31. Mazrooyisebdani, M.; Nair, V.A.; Garcia-Ramos, C.; Mohanty, R.; Meyerand, E.; Hermann, B.; Prabhakaran, V.; Ahmed, R. Graph Theory Analysis of Functional Connectivity Combined with Machine Learning Approaches Demonstrates Widespread Network Differences and Predicts Clinical Variables in Temporal Lobe Epilepsy. *Brain Connect.* **2020**, *10*, 39–50. [CrossRef]
32. Royer, J.; Bernhardt, B.C.; Larivière, S.; Gleichgerrcht, E.; Vorderwülbecke, B.J.; Vulliémoz, S.; Bonilha, L. Epilepsy and Brain Network Hubs. *Epilepsia* **2022**, *63*, 537–550. [CrossRef] [PubMed]
33. Gkiatis, K.; Garganis, K.; Benjamin, C.F.; Karanasiou, I.; Kondylidis, N.; Harushukuri, J.; Matsopoulos, G.K. Standardization of Presurgical Language fMRI in Greek Population: Mapping of Six Critical Regions. *Brain Behav.* **2022**, *12*, e2609. [CrossRef] [PubMed]
34. Dachena, C.; Casu, S.; Fanti, A.; Lodi, M.B.; Mazzarella, G. Combined Use of MRI, fMRIand Cognitive Data for Alzheimer's Disease: Preliminary Results. *Appl. Sci.* **2019**, *9*, 3156. [CrossRef]
35. Jenkinson, M.; Beckmann, C.F.; Behrens, T.E.J.; Woolrich, M.W.; Smith, S.M. FSL. *NeuroImage* **2012**, *62*, 782–790. [CrossRef] [PubMed]
36. Greve, D.N.; Fischl, B. Accurate and Robust Brain Image Alignment Using Boundary-Based Registration. *NeuroImage* **2009**, *48*, 63–72. [CrossRef]
37. Jenkinson, M.; Bannister, P.; Brady, M.; Smith, S. Improved Optimization for the Robust and Accurate Linear Registration and Motion Correction of Brain Images. *NeuroImage* **2002**, *17*, 825–841. [CrossRef]
38. Beckmann, C.F.; Smith, S.M. Probabilistic Independent Component Analysis for Functional Magnetic Resonance Imaging. *IEEE Trans. Med. Imaging* **2004**, *23*, 137–152. [CrossRef]
39. Gkiatis, K.; Garganis, K.; Karanasiou, I.; Chatzisotiriou, A.; Zountsas, B.; Kondylidis, N.; Matsopoulos, G.K. Independent Component Analysis: A Reliable Alternative to General Linear Model for Task-Based fMRI. *Front. Psychiatry* **2023**, *14*, 1214067. [CrossRef]
40. Destrieux, C.; Fischl, B.; Dale, A.; Halgren, E. Automatic Parcellation of Human Cortical Gyri and Sulci Using Standard Anatomical Nomenclature. *Neuroimage* **2010**, *53*, 1–15. [CrossRef]
41. Fischl, B. FreeSurfer. *Neuroimage* **2012**, *62*, 774–781. [CrossRef]
42. Benesty, J.; Chen, J.; Huang, Y.; Cohen, I. Pearson Correlation Coefficient. In *Noise Reduction in Speech Processing*; Cohen, I., Huang, Y., Chen, J., Benesty, J., Eds.; Springer: Berlin/Heidelberg, Germany, 2009; pp. 1–4. ISBN 978-3-642-00296-0.
43. Kraskov, A.; Stögbauer, H.; Grassberger, P. Estimating Mutual Information. *Phys. Rev. E* **2004**, *69*, 066138. [CrossRef] [PubMed]
44. Granger, C.W.J. Investigating Causal Relations by Econometric Models and Cross-Spectral Methods. *Econometrica* **1969**, *37*, 424–438. [CrossRef]
45. Garrison, K.A.; Scheinost, D.; Finn, E.S.; Shen, X.; Constable, R.T. The (in)Stability of Functional Brain Network Measures across Thresholds. *NeuroImage* **2015**, *118*, 651–661. [CrossRef] [PubMed]
46. Newman, M.E.J. Assortative Mixing in Networks. *Phys. Rev. Lett.* **2002**, *89*, 208701. [CrossRef]
47. Latora, V.; Marchiori, M. Efficient Behavior of Small-World Networks. *Phys. Rev. Lett.* **2001**, *87*, 198701. [CrossRef]
48. Watts, D.J.; Strogatz, S.H. Collective Dynamics of 'Small-World' Networks. *Nature* **1998**, *393*, 440–442. [CrossRef]
49. Nesterov, A.I. On Clustering Coefficients in Complex Networks 2024. *arXiv* **2024**, arXiv:2401.02999.
50. Freeman, L.C. A Set of Measures of Centrality Based on Betweenness. *Sociometry* **1977**, *40*, 35–41. [CrossRef]
51. Brandes, U. A Faster Algorithm for Betweenness Centrality. *J. Math. Sociol.* **2001**, *25*, 163–177. [CrossRef]
52. Newman, M.E.J. Finding Community Structure in Networks Using the Eigenvectors of Matrices. *Phys. Rev. E* **2006**, *74*, 036104. [CrossRef]
53. Liao, W.; Zhang, Z.; Pan, Z.; Mantini, D.; Ding, J.; Duan, X.; Luo, C.; Lu, G.; Chen, H. Altered Functional Connectivity and Small-World in Mesial Temporal Lobe Epilepsy. *PLoS ONE* **2010**, *5*, e8525. [CrossRef] [PubMed]
54. Stanley, M.L.; Moussa, M.N.; Paolini, B.; Lyday, R.G.; Burdette, J.H.; Laurienti, P.J. Defining Nodes in Complex Brain Networks. *Front. Comput. Neurosci.* **2013**, *7*, 169. [CrossRef] [PubMed]
55. Vaessen, M.J.; Braakman, H.M.H.; Heerink, J.S.; Jansen, J.F.A.; Debeij-van Hall, M.H.J.A.; Hofman, P.A.M.; Aldenkamp, A.P.; Backes, W.H. Abnormal Modular Organization of Functional Networks in Cognitively Impaired Children with Frontal Lobe Epilepsy. *Cereb. Cortex* **2013**, *23*, 1997–2006. [CrossRef] [PubMed]
56. Pedersen, M.; Omidvarnia, A.H.; Walz, J.M.; Jackson, G.D. Increased Segregation of Brain Networks in Focal Epilepsy: An fMRI Graph Theory Finding. *NeuroImage Clin.* **2015**, *8*, 536–542. [CrossRef]
57. Larivière, S.; Weng, Y.; Vos De Wael, R.; Royer, J.; Frauscher, B.; Wang, Z.; Bernasconi, A.; Bernasconi, N.; Schrader, D.V.; Zhang, Z.; et al. Functional Connectome Contractions in Temporal Lobe Epilepsy: Microstructural Underpinnings and Predictors of Surgical Outcome. *Epilepsia* **2020**, *61*, 1221–1233. [CrossRef]
58. Prando, G.; Zorzi, M.; Bertoldo, A.; Corbetta, M.; Zorzi, M.; Chiuso, A. Sparse DCM for Whole-Brain Effective Connectivity from Resting-State fMRI Data. *NeuroImage* **2020**, *208*, 116367. [CrossRef]

59. Al Musawi, A.F.; Roy, S.; Ghosh, P. Examining Indicators of Complex Network Vulnerability across Diverse Attack Scenarios. *Sci. Rep.* **2023**, *13*, 18208. [CrossRef]
60. Englot, D.J.; Konrad, P.E.; Morgan, V.L. Regional and Global Connectivity Disturbances in Focal Epilepsy, Related Neurocognitive Sequelae, and Potential Mechanistic Underpinnings. *Epilepsia* **2016**, *57*, 1546–1557. [CrossRef]
61. Park, K.M.; Cho, K.H.; Lee, H.-J.; Heo, K.; Lee, B.I.; Kim, S.E. Predicting the Antiepileptic Drug Response by Brain Connectivity in Newly Diagnosed Focal Epilepsy. *J. Neurol.* **2020**, *267*, 1179–1187. [CrossRef]
62. Yu, Y.; Qiu, M.; Zou, W.; Zhao, Y.; Tang, Y.; Tian, J.; Chen, X.; Qiu, W. Impaired Rich-Club Connectivity in Childhood Absence Epilepsy. *Front. Neurol.* **2023**, *14*, 1135305. [CrossRef]
63. Oldham, S.; Fornito, A. The Development of Brain Network Hubs. *Dev. Cogn. Neurosci.* **2019**, *36*, 100607. [CrossRef] [PubMed]
64. van den Heuvel, M.P.; Sporns, O. Network Hubs in the Human Brain. *Trends Cogn. Sci.* **2013**, *17*, 683–696. [CrossRef] [PubMed]
65. Lin, Q.; Li, W.; Li, Y.; Liu, P.; Zhang, Y.; Gong, Q.; Zhou, D.; An, D. Aberrant Structural Rich Club Organization in Temporal Lobe Epilepsy with Focal to Bilateral Tonic–Clonic Seizures. *NeuroImage Clin.* **2023**, *40*, 103536. [CrossRef] [PubMed]
66. Zhang, L.; Zhuang, B.; Wang, M.; Zhu, J.; Chen, T.; Yang, Y.; Shi, H.; Zhu, X.; Ma, L. Delineating Abnormal Individual Structural Covariance Brain Network Organization in Pediatric Epilepsy with Unilateral Resection of Visual Cortex. *Epilepsy Behav. Rep.* **2024**, *27*, 100676. [CrossRef] [PubMed]
67. Galovic, M.; van Dooren, V.Q.H.; Postma, T.S.; Vos, S.B.; Caciagli, L.; Borzì, G.; Cueva Rosillo, J.; Vuong, K.A.; de Tisi, J.; Nachev, P.; et al. Progressive Cortical Thinning in Patients With Focal Epilepsy. *JAMA Neurol.* **2019**, *76*, 1230–1239. [CrossRef]
68. Zhao, B.; Yang, B.; Tan, Z.; Hu, W.; Sang, L.; Zhang, C.; Wang, X.; Wang, Y.; Liu, C.; Mo, J.; et al. Intrinsic Brain Activity Changes in Temporal Lobe Epilepsy Patients Revealed by Regional Homogeneity Analysis. *Seizure* **2020**, *81*, 117–122. [CrossRef]
69. Ke, M.; Hou, Y.; Zhang, L.; Liu, G. Brain Functional Network Changes in Patients with Juvenile Myoclonic Epilepsy: A Study Based on Graph Theory and Granger Causality Analysis. *Front. Neurosci.* **2024**, *18*, 1363255. [CrossRef]
70. Wang, J.; Qiu, S.; Xu, Y.; Liu, Z.; Wen, X.; Hu, X.; Zhang, R.; Li, M.; Wang, W.; Huang, R. Graph Theoretical Analysis Reveals Disrupted Topological Properties of Whole Brain Functional Networks in Temporal Lobe Epilepsy. *Clin. Neurophysiol.* **2014**, *125*, 1744–1756. [CrossRef]
71. Bell, B.; Lin, J.J.; Seidenberg, M.; Hermann, B. The Neurobiology of Cognitive Disorders in Temporal Lobe Epilepsy. *Nat. Rev. Neurol.* **2011**, *7*, 154–164. [CrossRef]
72. McCormick, C.; Protzner, A.B.; Barnett, A.J.; Cohn, M.; Valiante, T.A.; McAndrews, M.P. Linking DMN Connectivity to Episodic Memory Capacity: What Can We Learn from Patients with Medial Temporal Lobe Damage? *NeuroImage Clin.* **2014**, *5*, 188–196. [CrossRef]
73. Seghier, M.L. Multiple Functions of the Angular Gyrus at High Temporal Resolution. *Brain Struct. Funct.* **2023**, *228*, 7–46. [CrossRef] [PubMed]
74. Hwang, K.; Bertolero, M.A.; Liu, W.B.; D'Esposito, M. The Human Thalamus Is an Integrative Hub for Functional Brain Networks. *J. Neurosci.* **2017**, *37*, 5594–5607. [CrossRef] [PubMed]
75. Blumenfeld, H.; Varghese, G.I.; Purcaro, M.J.; Motelow, J.E.; Enev, M.; McNally, K.A.; Levin, A.R.; Hirsch, L.J.; Tikofsky, R.; Zubal, I.G.; et al. Cortical and Subcortical Networks in Human Secondarily Generalized Tonic–Clonic Seizures. *Brain* **2009**, *132*, 999–1012. [CrossRef] [PubMed]
76. Brodovskaya, A.; Kapur, J. Circuits Generating Secondarily Generalized Seizures. *Epilepsy Behav.* **2019**, *101*, 106474. [CrossRef] [PubMed]
77. Labate, A.; Cerasa, A.; Gambardella, A.; Aguglia, U.; Quattrone, A. Hippocampal and Thalamic Atrophy in Mild Temporal Lobe Epilepsy: A VBM Study. *Neurology* **2008**, *71*, 1094–1101. [CrossRef]
78. He, X.; Doucet, G.E.; Pustina, D.; Sperling, M.R.; Sharan, A.D.; Tracy, J.I. Presurgical Thalamic "Hubness" Predicts Surgical Outcome in Temporal Lobe Epilepsy. *Neurology* **2017**, *88*, 2285–2293. [CrossRef]
79. Park, K.M.; Lee, B.I.; Shin, K.J.; Ha, S.Y.; Park, J.; Kim, S.E.; Kim, S.E. Pivotal Role of Subcortical Structures as a Network Hub in Focal Epilepsy: Evidence from Graph Theoretical Analysis Based on Diffusion-Tensor Imaging. *J. Clin. Neurol.* **2019**, *15*, 68–76. [CrossRef]
80. Chen, M.; Guo, D.; Li, M.; Ma, T.; Wu, S.; Ma, J.; Cui, Y.; Xia, Y.; Xu, P.; Yao, D. Critical Roles of the Direct GABAergic Pallido-Cortical Pathway in Controlling Absence Seizures. *PLOS Comput. Biol.* **2015**, *11*, e1004539. [CrossRef]
81. Moazeni, O.; Northoff, G.; Batouli, S.A.H. The Subcortical Brain Regions Influence the Cortical Areas during Resting-State: An fMRI Study. *Front. Hum. Neurosci.* **2024**, *18*, 1363125. [CrossRef]
82. Rodriguez-Sabate, C.; Gonzalez, A.; Perez-Darias, J.C.; Morales, I.; Sole-Sabater, M.; Rodriguez, M. Causality Methods to Study the Functional Connectivity in Brain Networks: The Basal Ganglia—Thalamus Causal Interactions. *Brain Imaging Behav.* **2024**, *18*, 1–18. [CrossRef]
83. Herbet, G.; Duffau, H. Revisiting the Functional Anatomy of the Human Brain: Toward a Meta-Networking Theory of Cerebral Functions. *Physiol. Rev.* **2020**, *100*, 1181–1228. [CrossRef] [PubMed]
84. van den Heuvel, M.P.; Sporns, O. Rich-Club Organization of the Human Connectome. *J. Neurosci.* **2011**, *31*, 15775–15786. [CrossRef] [PubMed]
85. Liu, M.; Chen, Z.; Beaulieu, C.; Gross, D.W. Disrupted Anatomic White Matter Network in Left Mesial Temporal Lobe Epilepsy. *Epilepsia* **2014**, *55*, 674–682. [CrossRef] [PubMed]

86. Mijalkov, M.; Volpe, G.; Pereira, J.B. Directed Brain Connectivity Identifies Widespread Functional Network Abnormalities in Parkinson's Disease. *Cereb. Cortex* **2022**, *32*, 593–607. [CrossRef] [PubMed]
87. Freedman, D.; Diaconis, P. On the Histogram as a Density Estimator:L 2 Theory. *Zeitschrift für Wahrscheinlichkeitstheorie und Verwandte Gebiete* **1981**, *57*, 453–476. [CrossRef]
88. McHugh, M.L. The Chi-Square Test of Independence. *Biochem. Med.* **2013**, *23*, 143–149. [CrossRef]

applied sciences

MDPI

Article

Differences in EEG Functional Connectivity in the Dorsal and Ventral Attentional and Salience Networks Across Multiple Subtypes of Depression

Ian D. Evans [1], Christopher F. Sharpley [1,*], Vicki Bitsika [1], Kirstan A. Vessey [1], Rebecca J. Williams [1], Emmanuel Jesulola [1,2] and Linda L. Agnew [1,3]

1 Brain-Behaviour Research Group, School of Science & Technology, University of New England, Armidale, NSW 2351, Australia; ievans3@une.edu.au (I.D.E.); vicki.bitsika@une.edu.au (V.B.); kvessey@une.edu.au (K.A.V.); rwilli90@une.edu.au (R.J.W.); doctorseasept@yahoo.com (E.J.); linda.agnew@griffith.edu.au (L.L.A.)

2 Department of Neurosurgery, The Alfred Hospital, Melbourne, VIC 3004, Australia

3 Griffith Health Group, Griffith University, Southport, QLD 4222, Australia

* Correspondence: csharpl3@une.edu.au

Abstract: Depression remains one of the most widespread and costly mental disorders, with the current first-line treatment efficacy of about a third, possibly due to its heterogeneous nature. Consequently, there is a need to identify reliable biomarkers for specific subtypes of depression, particularly neurological signatures that may help with targeted treatments. This study aimed to explore the connectivity between two important networks in the brain: the dorsal and ventral attention networks and the salience network, to determine their potential as biomarkers of depression subtypes. From resting electroencephalogram (EEG) data collected on 54 males and 46 females aged between 18 and 75 yr (M = 33 yr), functional network connectivity data were examined for their relationships with four depression subtypes. Beta and gamma wave connectivity was significantly associated with Anhedonia and Cognitive depression subtypes across and within all three networks while no significant results were found for alpha wave activity connectivity, and only one result was found for either the Mood or Somatic depression subtypes. In conclusion, these results provide further support for the concept of depression as heterogeneous rather than homogeneous and identify the novel neurophysiological signatures of two depression subtypes.

Keywords: depression; networks; functional connectivity; attention

1. Introduction

Despite carrying a major disease burden [1] and high risk for all-cause mortality [2], and bringing severe personal distress [3], depression has also been defined as an adaptive response to uncontrollable stress [4,5]. The 'adaptive' nature of depression is defined by the shaping effects of negative reinforcers (e.g., social withdrawal, which helps the individual avoid unpleasant events and/or escape from threat) and positive reinforcers (e.g., receiving sympathy or relaxing medication) that follow depressive behaviours [6–9]. The longer-term adaptive outcome of these consequences of depressive behaviour was noted some time ago by Charles Darwin, who commented that "Pain or suffering of any kind, if long continued, causes depression and lessens the power of action; yet, it is well adapted to make a creature guard itself against any great or sudden evil" [10], p. 51.

As such, it has been hypothesised that depression (or depressive behaviour) may be profitably analysed for its functionality, or how the depressive behaviour helps to protect the depressed individual from further discomfort [11]. Arising from this model of depression, one therapeutic approach applies behavioural activation strategies to address the depressed individual's disinclination to confront their behavioural withdrawal from the uncontrollable stressors that have led them to feel depressed. This approach to treatment has met with some success, as demonstrated in a meta-analysis of 28 research studies including 1853 participants [12], and providing support for the adaptive model for depressive behaviour.

However, although most studies of depression utilise a unitary model and metric of depression, such as Major Depressive Disorder (MDD), this disorder may occur in almost 1500 different combinations of the established diagnostic criteria and associated features for MDD [13], indicating that it is a heterogeneous entity. Consequently, some attention has been given to the identification of particular depression symptoms and how these may group together, often referred to as 'depression subtypes'. Several examples of these subtypes have been described, including melancholia, psychotic depression, atypical depression, and anxious depression [14]. Another model that has received some research attention focuses on subtypes of depression known as depressed Mood, Anhedonic depression, Cognitive depression, and Somatic depression [15,16]. Both of these models were developed using procedures that grouped MDD symptoms according to their clinical coherence, and then tested for significant associations between the defined subtype and other variables.

One particularly relevant set of variables that are valuable to investigate for their association with MDD subtypes are those generated by electroencephalography (EEG). EEG enables the identification of the relative electrical activity in specific areas of the brain, as well as the connections between those brain areas. The associations between this kind of information and MDD symptoms can make an obvious contribution to understanding the particular ways in which brain activity is related to specific MDD symptoms or specific depression subtypes, and thence to the eventual development of individualised treatment of depression based upon that brain activation–MDD symptomatology relationship [17–19]. Understanding depression from a neurological perspective has been a major and developing theme in the research literature [20–22], but mostly via the studies of depression as a unitary construct such as MDD. Although relatively easy to measure, depression as a unitary phenomenon may ignore variations in the way that EEG data relate to groups (i.e., subtypes) of depression symptoms.

There are many sites on the skull that can be used to measure an electrical signal representing brain activity, but it is worthwhile focusing on those which have a theoretical association with aspects of MDD. In particular, the attentional (DAN/VAN) and salience (SN) networks represent brain activities that are of relevance to depression when it is conceived as behavioural withdrawal. The dorsal and ventral attention networks (DAN/VAN) have been described as having multiple nodes located throughout the grey matter, each of which is related to specific aspects of attention [23]. These are (1) alerting (increases and maintains response readiness in response to a specific stimulus; located in the frontal and parietal regions, mostly in the right hemisphere); (2) orienting (selecting pertinent information; located in the pulvinar, superior colliculus, superior parietal lobe, temporoparietal junction, superior temporal lobe, and frontal eye fields); and (3) executive attention (the monitoring and resolution of conflict between information from different neural regions; the dorsal anterior cingulate cortex (ACC) in cognitive conflict tasks, and activation in the rostral ACC after the commission of errors. Whether the ACC monitors or resolves conflict is not clear, but the ACC function seems to preferentially relate to conflict at the response level [24]).

The salience network (SN; [24,25]) is mostly located in the cortical and subcortical prominent nodes of the anterior cingulate cortex, as well as the rostral prefrontal cortex and the supramarginal gyrus. The SN has a major role in emotional control [26]. Altered SN function has been associated with reduced motivation to engage with the external environments of depressed patients [27]. It has been suggested that the primary function of the SN is to identify the most relevant stimuli that reach the brain, including both the cognitive and affective aspects of the stimulus [28]. If these were associated with the emotional and cognitive aspects of MDD, then the SN may well be conceptualised as helping the depressed patient identify which stimuli held the most emotional and/or cognitive valence vis à vis the MDD symptomatology and the individual's response to those stimuli. With relevance to the attentional networks, the SN may assist in deciding if the information identified by the DAN or VAN requires a behavioural response from the individual. That response may differ across depressed and non-depressed individuals and reflect the underlying propensity towards behavioural withdrawal from stressful stimuli that is a hallmark of MDD. Connectivity between the DAN, VAN, and SN might provide a confirmation of the behavioural withdrawal model of MDD, and assist in the development of effective treatment models for depression or its subtypes.

Therefore, the aim of the study was to examine the DAN, VAN, and SN network activity (and their connectivity with each other), and their association with MDD as well as the four MDD subtypes (Anhedonic, Cognitive, Somatic, and Mood) described by Sharpley and Bitsika [15]. Although the lack of previous conclusive research findings on these issues prevented the raising of specific directional hypotheses, it was expected that high depression scores (characterised by behavioural withdrawal in the form of MDD symptoms) would be associated with lowered activity in frequency bands associated with cognitive activation such as beta and gamma. That is, the depressed individual would exert neurological energy on the environmental cues that contributed to their depression but fail to find them sufficiently relevant to initiate behavioural activity.

2. Materials and Methods

2.1. Participants

A sample of 54 males and 46 females (M age = 32.53 yr, SD = 14.13 yr, and range = 18–75 yr) was drawn from a larger study undertaken in the New England region of New South Wales, Australia [29]. The participants were selected on the basis of not having a medical history of severe physical brain injury, brain surgery, or history of epilepsy or seizure disorder.

2.2. Depression Scales

The Zung Self-Rating Depression Scale (SDS) [30] comprises ten positively worded and ten negatively worded statements based on the Diagnostic Criteria and Associated Features of MDD [3]. Participants are required to state their frequency of these 20 items during the last two weeks on a four-point scale (None or a little of the time = 1, Some of the time = 2, Good part of the time = 3, and Most or all of the time = 4) so that the total raw scores range from 20 to 80 [30,31]. SDS raw scores of 40 or above indicate "clinically significant depression" [31], p. 335, a threshold that was met by 34 members of the overall sample. Split-half reliability for the SDS has been reported as 0.81, [30], 0.79 [32], and 0.94 [33], and internal consistency (Cronbach alpha) of 0.88 for depressed patients and 0.93 for non-depressed patients [34]. The four depression subtypes described by these data were derived from the 20 SDS items; for more information on each depression subtype see [15]. Rather than divide the sample into depressed and non-depressed groups, the depression score for the overall MDD and each subtype was calculated for each individual to be used for the subsequent correlation analyses.

2.3. EEG Data Processing

A 40-channel Neuroscan QuikCap EEG machine (Compumedics USA 144 Ltd., El Paso, TX, USA) with sintered Ag/AgCl electrodes was used to record EEG signals with electrode arrangement in consonance with the international 10/20 system and aligned with the anatomical inion and nasion points. A NuAmps digital amplifier (Compumedics USA Ltd., El Paso, TX, USA) was used for signal acquisition and digitalization at a sampling rate of 1000 Hz and passed through a bandpass filter of DC to 250 Hz. The amplifier was connected to the Curry 7 Acquisition software (Compumedics USA Ltd., El Paso, TX, USA) running on a Dell Optiflex 9020 desktop PC.

EEG recordings were initially referenced to the average of the A1-A2 earlobe electrodes, and subsequently converted to a common average reference offline. By using the four electrodes, EOG data were collected. Specifically, two electrodes were placed above and below the left eye to measure vertical eye movement, and the other two electrodes were located on the outside left and right canthus to measure horizontal eye movement. Prior to the commencement of EEG recording, the electrode impedance values were adjusted to <5 kΩ at all the electrodes to ensure the quality of signal acquisition. The EEG signals were processed using a 1–45 Hz 2nd order Butterworth bandpass filter, then re-referenced to a common average. Data tapering was performed using a Hann window with a 10% width to prevent data loss. This was then followed by a visual examination of the EEG signals to identify artefacts (eye movements, muscle movements, spontaneous discharges or electrode pops etc.), which were then removed from the recorded data. Bad block and eye blink detection (using the magnitude of eye blink deflections as a set threshold criterion to detect artefacts) was undertaken by three automated methods (subtraction, covariance and principal component analysis) to produce clean EEG data.

From the cleaned EEG data generated, 2 secs back-to-back epochs were then created and re-examined such that epochs with bad blocks were removed from averaged data; in this study, all the participants had a minimum of 87% usable epochs in the Eyes Open condition. By using the Key Institute eLORETA (exact low-resolution brain electromagnetic tomography) [35] software, the functional lagged linear connectivity (also known as coherence) estimates of four EEG frequency band activity were obtained, i.e., theta (4.05–8 Hz), alpha (8.05–12 Hz), beta (12.05–30 Hz), and gamma (30.05–45 Hz). This technique provides a single weighted minimum norm solution to the inverse problem and has been demonstrated to provide zero error (despite low spatial resolution) in localising cortical grey matter test sources [36,37]. The weights utilised by eLORETA (based on the EEG montage used in the recording) are used to calculate the current source density throughout the grey matter of a standardised realistic head model [38] based on the MNI template [39]. The resulting current density distribution is used to calculate measures of linear dependence between 'virtual' electrodes placed at regions of interest (ROIs) within the grey matter; for a comprehensive description of the mathematics underlying the eLORETA methodology and how the weighted norms are calculated, see [35,40]. ROIs were selected using commonly identified grey matter nodes in the DAN, VAN, and SN based on MNI coordinates identified in previous studies [41,42], with all the grey matter tissue within 10 mm of the identified source included as part of that node. This resulted in 15 ROIs being selected (see Table 1). In Table 1, 'Location' refers to the generally identified brain region. However, MNI coordinates locate each region more specifically. MNI X describes the brain region according to its location from left (negative values) to right (positive values); MNI Y coordinates refer to the front (positive values) to rear (negative values); and MNI Z coordinates position a brain site on the transverse plane of the brain, from top (positive values) to the bottom (negative values). Some regions listed in Table 1 are widespread, and thus occur in more than one network. In those cases, the regions may be more precisely identified by their MNI coordinates.

Table 1. Neural networks, brain sites, and MNI coordinates of centroid for each ROI.

Network	Location	MNI (X, Y, Z)
Dorsal Attention Network	left frontal eye field	−25, 12, 55
	right frontal eye field	28, −10, 53
	left posterior intraparietal sulcus	−22, −68, 46
	right posterior intraparietal sulcus	20, −67, 51
Ventral Attention Network	left middle frontal gyrus	−47, 14, 32
	right middle frontal gyrus	47, 14, 32
	left supramarginal gyrus	−57, −43, 34
	right supramarginal gyrus	57, −43, 34
Salience Network	dorsal anterior cingulate	0, −21, 36
	left anterior prefrontal cortex	−35, 45, 30
	right anterior prefrontal cortex	32, 45, 30
	left insula	−41, 3, 6
	right insula	41, 3, 6
	left lateral parietal lobule	−62, −45, 30
	right lateral parietal lobule	62, −45, 30

2.4. Procedure

The participants signed a consent form, filled out a background questionnaire (age and sex) and the SDS, and had their scalps prepared and the electrode cap fitted. Headphones for the participants were used to minimise the effect of external stimuli. Following 15 min of sitting still (adaptation), the audio-recorded experimental protocol (not used in this study) was presented via headphones to ensure consistency across the participants. Ethics approval was received from the Human Research Ethics Committee of the University of New England, Australia (Approval No. HE14-051).

3. Results

3.1. Age, Sex, and SDS Scores

SDS internal consistency (Cronbach's alpha) was 0.905. The normal Q-Q plots for the SDS revealed an almost completely straight line, indicative of normality. The mean SDS score was 36.70 (SD = 11.25), minimum = 21, and maximum = 66. Using Zung's [31] cutoff score of 40, there were 33 participants who had clinically significant depression (mean SDS score = 50.39, SD = 7.43) and 67 participants whose SDS scores did not meet this category (mean SDS score = 29.95, SD = 4.83; F (1,99) = 273.729, $p < 0.001$, $\eta_p{}^2 = 0.736$). There were no significant correlations between age or sex and SDS total score, and any of the four SDS subtype scores.

3.2. MDD Subtypes

Mean scores for the four MDD subtypes were depressed Mood = 1.736 (SD = 0.653; range = 1.00 to 3.67), Anhedonia = 1.850 (SD = 0.657; range = 1.00 to 3.75), Cognitive depression = 2.183 (SD = 0.810; range = 1.00 to 4.00), and Somatic depression = 1.618 (SD = 0.511; range = 1.00 to 3.50) using the 4-point scale described above. These means were used as cutoffs to identify the most depressed from the least depressed participants in each SDS subtype. All four SDS subtype scores were significantly correlated with each other (all $\rho \geq 0.646$, $p < 0.001$).

3.2.1. Theta Band

Correlation analyses were conducted in eLORETA using each depression score (MDD and the four subtypes) and functional connectivity between all the ROIs. Significant positive correlations were found between theta coherence and the Anhedonia scores. Two distinct neural systems were observed: firstly, connectivity between the right middle frontal gyrus (VAN) and both the left and right posterior IPS (DAN). Secondly, a four-node ring-like system including the left FEF (DAN), the left middle frontal gyrus (VAN), the dorsal ACC (SN), and the left insula (SN); significant r-values from 0.311 to 0.342 and p-values from 0.037 to 0.082 (see Figure 1A and Table 2).

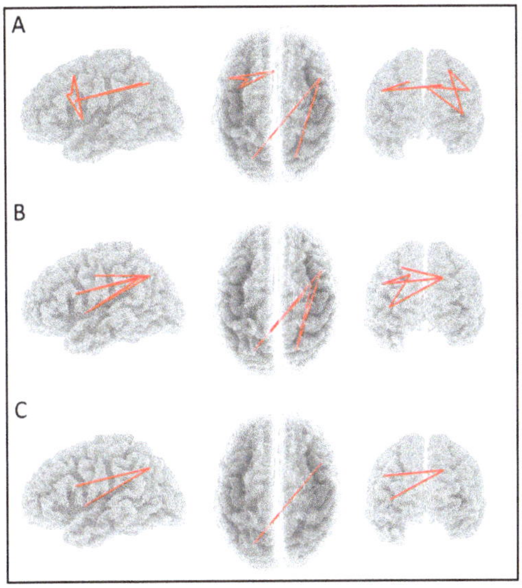

Figure 1. Significant positive correlations ($p < 0.1$) between theta coherence and the (**A**) Anhedonic depression scores, (**B**) Cognitive depression scores, and (**C**) overall MDD depression scores. Views are from the left, top, and front, respectively.

Table 2. Pearson's r- and p-values for each significant connection in the theta band Anhedonic depression condition.

ROI	Network	r	p
left frontal eye field right middle frontal gyrus	DAN VAN	0.311	0.082
right middle frontal gyrus dorsal anterior cingulate	VAN SN	0.342	0.037
dorsal anterior cingulate left insula	SN SN	0.323	0.064
left insula left frontal eye field	SN DAN	0.325	0.060
left posterior IPS left middle frontal gyrus	DAN VAN	0.323	0.063
left middle frontal gyrus right posterior IPS	VAN DAN	0.326	0.059

Significant positive correlations were also found between theta coherence and the Cognitive scores, with a five-node system including the right FEF (DAN), both the left and right posterior IPS (DAN), the right middle frontal gyrus (VAN), and the right insula (SN); significant r-values from 0.289 to 0.358 and p-values from 0.009 to 0.089 (see Figure 1B and Table 3).

Table 3. Pearson's r- and p-values for each significant connection in the theta band Cognitive depression condition.

ROI	Network	r	p
right insula left posterior IPS	SN DAN	0.309	0.049
left posterior IPS left middle frontal gyrus	DAN VAN	0.358	0.009
left middle frontal gyrus right posterior IPS	VAN DAN	0.330	0.022
right posterior IPS right insula	DAN SN	0.310	0.047
right frontal eye field left posterior IPS	DAN DAN	0.289	0.089

Significant positive correlations were found between theta band coherence and the SDS scores in a small system consisting of the right insula (SN), the left posterior IPS (DAN), and the right middle frontal gyrus (VAN); significant r-values from 0.294 to 0.300 and p-values from 0.082 to 0.096 (see Figure 1C and Table 4).

Table 4. Pearson's r- and p-values for each significant connection in the theta band overall SDS depression condition.

ROI	Network	r	p
right insula left posterior IPS	SN DAN	0.294	0.096
left posterior IPS left middle frontal gyrus	DAN VAN	0.300	0.082

3.2.2. Alpha Band

No significant results were found in the alpha band across all the subtypes and the overall SDS scores.

3.2.3. Beta Band

Significant negative correlations were found between beta band coherence and the Cognitive scores. This connectivity system consisted of a four-node chain involving the left middle frontal gyrus (VAN), the left lateral parietal lobule (SN), the left insula (SN), and the left supramarginal gyrus (VAN); significant r-values from −0.253 to −0.279 and p-values from 0.037 to 0.092 (see Figure 2A and Table 5).

Significant negative correlations were also found between beta band coherence and the SDS scores between the left supramarginal gyrus (VAN), the left insula (SN), and the left lateral parietal lobule (SN); $r = -0.239$, p-values from 0.093 to 0.097 (see Figure 2B and Table 6). Negative correlations between beta band coherence and the Anhedonic scores were also found in the same connectivity pattern; significant r-values from −0.266 to −0.271 and p-values from 0.032 to 0.040 (see Table 6).

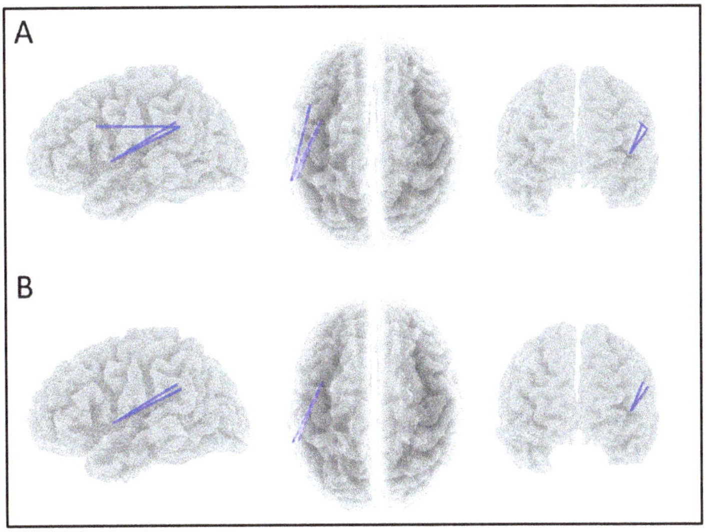

Figure 2. Significant negative correlations ($p < 0.1$) between beta coherence and (**A**) the Cognitive depression scores and (**B**) both the overall MDD depression scores and Anhedonic depression scores. Views are from the left, top, and front, respectively.

Table 5. Pearson's r- and p-values for each significant connection in the beta band Cognitive depression condition.

ROI	Network	r	p
right supramarginal gyrus left insula	VAN SN	−0.273	0.046
left insula left lateral parietal lobule	SN SN	−0.279	0.037
left lateral parietal lobule right middle frontal gyrus	SN VAN	−0.253	0.092

Table 6. Pearson's r- and p-values for each significant connection in the beta band Anhedonic and SDS depression conditions.

		Anhedonic		SDS	
ROI	Network	r	p	r	p
right supramarginal gyrus left insula	VAN SN	−0.266	0.040	−0.239	0.097
left insula left lateral parietal lobule	SN SN	−0.271	0.032	−0.239	0.093

3.2.4. Gamma Band

Significant negative correlations were also found between gamma band coherence and the SDS scores in connectivity between the left supramarginal gyrus (VAN) and the left insula (SN); $r = -0.275$, $p = 0.024$ (see Figure 3). Negative correlations between gamma band coherence and the Anhedonia ($r = -0.261$, $p = 0.049$), Cognitive ($r = -0.301$, $p = 0.010$), and Mood subtype scores ($r = -0.247$, $p = 0.065$) were also found in the same connectivity pattern.

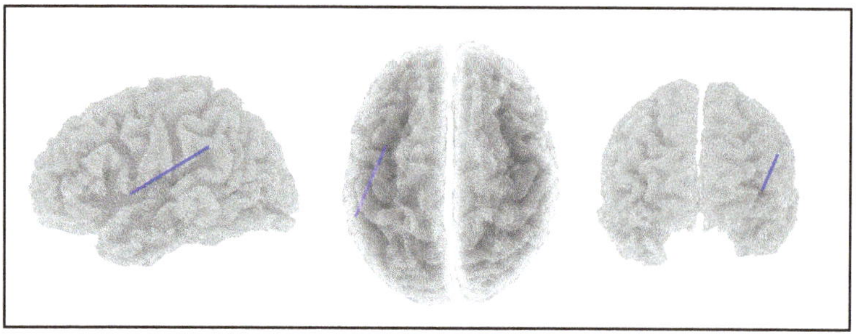

Figure 3. Significant negative correlations ($p < 0.1$) between gamma coherence and the Anhedonic, Cognitive, Mood, and overall MDD depression scores. Views are from the left, top, and front, respectively.

4. Discussion

The primary aim of this study was to investigate and describe any differences in the attentional and salience network connectivity in the four MDD subtypes as defined by the scores on groups of items from the SDS obtained by the participants regardless of whether or not they met the criteria for clinical depression.

Theta coherence increased along with the overall SDS scores in a short three-node system typically involved in orienting visual attention across all the networks used in this study [43,44]. As increases in theta activity are typically associated with drowsiness or similar low-energy states, this indicates that people with MDD are increasingly likely to have less interest in monitoring their surroundings, leading to a sense of withdrawal from the environment given that it is less likely that attentional networks will identify anything considered worth engaging with. This finding is congruent with the 'depression as behavioural withdrawal' hypothesis mentioned in the Introduction.

This situation is increasingly the case with the Cognitive and particularly the Anhedonic depression subtypes. The theta band coherence system associated with the Cognitive depression subtype includes all the nodes found in those listed in the broader SDS depression and two more nodes from the DAN (the right frontal eye field and right posterior IPS). Given that activation in the DAN is indicative of increased top-down processing effects, this indicates that Cognitive depression may also be affected by the top-down control of attention [45,46]. The theta results relating to Anhedonic depression are even broader in scope but mostly involve the DAN and VAN, indicating that Anhedonic depression may be the result of a more chronic cause. This view is backed by the inclusion of the dorsal ACC (as part of the SN) in the theta results, given the broad range of mental processes that the dACC is associated with [47,48].

As distinct from theta coherence, beta coherence decreased as the overall SDS scores increased in a short three-node system primarily in the SN. Similar results were also found with the Cognitive subtypes with the addition of the middle frontal gyrus node from the VAN. Each node here plays an important role in identifying relevant stimuli based on episodic memory, particularly the supramarginal gyrus and the lateral parietal lobule [49,50]. This function would still be active during resting states such as those used in this study because monitoring the environment is an ongoing process rather than a task-positive one. Since beta and gamma activation is typically associated with active cognitive processing, this would indicate that greater degrees of MDD are associated with a fall in cognitive resources available for engaging with the environment (perhaps contributing to the behavioural withdrawal noted above). Similar results were found in the gamma band as those in the beta band where gamma coherence between the left supramarginal gyrus and left insula decreased as Anhedonia, Cognitive, Mood, and overall SDS scores increased.

The overlap between the beta and gamma bands is likely the result of the same processes discussed above in the beta band results.

The Mood and Somatic depression subtypes are conspicuous in their absence from these results; apart from a single connectivity result in the gamma band, there was no relationship between either depression subtype and any measure of connectivity in the attentional networks. This can be explained by, or may be a product of, the more inward-focused nature of these depression subtypes; Mood-based depression is associated with negative emotional states (sadness; crying; and feeling hopeless, useless, or irritable) while Somatic-based depression is associated with negative bodily states (sleeping difficulties, weight loss, high pulse, and feeling tired and restless [15]). The absence of any relationship between the Mood or Somatic depression subtypes and attentional network activation indicates that these forms of depression are primarily driven by internal psychological and/or physical factors rather than a response to external stimuli.

4.1. Clinical Implications

This study was based on the assumption of MDD being heterogeneous, and this was confirmed by the different connectivity results across the four MDD subtypes used here. As such, the use of a homogeneous construct of MDD in clinical settings is challenged by these data. Instead, depressed patients might benefit from the consideration of their MDD subtype symptom profiles and a consequent therapeutic focus upon specific treatments for the subtypes they present with most severely. Those treatment choices could also be subsequently driven by the inclusion of the underlying cognitive processes inherent in the attentional and salience networks. This kind of diagnostic and therapeutic approach could move depression diagnosis and treatment from a 'one-size fits all' model, in which all depressions are treated similarly, towards a more personalised medicine approach [51,52]. The first step in that process is to recognise that, at a neurophysiological level (i.e., brain network connectivity), different groupings of MDD symptoms are more likely to be of benefit than a single score on an inventory or clinical interview that assumes a dichotomous classification is the best diagnosis available. The current study purposely examined only three major brain networks, and there are others that could also repay investigation such as task-positive networks.

4.2. Limitations

The Zung SDS is a creditable and well-justified method of measuring multiple MDD symptoms; however, it is not the only such instrument, and the use of alternative self-report instruments could add some greater validity to the overall results. Additionally, the DSM-5-TR Diagnostic Criteria are complemented by Diagnostic Features, but no current instrument exists that includes all of these. In fact, the tendency has been towards the development and use of briefer lists of MDD symptoms to diagnose MDD, such as the Personal Health Questionnaire-9 [53], but there are reasons from this study for moving towards more inclusive diagnostic instruments that measure all the aspects of depressive behaviour that contribute to the final diagnosis. This study used a 'snap-shot' research design, with data being collected at a single time point, whereas individuals' mood states vary across days or weeks, and so repeated sampling of mood states would enhance the generalizability of this research. The sample was composed of volunteers from a single cultural and geographical region; comparative data from other settings could help extend the current findings. EEG does produce precise measures of site-specific electrical activity in the brain, but might be profitably matched with fMRI (for example) to provide a more detailed account of neurocognitive activity. The method for identifying MDD subtypes used in this study might be described as 'a priori' because the subtypes were grouped according to the commonalities between specific MDD symptoms, performed by a panel of clinicians [15]. Another method of identifying subtypes is 'a posteriori' by applying cluster analysis to EEG data and then regressing those data upon MDD symptoms from a scale like the SDS. Neither method has yet shown superiority, and both represent valuable models for defining MDD subtypes. Although some of these limitations do limit generalizability, they do not invalidate these results. Finally, research is always strengthened by testing hypotheses based on previous findings, but that was not possible here due to the lack of studies in this field.

5. Conclusions

Although each of the four MDD subtypes did not show a specific set of inter-network connections, the fact that two of them did so (Anhedonia and Cognitive depression) provides strong support for the notion of MDD as heterogeneous, and also provides some basis for the subtype model tested here. It may be that Somatic depression and depressed Mood are also able to be identified by reference to alternative neurocognitive variables, and that work is being undertaken by the authors.

Author Contributions: I.D.E.: conceptualisation, methodology, software, formal analysis, writing (original draft), and visualisation. C.F.S.: conceptualisation, methodology, resources, and writing (review and editing). V.B.: conceptualisation and writing (review and editing). K.A.V.: writing (review and editing). R.J.W.: writing (review and editing). E.J.: investigation and data curation. L.L.A.: resources. All authors have read and agreed to the published version of the manuscript.

Funding: This research did not receive any specific grant from funding agencies in the public, commercial, or not-for-profit sectors.

Institutional Review Board Statement: Ethics approval for this study was provided by the Human Research Ethics Committee of the University of New England, Australia (Approval No. HE14-051), consistent with the Code of Ethics of the World Medical Association (Declaration of Helsinki).

Informed Consent Statement: Written and informed consent was obtained from all the subjects involved in the study.

Data Availability Statement: EEG and source localization data will be made available upon reasonable request.

Conflicts of Interest: The authors declare no conflicts of interest.

References

1. Liu, Q.; He, H.; Yang, J.; Feng, X.; Zhao, F.; Lyu, J. Changes in the global burden of depression from 1990 to 2017: Findings from the Global Burden of Disease study. *J. Psychiatr. Res.* **2020**, *126*, 134–140. [CrossRef] [PubMed]
2. Li, W.; Chen, D.; Ruan, W.; Peng, Y.; Lu, Z.; Wang, D. Associations of depression, sleep disorder with total and cause-specific mortality: A prospective cohort study. *J. Affect. Disord.* **2022**, *298*, 134–141. [CrossRef] [PubMed]
3. American Psychiatric Association. *Diagnostic and Statistical Manual of Mental Disorders*, 5th ed.; American Psychiatric Association: Washington DC, USA, 2022.
4. LeMoult, J. From stress to depression: Bringing together cognitive and biological science. *Curr. Dir. Psychol. Sci.* **2020**, *29*, 592–598. [CrossRef] [PubMed]
5. Ferster, C.B. A Functional Analysis of Depression. *Am. Psychol.* **1973**, *28*, 857–870. [CrossRef]
6. Dougher, M.J.; Hackbert, L. A Behavior-Analytic Account of Depression and a Case Report Using Acceptance-Based Procedures. *Behav. Anal.* **1994**, *17*, 321–334. [CrossRef]
7. Bolling, M.; Kohlenberg, R.; Parker, C. Behavior Analysis and Depression. In *Clinical Behavior Analysis*; Dougher, M.J., Ed.; Context Press: Reno, NV, USA, 1999; pp. 127–153.
8. Kanter, J.; Cautilli, J.; Busch, A.; Baruch, D. Towards a comprehensive functional analysis of depressive behavior: Five environmental factors and a possible sixth and seventh factor. *Int. J. Behav. Consult. Ther.* **2011**, *7*, 5–14. [CrossRef]
9. Santos, M.M.; Nagy, G.A.; Kanter, J.W.; López, S.R. Applying a Process-Oriented Model of Cultural Competence to Behavioral Activation for Depression. *Cogn. Behav. Pract.* **2021**, *28*, 127–146. [CrossRef]

10. Darwin, C. Recollections of the development of my mind and character. In *Autobiographies*; Penguin: London, UK, 2002.
11. Sturmey, P. *Functional Analysis in Clinical Psychology*; John Wiley & Sons: New York, NY, USA, 1996.
12. Stein, A.T.; Carl, E.; Cuijpers, P.; Karyotaki, E.; Smits, J.A. Looking beyond depression: A meta-analysis of the effect of behavioral activation on depression, anxiety, and activation. *Psychol. Med.* **2021**, *51*, 1491–1504. [CrossRef]
13. Ostergaard, S.; Jensen, S.; Bech, P. The heterogeneity of the depressive syndrome: When numbers get serious. *Acta Psychiatr. Scand.* **2011**, *124*, 495–496. [CrossRef]
14. Baumeister, H.; Parker, G. Meta-review of depressive subtyping models. *J. Affect. Disord.* **2012**, *139*, 126–140.
15. Sharpley, C.F.; Bitsika, V. Differences in neurobiological pathways of four "clinical content" subtypes of depression. *Behav. Brain Res.* **2013**, *256*, 368–376. [CrossRef] [PubMed]
16. Sharpley, C.F.; Bitsika, V. Validity, reliability and prevalence of four 'Clinical Content' subtypes of depression. *Behav. Brain Res.* **2014**, *259*, 9–15. [CrossRef] [PubMed]
17. Milaneschi, Y.; Lamers, F.; Berk, M.; Penninx, B. Depression Heterogeneity and Its Biological Underpinnings: Toward Immunometabolic Depression. *Biol. Psychiatry* **2020**, *88*, 369–380. [CrossRef] [PubMed]
18. Beijers, L.; Wardenaar, K.; van Loo, H.; Schoevers, R. Data-driven biological subtypes of depression: Systematic review of biological approaches to depression subtyping. *Mol. Psychiatry* **2019**, *24*, 888–900. [CrossRef]
19. Buch, A.; Liston, C. Dissecting diagnostic heterogeneity in depression by integrating neuroimaging and genetics. *Neuropsychopharmacology* **2021**, *46*, 156–175. [CrossRef]
20. Pandya, M.; Altinay, M.; Malone, D.A.; Anand, A. Where in the brain is depression? *Curr. Psychiatry Rep.* **2012**, *14*, 634–642. [CrossRef]
21. Drevets, W.C.; Price, J.L.; Furey, M.L. Brain structural and functional abnormalities in mood disorders: Implications for neurocircuitry models of depression. *Brain Struct. Funct.* **2008**, *213*, 93–118. [CrossRef]
22. Zhang, F.F.; Peng, W.; Sweeney, J.A.; Jia, Z.Y.; Gong, Q.Y. Brain structure alterations in depression: Psychoradiological evidence. *CNS Neurosci. Ther.* **2018**, *24*, 994–1003. [CrossRef]
23. Raz, A.; Buhle, J. Typologies of attentional networks. *Nat. Rev. Neurosci.* **2006**, *7*, 367–379. [CrossRef]
24. Huang, H.; Chen, C.; Rong, B.; Wan, Q.; Chen, J.; Liu, Z.; Zhou, Y.; Wang, G.; Wang, H. Resting-state functional connectivity of salience network in schizophrenia and depression. *Sci. Rep.* **2022**, *12*, 11204. [CrossRef]
25. Han, S.; Cui, Q.; Wang, X.; Li, L.; Li, D.; He, Z.; Guo, X.; Fan, Y.; Guo, J.; Sheng, W.; et al. Resting state functional network switching rate is differently altered in bipolar disorder and major depressive disorder. *Hum. Brain Mapp.* **2020**, *41*, 3295–3304. [CrossRef] [PubMed]
26. Goulden, N.; Khusnulina, A.; Davis, N.J.; Bracewell, R.M.; Bokde, A.L.; McNulty, J.P.; Mullins, P.G. The salience network is responsible for switching between the default mode network and the central executive network: Replication from DCM. *Neuroimage* **2014**, *99*, 180–190. [CrossRef] [PubMed]
27. Nolen-Hoeksema, S.; Wisco, B.E.; Lyubomirsky, S. Rethinking rumination. *Perspect. Psychol. Sci.* **2008**, *3*, 400–424. [CrossRef] [PubMed]
28. Menon, V.; Uddin, L.Q. Saliency, switching, attention and control: A network model of insula function. *Brain Struct. Funct.* **2010**, *214*, 655–667. [CrossRef]
29. Sharpley, C.F.; Agnew, L.L. *The New England Mental Health Study*, 1st ed.; University of New England: Amidale, NSW, Australia, 2016.
30. Zung, W. A self-rating depression scale. *Arch. Gen. Psychiatry* **1965**, *12*, 63–70. [CrossRef]
31. Zung, W. From art to science: The diagnosis and treatment of depression. *Arch. Gen. Psychiatry* **1973**, *29*, 328–337. [CrossRef]
32. DeJonge, J.; Baneke, J. The Zung Self-rating Depression Scale: A replication study on reliability, validity and prediction. *Psychol. Rep.* **1989**, *64*, 833–834. [CrossRef]
33. Gabrys, J.; Peters, K. Reliability, discriminant and predictive validity of the Zung Self-Rating Depression Scale. *Psychol. Rep.* **1985**, *57*, 1091–1096. [CrossRef]
34. Schaefer, A.; Brown, J.; Watson, C.; Plenel, D.; DeMotts, J.; Howard, M.; Norman, P.; Balleweg, B.J.; Douglas, A. Comparison of the validities of the Beck, Zung and MMPI depression scales. *J. Consult. Clin. Psychol.* **1985**, *53*, 415–418. [CrossRef]
35. Pascual-Marqui, R.D.; Lehmann, D.; Koukkou, M.; Kochi, K.; Anderer, P.; Saletu, B.; Tanaka, H.; Hirata, K.; John, E.R.; Prichep, L.; et al. Assessing interactions in the brain with exact low-resolution electromagnetic tomography. *Philos. Trans. R. Soc. A Math. Phys. Eng. Sci.* **2011**, *369*, 3768–3784. [CrossRef]
36. Marco-Pallarés, J.; Grau, C.; Ruffini, G. Combined ICA-LORETA analysis of mismatch negativity. *Neuroimage* **2005**, *25*, 471–477. [CrossRef] [PubMed]
37. Pascual-Marqui, R.D.; Esslen, M.; Kochi, K.; Lehmann, D. Functional imaging with low-resolution brain electromagnetic tomography (LORETA): A review. *Methods Find. Exp. Clin. Pharmacol.* **2002**, *24* (Suppl. C), 91–95. [PubMed]
38. Fuchs, M.; Kastner, J.; Wagner, M.; Hawes, S.; Ebersole, J.S. A standardized boundary element method volume conductor model. *Clin. Neurophysiol.* **2002**, *113*, 702–712. [CrossRef] [PubMed]

39. Mazziotta, J.; Toga, A.; Evans, A.; Fox, P.; Lancaster, J.; Zilles, K.; Woods, R.; Paus, T.; Simpson, G.; Pike, B.; et al. A probabilistic atlas and reference system for the human brain: International Consortium for Brain Mapping (ICBM). *Philos. Trans. R. Soc. Lond. Ser. B Biol. Sci.* **2001**, *356*, 1293–1322. [CrossRef]
40. Pascual-Marqui, R.D. Discrete, 3D distributed, linear imaging methods of electric neuronal activity. Part 1: Exact, zero error localization. *arXiv* **2007**, arXiv:07103341.
41. Fox, M.D.; Corbetta, M.; Snyder, A.Z.; Vincent, J.L.; Raichle, M.E. Spontaneous neuronal activity distinguishes human dorsal and ventral attention systems. *Proc. Natl. Acad. Sci. USA* **2006**, *103*, 10046–10051. [CrossRef]
42. Raichle, M.E. The restless brain. *Brain Connect.* **2011**, *1*, 3–12. [CrossRef]
43. Japee, S.; Holiday, K.; Satyshur, M.D.; Mukai, I.; Ungerleider, L.G. A role of right middle frontal gyrus in reorienting of attention: A case study. *Front. Syst. Neurosci.* **2015**, *9*, 23. [CrossRef]
44. Suo, X.; Ding, H.; Li, X.; Zhang, Y.; Liang, M.; Zhang, Y.; Yu, C.; Qin, W. Anatomical and functional coupling between the dorsal and ventral attention networks. *Neuroimage* **2021**, *232*, 117868. [CrossRef]
45. Astafiev, S.V.; Shulman, G.L.; Corbetta, M. Visuospatial reorienting signals in the human temporo-parietal junction are independent of response selection. *Eur. J. Neurosci.* **2006**, *23*, 591–596. [CrossRef]
46. Corbetta, M.; Shulman, G.L. Control of goal-directed and stimulus-driven attention in the brain. *Nat. Rev. Neurosci.* **2002**, *3*, 201–215. [CrossRef] [PubMed]
47. Margulies, D.S.; Kelly, A.C.; Uddin, L.Q.; Biswal, B.B.; Castellanos, F.X.; Milham, M.P. Mapping the functional connectivity of anterior cingulate cortex. *Neuroimage* **2007**, *37*, 579–588. [CrossRef] [PubMed]
48. Posner, M.I.; Rothbart, M.K.; Sheese, B.E.; Tang, Y. The anterior cingulate gyrus and the mechanism of self-regulation. *Cogn. Affect. Behav. Neurosci.* **2007**, *7*, 391–395. [CrossRef] [PubMed]
49. Menon, V.; D'Esposito, M. The role of PFC networks in cognitive control and executive function. *Neuropsychopharmacology* **2022**, *47*, 90–103. [CrossRef] [PubMed]
50. Rubinstein, D.Y.; Camarillo-Rodriguez, L.; Serruya, M.D.; Herweg, N.A.; Waldman, Z.J.; Wanda, P.A.; Sharan, A.D.; Weiss, S.A.; Sperling, M.R. Contribution of left supramarginal and angular gyri to episodic memory encoding: An intracranial EEG study. *Neuroimage* **2021**, *225*, 117514. [CrossRef]
51. Kambeitz-Ilankovic, L.; Koutsouleris, N.; Upthegrove, R. The potential of precision psychiatry: What is in reach? *Br. J. Psychiatry* **2022**, *220*, 175–178. [CrossRef]
52. Tozzi, L.; Zhang, X.; Pines, A.; Olmsted, A.M.; Zhai, E.S.; Anene, E.T.; Chesnut, M.; Holt-Gosselin, B.; Chang, S.; Stetz, P.C.; et al. Personalized brain circuit scores identify clinically distinct biotypes in depression and anxiety. *Nat. Med.* **2024**, *30*, 2076–2087. [CrossRef]
53. Kroenke, K.; Spitzer, R.; Williams, J. The PHQ-9: Validity of a brief depression severity measure. *J. Gen. Intern. Med.* **2001**, *16*, 606–613. [CrossRef]

applied sciences

MDPI

Article

Neural and Cardio-Respiratory Responses During Maximal Self-Paced and Controlled-Intensity Protocols at Similar Perceived Exertion Levels: A Pilot Study

Luc Poinsard [1,*], Florent Palacin [1], Iraj Said Hashemi [2] and Véronique Billat [1,3]

1 EA 4445—Movement, Balance, Performance, and Health Laboratory, Université de Pau et des Pays de l'Adour, 65000 Tarbes, France; palacinflorent@gmail.com (F.P.); veronique.billat@univ-evry.fr (V.B.)
2 Laboratory of Neurophysiology and Movement Biomechanics, Faculté Des Sciences de La Motricité, ULB Neuroscience Institute, Université Libre de Bruxelles, 1070 Brussels, Belgium; said-iraj.hashemi@ulb.be
3 Faculty of Sport Science, Université Évry Paris-Saclay, 91000 Évry-Courcouronnes, France
* Correspondence: lpoinsard@univ-pau.fr

Abstract: Self-paced exercise protocols have gained attention for their potential to optimize performance and manage fatigue by allowing individuals to regulate their efforts based on perceived exertion. This pilot study aimed to investigate the neural and physiological responses during a self-paced $\dot{V}O_2max$ (SPV) and incremental exercise tests (IET). Six trained male cyclists (mean age 39.2 ± 13.3 years; $\dot{V}O_2max$ 54.3 ± 8.2 mL·kg^{-1}·min^{-1}) performed both tests while recording their brain activity using electroencephalography (EEG). The IET protocol involved increasing the power every 3 min relative to body weight, while the SPV allowed participants to self-regulate the intensity using ratings of perceived exertion (RPE). Gas exchange, EEG, heart rate (HR), stroke volume (SV), and power output were continuously monitored. Statistical analyses included a two-way repeated measures ANOVA and Wilcoxon signed-rank tests to assess differences in alpha and beta power spectral densities (PSDs) and the EEG/$\dot{V}O_2$ ratio. Our results showed that during the SPV test, the beta PSD initially increased but stabilized at around 80% of the test duration, suggesting effective management of effort without further neural strain. In contrast, the IET showed a continuous increase in beta activity, indicating greater neural demand and potentially leading to an earlier onset of fatigue. Additionally, participants maintained similar cardiorespiratory parameters ($\dot{V}O_2$, HR, SV, respiratory frequency, etc.) across both protocols, reinforcing the reliability of the RPE scale in guiding exercise intensity. These findings suggest that SPV better optimizes neural efficiency and delays fatigue compared to fixed protocols and that individuals can accurately control exercise intensity based on perceived exertion. Despite the small sample size, the results provide valuable insights into the potential benefits of self-paced exercise for improving adherence to exercise programs and optimizing performance across different populations.

Keywords: brain activity; incremental exercise test; $\dot{V}O_2max$; self-regulation in exercise; aerobic capacity; exercise physiology; cycling performance; effort perception

1. Introduction

Self-pacing exercise allows individuals to regulate their exercise efforts according to subjective sensations and physiological states. This approach has been shown to improve adherence to exercise programs, particularly in certain populations, such as overweight and obese populations [1,2], and this trend also extends to normal-weight individuals [3,4]. The improved feelings of autonomy and enhanced positive affective responses contribute to exercise adherence. In addition, self-paced maximal exercise tests—where individuals regulate their pace during maximal exertion—have been shown to produce similar [5,6] or even higher maximal oxygen uptake ($\dot{V}O_2max$) values compared to traditional fixed incremental exercise tests (IETs). This suggests that giving individuals control over their pacing can optimize performance outcomes by allowing better management of effort and fatigue throughout the exercise [7–9].

However, despite these advantages, self-pacing requires careful application, particularly when working with less experienced athletes or when external feedback is limited. Ensuring that athletes maintain appropriate intensities is important to avoid potential risks of under- or overtraining [10]. With proper guidance and feedback, self-pacing can help athletes benefit from individualized intensity regulation while minimizing these risks.

The Self-Paced $\dot{V}O_2$max (SPV) protocol employs a progressive exercise design, where intensity increases are guided by the limits of ratings of perceived exertion (RPE), allowing individuals to self-regulate their pace. The RPE, developed by Borg, is a widely used subjective measure allowing individuals to rate their exertion levels during physical activity on a scale from 6 to 20 [11]. The RPE scale correlates well with physiological markers, such as heart rate (HR) and oxygen uptake ($\dot{V}O_2$), in healthy subjects, making it a reliable tool for prescribing and regulating exercise intensity [11,12]. Recent research has further highlighted the utility of self-paced exercise, demonstrating that the RPE scale can effectively guide exercise intensity in various settings, including prolonged endurance events, like marathons [13]. This scale is an integral part of self-pacing, as it enables individuals to modulate their exercise intensity according to perceived exertion, thus aligning their physical effort with their subjective experience [14,15].

The Central Governor Model proposed by Noakes and Gibson suggests that the brain plays a crucial role in regulating exercise performance by modulating effort perception and fatigue, further underscoring the importance of subjective measures like RPE in exercise science [16–18]. This theory is supported by recent findings showing early modifications in brain activity during prolonged exercise, even when cardiorespiratory responses remain steady [13]. This highlights the brain's critical role in managing exertion and preventing overexertion during endurance events, as indicated by the electroencephalography (EEG) and $\dot{V}O_2$ relationship, where EEG changes can precede observable drops in $\dot{V}O_2$, suggesting central regulation of fatigue.

However, the literature still lacks proof of concept concerning the direct link between RPE and brain response during exhaustive exercise. Consequently, whether the brain acts as a limitation or, conversely, as a super controller for power optimization at the same RPE remains unknown.

EEG is an effective method for assessing brain activity during exercise, providing high temporal resolution and enabling real-time monitoring of neural responses. Technological advancements have facilitated the use of EEG in dynamic exercise settings, reducing movement artifacts and enabling detailed analysis of cortical activity [19–24]. EEG allows for the examination of how perceived exertion translates into measurable brain activity, linking subjective sensations with physiological responses [25,26]. Key frequency bands studied in exercise research include alpha (8–13 Hz) and beta (13–30 Hz) waves, which are associated with relaxation, cognitive processing, and motor control [25,27]. Studies indicate that both alpha and beta wave activity increase with exercise intensity, reflecting enhanced neural synchronization and motor planning [24,28]. However, conflicting evidence exists regarding the patterns of cortical activity during IET, with some studies reporting linear increases, while others observe plateaus or declines in neural activity as fatigue sets in [29,30].

To our knowledge, only one study, by Dykstra et al. (2019), has directly compared EEG responses between IET and SPV tests, using RPE as both a dependent and an independent variable [31]. Dykstra found significant differences between the two protocols, with the SPV test showing continuous increases in alpha and beta activity, while the IET exhibited a peak followed by a decline, suggesting an earlier onset of fatigue.

Maceri et al. (2019) also employed RPE as an independent variable and used EEG to assess brain activity during the SPV test [28]. Their study, which investigated EEG responses in younger and middle-aged adults, found that alpha and beta wave activity increased with exercise intensity regardless of age, highlighting the effectiveness of RPE in regulating exercise intensity. However, unlike Dykstra's study, Maceri did not compare different exercise protocols but rather examined EEG variations within a single self-paced protocol across different age groups.

Both studies imposed RPE levels (11, 13, 15, 17, and 20 on Borg's scale) rather than basing subsequent test steps on the RPE given by subjects at the end of each step [28,31]. Imposing a standard RPE on each subject does not account for individual variations in response. For instance, at an RPE of 15, subjects may be at different percentages of their ventilatory threshold or $\dot{V}O_2max$. Our study aims to fill this gap by defining the steps of the second test based on the RPE indicated by subjects at the end of each step of the first test.

The primary objectives of this study are (1) to investigate neural responses to self-paced and externally controlled incremental exercise using EEG and (2) to assess whether individuals can effectively regulate their effort based solely on perceived exertion, as indicated by the RPE scale. We hypothesize that alpha and beta power spectral density (PSD) will increase with exercise intensity in both the IET and SPV tests, but this increase will be more pronounced in the incremental exercise test due to higher levels of perceived exertion and physiological strain. Additionally, we expect that a decline in the EEG/$\dot{V}O_2$ ratio will be observed as exercise intensity increases, particularly in the IET, suggesting that neural activity decreases relative to physiological demand, and potentially serving as a marker of central fatigue. Furthermore, we hypothesize that subjects will demonstrate a capacity to control their exercise intensity accurately using the RPE scale, reflected by similar cardiovascular parameters ($\dot{V}O_2$, respiratory rate (Rf), HR, power output) between the IET and SPV tests.

2. Materials and Methods

2.1. Subjects

Six non-elite male cyclists with the following characteristics were recruited from local sports clubs to participate in the study: mean age ± SD 39.2 ± 13.3 years; height 179.8 ± 9.0 cm; weight 70.8 ± 9.7 kg; and body mass index 21.9 ± 2.4 kg/m^2. Eligible participants met the following inclusion criteria: (1) non-smokers, (2) performed in at least 8 h of cycling training per week, and (3) no existing health issues. Although classified as "non-elite", the participants demonstrated a good level of aerobic fitness, as indicated by their $\dot{V}O_2peak$ values (mean $\dot{V}O_2peak$ = 54.3 ± 8.2 mL·kg^{-1}·min^{-1}), which are consistent with cyclists competing at a departmental or regional level.

The study was approved by the ethics committee of the University Hospital of Brugmann (Brussels, Belgium; reference: B0772022000014). All subjects provided informed consent prior to participation in the study.

2.2. Measurement Protocols

Each participant completed two separate exercise test sessions: the IET and the SPV test. The IET gradually increases intensity until exhaustion is reached, while the SPV allows participants to regulate their effort based on perceived exertion (RPE) [11]. Both tests were conducted in a laboratory setting, with two days of rest between sessions to ensure full recovery. The experimental sessions are presented in Figure 1.

2.2.1. Incremental Exercise Test

The IET was designed to assess maximal aerobic power (MAP) and $\dot{V}O_2max$ [32]. The test started at a power output of 1.5 W per kilogram of body mass (i.e., 105 W for a 70 kg participant), and the intensity increased by 0.5 W per kilogram every three minutes. Participants were instructed to maintain a cadence of at least 60 rotations per minute (rpm) until exhaustion. At random intervals within a 30 s window before the end of each step, the experimenter prompted the subject to rate their perceived exertion using the Borg 6–20 scale [11]. The test was terminated if the participant's cadence dropped below 60 rpm for five consecutive seconds.

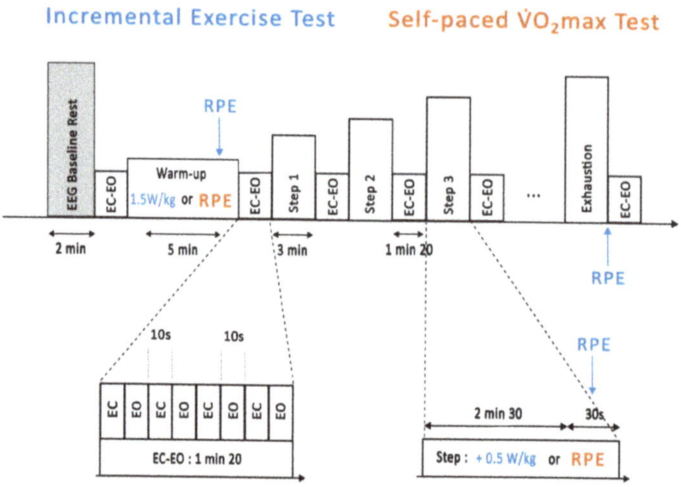

Figure 1. Schematic representation of the incremental exercise test (IET) and the self-paced $\dot{V}O_2max$ test (SPV). During the IET, intensity increases progressively by 0.5 W/kg every 3 min until exhaustion, while the SPV test allows participants to adjust their pace based on perceived exertion. Between each step, there was an alternation between eyes open (EO) and eyes closed (EC) at rest. In the figure, elements common to both tests are shown in black, IET-specific elements are shown in blue, and SPV-specific elements are shown in orange. Abbreviations: EEG = electroencephalogram; RPE = rate of perceived exertion; EC-EO = eyes closed eyes open phase.

2.2.2. Self-Paced $\dot{V}O_2max$ Test

The SPV test followed a similar structure to the IET but replaced power increments with RPE regulation. Participants self-regulated their exercise intensity based on their perceived exertion (using the RPE scale) [11], adjusting their cadence and/or resistance to maintain the target RPE for each step. The RPE values obtained during the IET were used to guide the SPV.

For both tests, participants were seated on a cycling ergometer (Cyclus2, RBM elektronik-automation GmbH, Leipzig, Germany) and underwent a two-minute baseline EEG recording while resting with their eyes open. The tests then proceeded according to the measurement protocols, with real-time data collected from all devices, including the EEG, metabolic cart, heart rate monitor, and cycling ergometer.

Participants alternated between eyes open and eyes closed phases at each exercise stage, lasting 80 s in total, to allow for more accurate EEG data collection by reducing movement artifacts. No information regarding power output, cadence, or time remaining was provided to the participants, and no verbal encouragement was given to ensure that effort regulation was based solely on perceived exertion.

2.3. Measurement Tools

To capture detailed physiological and neural data, several measurement tools were employed during both exercise tests.

2.3.1. Gas Exchange Measurements

Gas exchange was measured breath by breath using a facemask connected to a metabolic cart (Quark CPET, Cosmed, Rome, Italy) [33,34]. Flow and gas calibrations were performed approximately 10 min before each test, following the manufacturer's guidelines to ensure accuracy. The data were processed using Omnia Software (version 2.2, Cosmed, Rome, Italy), which calculated the Rf, tidal volume (Vt), ventilation rate ($\dot{V}E$), $\dot{V}O_2$, and carbon dioxide output ($\dot{V}CO_2$). The software also synchronized gas exchange data with HR measurements (HRM-Run, Garmin, KS, USA) and ergometer data, including power output and cadence.

2.3.2. Cardiac Output and Heart Rate Monitoring

HR was continuously monitored using a chest strap (HRM-Run, Garmin, KS, USA) [35]. Hemodynamic function was assessed using the Physioflow® PF07 Enduro (Manatec Biomedical, Poissy, France), an impedance cardiograph device that measures changes in transthoracic impedance during the cardiac cycle. This method allows for the calculation of HR and stroke volume (SV), as well as the estimation of cardiac output (CO). It has been validated in both resting and exercise conditions, including up to $\dot{V}O_2max$, and it is considered reliable for continuous hemodynamic monitoring [36,37].

2.3.3. Electroencephalography (EEG) Recording

Brain activity was recorded using a 64-channel ActiCap system (actiCHamp Plus, Brain Products, Gilching, Germany), which uses high-quality Ag/AgCl active electrodes. Conductive gel (SuperVisc, EasyCap GmbH, Wörthsee, Germany) was applied to enhance electrode–skin contact and to maintain low impedances. Active electrodes were chosen due to their reliability in capturing electro-cortical activity during intense exercise while minimizing movement artifacts [29].

Data were recorded from all 64 electrode sites, but the analysis focused on 15 sites in the extended 10–20 system: F3, F1, Fz, F2, F4, C3, C1, Cz, C2, C4, P3, P1, Pz, P2, and P4. These sites were chosen based on both the quality of the recorded signal and their relevance to the motor, sensory, and cognitive processes associated with exercise [21,38]. Some other electrode sites were excluded due to poor signal quality or noise artifacts. The ground electrode was positioned at AFz, and all electrodes were referenced to FCz. Impedances were maintained below 5 kΩ for all sensors to ensure high-quality signal acquisition [39]. Data were amplified using the BrainVision amplifier, sampled at 512 Hz, and recorded with BrainVision Recorder software (version 1.25.0204, Brain Products GmbH, Gilching, Germany).

2.4. Data Analysis

2.4.1. Data Synchronization

Markers were systematically placed at the start and end of each step, as well as during the transitions between eyes open and eyes closed conditions, to ensure the precise synchronization of data collected from the different measurement devices. This allowed for accurate alignment of the physiological and neural data streams.

2.4.2. EEG Analysis

EEG data were processed using MATLAB R2024a-based software EEGLAB (version 2023.4) [40]. After reducing the sampling rate to 256 Hz, high- and low-pass filters were applied to retain a frequency range from 0.5 to 80 Hz.

For the analysis of brain activity recorded during both tests, only the last 90 s of every 3-min step were used to minimize the possible influence of adaptation processes due to changes in load. A 50 Hz notch filter was applied, followed by an independent component analysis [41]. Recurring artefacts, such as those from eye blinks, eye movement, and muscular activity, were identified and removed. The cleaned data were then transformed into power spectra using a fast Fourier transform, and power spectral densities were calculated for two key frequency bands: alpha (α: 8–12 Hz) and beta (β: 12–30 Hz).

In addition, the power spectrum for each electrode was averaged across specific scalp zones. The frontal zone included electrodes F3, F1, Fz, F2, and F4; the central zone included C3, C1, Cz, C2, and C4; and the parietal zone included P3, P1, Pz, P2, and P4. To obtain an overall measure of brain activity, the data from all electrodes were also averaged to calculate the total brain power (total power, TP). Finally, exercise data were expressed as percentage changes from baseline PSD measures to account for day to day and between-subject variability, as reported in a previous study [25,42].

2.4.3. Statistical Analysis

All collected data were compiled and analyzed using XLSTAT 2023 (version 2.0, Paris, France). Descriptive statistics were calculated for all variables, and the results are presented as mean ± standard deviation (SD). Given the small sample size ($n = 6$), the normality of the data was tested using the Shapiro–Wilk test. For variables where normality was confirmed ($p > 0.05$), a two-way repeated measures ANOVA was conducted to assess the main effects of the condition (IET vs. SPV), the time, and the interaction between the condition and the time on alpha and beta PSD, as well as the EEG/$\dot{V}O_2$ ratio. In cases where the sphericity assumption was violated, the Greenhouse–Geisser correction was applied to adjust the degrees of freedom. Additionally, a post hoc Wilcoxon signed-rank test was used to compare the two conditions at each time point. For variables where normality was not confirmed ($p \leq 0.05$), a Friedman test was used to assess the main effects of the condition and the time. The Wilcoxon signed-rank test was used as a post hoc test to compare the two conditions at each time point for non-normally distributed variables. Statistical significance was set at $p < 0.05$ for all analyses.

3. Results

3.1. Physiological and Power Variables

Table 1 provides the maximal values for physiological and power variables during the IET and SPV tests. Across both protocols, most variables showed no significant differences, except for cadence, which was significantly higher in the SPV test compared to the IET test ($p = 0.031$).

Table 1. Maximal values for physiological and power variables during the incremental exercise test and the self-paced $\dot{V}O_2$max test.

Variable	IET	SPV	p
Rf (1/min)	61.1 ± 5.1	61.7 ± 8.1	0.844
Vt (L(btps))	2.9 ± 0.6	2.8 ± 0.5	0.131
$\dot{V}E$ (L/min)	147.8 ± 27.4	144.7 ± 31.4	0.844
RER	1.2 ± 0.1	1.1 ± 0.04	0.674
Relative $\dot{V}O_2$ (mL·kg^{-1}·min^{-1})	54.3 ± 8.2	53.7 ± 7.7	0.313
Absolute $\dot{V}O_2$ (mL·min^{-1})	3832.0 ± 648.3	3792.1 ± 630.3	0.313
Relative $\dot{V}CO_2$ (mL·kg^{-1}·min^{-1})	58.0 ± 9.6	57.2 ± 8.6	0.438
Absolute $\dot{V}CO_2$ (mL·min^{-1})	4088.3 ± 721.9	4030.2 ± 673.0	0.438
HR (bpm)	171.8 ± 14.7	173.0 ± 15.5	0.498
SV (mL)	145.1 ± 8.4	145.8 ± 11.8	0.813
CO (L/min)	24.4 ± 2.2	24.4 ± 3.7	0.813
MAP (W)	326.5 ± 67.3	331.7 ± 67.1	0.563
Pmax (W)	358.3 ± 74.3	374.8 ± 66.3	0.343
Cadence (rpm)	89.16 ± 4.5	96.65 ± 4.4	**0.031**

Note: the value in bold indicates a statistically significant difference between the two tests ($p < 0.05$). Abbreviations: IET = incremental exercise test, SPV = self-paced $\dot{V}O_2$max test, Rf = respiratory frequency, Vt = tidal volume, $\dot{V}E$ = ventilatory flow, RER = respiratory exchange ratio, $\dot{V}O_2$ = oxygen uptake, $\dot{V}CO_2$ = carbon dioxide output, HR = heart rate, SV = stroke volume, CO = cardiac output, MAP = maximal aerobic power, Pmax = maximal power output, rpm = rotations per minute.

3.2. EEG

3.2.1. Alpha PSD

There were no significant differences in the alpha frequency band between the two protocols in the central, frontal, and parietal scalp zones, nor for TP (all $p > 0.05$) (Table 2). However, a significant main effect of the time was observed in the parietal scalp zone ($F = 1.71$, $p = 0.043$), indicating a time-related increase in alpha power in this region. No significant effects were found for the condition ($F = 0.24$, $p = 0.623$) or the time x condition interaction ($F = 0.54$, $p = 0.898$). Additionally, in the central scalp zone, there was a trend towards a time-related effect, but it did not reach significance ($F = 1.65$, $p = 0.053$).

Table 2. Comparison of EEG alpha and beta power spectral densities between the incremental exercise test and the self-paced $\dot{V}O_2$max test.

Variable	IET	SPV	*p*
Central α	2.03 ± 1.6	1.92 ± 1.1	0.856
Frontal α	1.80 ± 1.2	1.72 ± 1.0	0.856
Parietal α	1.83 ± 1.3	1.74 ± 0.8	0.579
TP α	1.83 ± 1.3	1.75 ± 0.9	0.587
Central β	1.65 ± 0.9	1.33 ± 0.8	**<0.001**
Frontal β	1.54 ± 0.7	1.49 ± 0.7	0.309
Parietal β	1.63 ± 0.8	1.57 ± 0.8	0.274
TP β	1.58 ± 0.8	1.46 ± 0.7	**0.027**
Central α/β ratio	0.99 ± 0.7	0.87 ± 0.9	0.122
Frontal α/β ratio	1.16 ± 0.6	1.12 ± 0.6	0.422
Parietal α/β ratio	0.81 ± 0.5	0.84 ± 0.6	0.618
TP α/β ratio	1.01 ± 0.6	0.96 ± 0.5	0.310

Note: central, frontal, and parietal scalp zones are assessed for both alpha and beta bands, along with the alpha/beta (α/β) ratio and total power (TP). Values are expressed as a percentage of baseline EEG measures. The value in bold indicates a statistically significant difference between the two tests ($p < 0.05$). Abbreviations: IET = incremental exercise test, SPV = self-paced $\dot{V}O_2$max test, α = alpha, β = beta, α/β = alpha/beta ratio, TP = total power.

3.2.2. Beta PSD

Significant differences in beta PSD were observed between the two protocols in the central scalp zone ($p < 0.001$) and TP ($p = 0.027$), but not in the frontal or parietal zones (all $p > 0.05$) (Table 2). Protocol differences occurred at 80% ($p = 0.028$) and 100% ($p = 0.044$) of the test duration (Figure 2). Within the IET protocol, beta power increased significantly from 60% to 80% ($p = 0.006$), with no significant change from 80% to 100% ($p = 0.069$). Similarly, within the SPV protocol, beta power differed from 60% to 80% ($p = 0.048$), but not from 80% to 100% ($p = 0.733$) (Figure 2).

Unlike cadence, which was consistently higher in the SPV protocol than in the IET protocol, EEG responses were greater in the IET only towards the end of the tests (after 80% of the time to exhaustion) (Figure 2). Additionally, a significant main effect was observed for the time ($F = 7.73$, $p < 0.001$) and the condition ($F = 6.32$, $p = 0.014$) in the central scalp zone, although no significant interaction between the time and the condition was found.

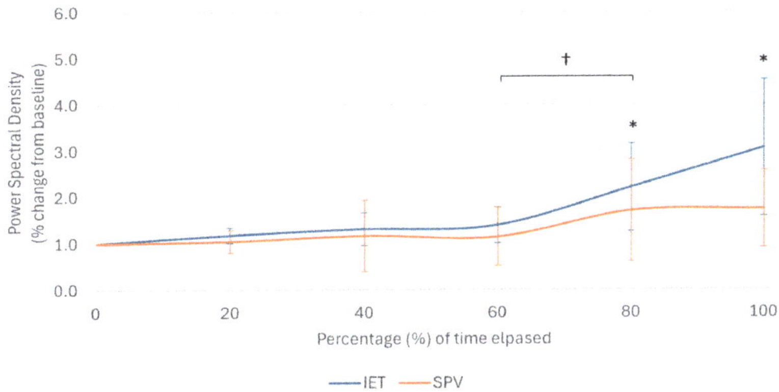

Figure 2. Comparison of changes in beta power spectral density in the central brain scalp zone during the incremental exercise test (IET) and the self-paced $\dot{V}O_2$max (SPV) test. The IET data are represented by the blue curve, and the SPV data are shown by the orange curve. The error bars are color-coded to match the corresponding condition (blue for IET and orange for SPV) to enhance the visual distinction between the two tests. Significant differences ($p < 0.05$) between the two tests are marked with asterisks (*), and differences between previous steps within the same test are indicated by daggers (†).

3.2.3. Alpha/Beta Ratio

No significant differences were observed for the alpha/beta ratio between the two protocols in the central, frontal, or parietal scalp zones, nor for TP (all $p > 0.05$) (Table 2). Additionally, there were no significant main effects of the time, the condition, or the time x the condition in any of the scalp zones or for TP.

3.3. EEG/$\dot{V}O_2$ Ratio

The EEG Alpha/$\dot{V}O_2$ and Beta/$\dot{V}O_2$ ratio graphs (Figure 3) show a sharp initial decline in both ratios at the beginning of the exercise for both the IET and SPV tests. Following this initial drop, the ratios stabilize and remain relatively flat for the majority of the test duration. When around 80% of the time has elapsed, a slight increase is observed in both the Alpha/$\dot{V}O_2$ and Beta/$\dot{V}O_2$ ratios, with this increase being slightly more pronounced in the IET compared to the SPV. However, it is important to note that these changes were not statistically significant (Alpha/$\dot{V}O_2$: $p = 0.636$; Beta/$\dot{V}O_2$: $p = 0.977$).

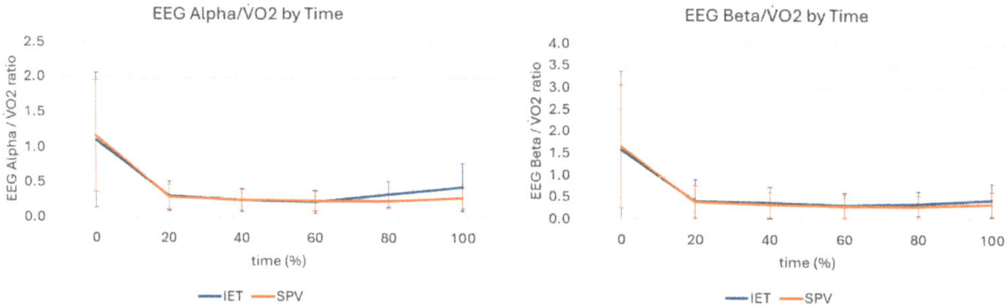

Figure 3. Comparison of EEG Alpha/$\dot{V}O_2$ and Beta/$\dot{V}O_2$ ratios between the incremental exercise test (IET, blue line) and the self-paced $\dot{V}O_2$max (SPV, orange line) test. The error bars are color-coded to match the corresponding condition (blue for IET and orange for SPV) to enhance the visual distinction between the two tests.

3.3.1. EEG Alpha/$\dot{V}O_2$ Ratio

There was a significant main effect of the time for the Alpha/$\dot{V}O_2$ ratio in the central ($F = 6.37$, $p < 0.001$), frontal ($F = 19.4$, $p < 0.001$), and parietal ($F = 8.86$, $p < 0.001$) scalp zones, as well as for TP ($F = 13.49$, $p < 0.001$). However, no significant effects were observed for the condition or the time x condition interaction across any of the scalp zones or TP (Table 3).

Table 3. Multi-factor ANOVA outcomes for the EEG Alpha/$\dot{V}O_2$ ratio analyzed across various brain scalp zones (central, frontal, parietal) and total power (TP) during the incremental exercise test and the self-paced $\dot{V}O_2$max test.

				Condition		Time (%)		Condition × Time (%)	
	R^2	F	Pr > F	F	Pr > F	F	Pr > F	F	Pr > F
Central Alpha/$\dot{V}O_2$	0.275	3.035	0.002	0.954	0.331	6.374	<0.001	0.145	0.981
Frontal Alpha/$\dot{V}O_2$	0.528	8.952	<0.001	0.384	0.537	19.399	<0.001	0.232	0.947
Parietal Alpha/$\dot{V}O_2$	0.343	4.177	<0.001	0.090	0.764	8.861	<0.001	0.322	0.899
TP Alpha/$\dot{V}O_2$	0.438	6.237	<0.001	0.185	0.668	13.486	<0.001	0.198	0.962

3.3.2. EEG Beta/$\dot{V}O_2$ Ratio

A significant main effect of the time was observed for the Beta/$\dot{V}O_2$ ratio in the central (F = 13.43, p < 0.001), frontal (F = 10.87, p <0.001), and parietal (F = 4.87, p = 0.001) scalp zones, as well as for TP (F = 8.68, p < 0.001). However, there were no significant effects for the condition or the time x condition interaction across any of the scalp zones or TP (Table 4).

Table 4. Multi-factor ANOVA outcomes for the EEG Beta/$\dot{V}O_2$ ratio analyzed across various brain scalp zones (central, frontal, parietal) and total power (TP) during the incremental exercise test and the self-paced $\dot{V}O_2$max test.

				Condition		Time (%)		Condition × Time (%)	
	R^2	F	Pr > F	F	Pr > F	F	Pr > F	F	Pr > F
Central Beta/$\dot{V}O_2$	0.466	6.810	<0.001	2.415	0.124	13.439	<0.001	1.250	0.293
Frontal Beta/$\dot{V}O_2$	0.389	4.969	<0.001	0.052	0.819	10.867	<0.001	0.054	0.998
Parietal Beta/$\dot{V}O_2$	0.225	2.270	0.017	0.475	0.493	4.868	0.001	0.041	0.999
TP Beta/$\dot{V}O_2$	0.336	3.964	<0.001	0.030	0.863	8.682	<0.001	0.032	0.999

4. Discussion

The objective of this study was to investigate neural responses to both IET and SPV tests using EEG and to determine whether individuals can regulate their effort based solely on perceived exertion, as indicated by the RPE scale. The results yield substantial insights into the neural mechanisms governing exercise regulation and the effectiveness of the RPE scale for guiding exercise intensity.

Our results showed that during the SPV test, the beta PSD initially increased but then stabilized at around 80% of the test duration. This suggests that participants reached a point where their perceived exertion allowed them to maintain their effort without requiring further increases in neural engagement. In contrast, the IET exhibited a continuous increase in beta activity throughout the test, reflecting a gradual rise in neural demand as participants neared exhaustion.

Our findings diverge from those reported by Dykstra et al. (2019). He observed sustained elevations in both alpha and beta wave activity during the SPV test and during the IET, and beta activity peaked before declining, indicating an earlier onset of fatigue [31]. The possible explanations for these discrepancies are several. Dykstra focused on the dorsolateral prefrontal cortex, which is associated with executive function and inhibitory control [43,44], whereas our study examined the central area of the brain [45,46], particularly motor-related cortical regions. These central regions may show different activation patterns due to their direct involvement in motor control and sensory feedback integration during intense physical exertion [46–48].

Finally, another possible explanation for the divergence in results is the methodological difference between the studies. Unlike Dykstra's study, which imposed specific RPE levels, our study allowed participants to self-regulate their exertion based on subjective feedback at each step of the exercise. This individualized approach likely provides a more accurate reflection of how individuals naturally regulate exertion, leading to more sustained neural engagement. However, it is worth noting that the advantages of self-paced protocols may be less pronounced in populations that are not as aware of their physiological cues or in those with less exercise experience [10,48], which could explain variations across different studies.

The stabilization of beta activity in the SPV test in our study could be interpreted in line with the Central Governor Model, as proposed by Noakes [18]. This model suggests that the brain regulates effort to prevent overexertion and maintain homeostasis, potentially explaining why beta activity plateaus during self-paced exercise [49]. In contrast, the continuous rise in beta activity during the IET reflects the higher neural strain imposed by the lack of self-regulation, which may have contributed to earlier fatigue onset [12,21,25].

Another notable finding in our study was the behavior of the EEG/$\dot{V}O_2$ ratio, which remained stable throughout most of the exercise but increased slightly around 80% of the test duration in both the alpha and beta bands. This increase, which was more pronounced in the IET, indicates a tendency towards heightened neural engagement relative to physiological demand as exercise intensity peaked. This suggests that the brain was allocating more resources to manage the escalating strain. The more pronounced rise in the IET may indicate that this protocol required greater cognitive and motor effort, potentially contributing to earlier fatigue compared to the SPV test. In contrast, the stable EEG/$\dot{V}O_2$ ratio in the SPV test suggests better exertion management, thus maintaining neural efficiency and delaying central fatigue.

This pattern of neural engagement aligns with the concept of the Estimated Time Limit (ETL), which posits that individuals subconsciously regulate their effort based on an internal estimate of how long they can sustain a given intensity before reaching exhaustion [50,51]. As exercise progresses, the brain continually reassesses this internal time limit, adjusting neural and physical effort to prevent premature fatigue. The stabilization of the EEG/$\dot{V}O_2$ ratio around 80% of the test duration across both protocols suggests that participants may have been tapping into this internal pacing strategy as they approached their perceived time limit. This regulatory mechanism is thought to involve both physiological signals, such as cardiovascular and metabolic feedback, as well as psychological factors, including motivation and perceived exertion, as suggested by previous research [10,52]. Although we did not directly assess psychological factors in this study, the observed increase in neural engagement towards the end of the exercise could reflect the brain's efforts to marshal additional resources to extend performance as participants neared their ETL, thereby delaying the onset of central fatigue and maintaining overall effort.

These findings regarding the EEG/$\dot{V}O_2$ ratio contrast with those of Billat et al. (2024), who observed a significant decline in the EEG/$\dot{V}O_2$ ratio during the IET, suggesting reduced neural engagement [53]. The divergence in results could be attributed to several factors. First, while both studies employed an IET, Billat's study used 2 min stage durations, whereas our protocol used 3 min stages. Research suggests that shorter stages, like those used by Billat, may lead to quicker attainment of peak values but may not allow sufficient time for physiological stabilization, particularly for $\dot{V}O_2$max and other physiological parameters. In contrast, the longer stages in our protocol likely promoted more stable physiological responses, which could explain the variation in the EEG/$\dot{V}O_2$ ratio between the two studies [54,55]. Additionally, our IET protocol included alternating eyes open and eyes closed phases between increments, potentially introducing brief moments of neural recovery that were absent in Billat's continuous protocol. These differences in stage duration and protocol design may account for the distinct neural and physiological responses observed in our study. Furthermore, the strictly controlled incremental protocol employed by Billat might have imposed a higher cognitive load and led to more pronounced central fatigue earlier in the exercise [10].

Moreover, the populations studied could also contribute to the differences in results. Billat's study included active males who participated in a variety of sports, such as judo and water polo, and not exclusively endurance sports. The diversity in training backgrounds certainly would have influenced their physiological responses and how they engaged neural resources during exercise. In contrast, our study involved trained cyclists, which might explain the more stable neural engagement observed in our participants. Additionally, variations in fitness levels [56–58], age [59,60], and training backgrounds [61,62] between the two studies may have significantly affected both physiological responses and participants' perceived exertion, potentially explaining the discrepancies between our findings and those of Billat [53].

The second major objective of our study was to assess whether individuals could effectively regulate their effort based on perceived exertion, as indicated by the RPE scale. Our findings revealed that participants maintained similar cardiovascular parameters ($\dot{V}O_2$, Rf, HR, power output, etc.) across both the SPV and the IET tests. This reinforces the reliability of the RPE scale as a tool for guiding exercise intensity, even when the pacing strategy differs. Previous studies support these findings, showing that when using RPE to regulate exercise intensity, participants can achieve similar physiological responses compared to when power output is externally controlled [5,8,63].

It is important to consider the role of cadence in these findings. During self-paced cycling, individuals tend to favor higher cadence over applying more force, which may optimize energy efficiency and delay the onset of fatigue. Research by Marsh and Martin (1993) demonstrated that self-selected cadences are generally higher, which can reduce the overall force required per pedal stroke, allowing for sustained effort over longer periods [64]. In our study, we found that at equal power outputs and physiological parameters (e.g., $\dot{V}O_2$, HR), the cadence was higher in the SPV test compared to the IET test. This could be explained by the self-regulation aspect of the SPV, where participants may subconsciously choose a higher cadence to maintain a steady perception of effort. Higher cadence at the same power output reduces the force required per pedal stroke, potentially minimizing muscle fatigue and maintaining comfort throughout the test. Research by Takaishi et al. (1996, 1998) supports this notion, suggesting that higher cadences are preferred in self-paced settings as they help sustain effort by reducing muscular strain [65,66]. This preference for higher cadence in self-paced exercise aligns with the idea that individuals naturally regulate their effort to optimize both comfort and performance, thus contributing to the similar cardiovascular outcomes observed across different exercise protocols.

The findings of this study suggest several practical applications for exercise prescription and training. The demonstrated reliability of the RPE scale in guiding exercise intensity across both self-paced and fixed protocols highlights its utility as a versatile tool in various settings, including clinical and athletic populations. Importantly, the higher cadence observed during self-paced exercise—which likely contributes to the similar cardiovascular outcomes despite different pacing strategies—suggests that encouraging self-regulated exercise may help optimize both comfort and performance. This has significant implications for exercise adherence, as explicit recommendations for self-paced exercise have been shown to improve adherence to exercise programs, particularly among populations less accustomed to structured training [2,67,68].

Additionally, recent research by Palacin et al. (2024) underscores the importance of self-paced exercise in long-duration endurance events, such as marathons, where maintaining an appropriate pacing strategy can significantly influence performance outcomes [13]. Their study demonstrated that brain activity and RPE can be effectively monitored to optimize pacing and delay the onset of fatigue during a marathon. These findings suggest that similar principles could be applied in various exercise contexts to enhance both performance and adherence. By allowing individuals to tailor their effort based on perceived exertion, self-paced exercise can enhance feelings of autonomy and enjoyment, key factors in promoting long-term adherence to exercise programs and improving overall health outcomes. Exercise adherence plays a significant role in reducing the risk of cardiovascular disease [69,70], which remains the leading cause of death in both Europe and the United States [71].

Furthermore, the observed stability and slight increases in the EEG/$\dot{V}O_2$ ratio suggest that monitoring neural activity relative to physiological demand could be a valuable tool in personalized training programs. This approach could be particularly beneficial in settings that require fine-tuning exercise intensity to manage cognitive and motor demands effectively, such as in rehabilitation or high-performance sports. The differences in neural engagement between self-paced and fixed protocols also highlight the importance of tailoring exercise strategies to individual needs, thus potentially reducing the risk of central fatigue and enhancing overall performance.

Despite the valuable insights provided by this pilot study, several limitations must be considered.

First, the small sample size of six trained male cyclists limits the generalizability of the results. This may reduce the external validity, and future research should consider larger, more diverse samples to ensure that findings are applicable across broader populations. Additionally, because the participants in this study were trained, it is important to investigate whether the relationships between neural effort, exercise regulation, and performance observed here remain consistent in less-trained cohorts. These individuals may not be as accustomed to the sensations of self-pacing, exhaustion, and fatigue, making it crucial for future research to examine how self-paced exercise protocols perform across a wider range of fitness levels.

Second, the lack of randomization in the order of the two exercise tests introduces the potential for a learning or familiarization effect. However, this design was necessary, as the SPV test required the RPE values from the IET to determine appropriate intensity levels. Future studies could explore alternative methods to address this limitation.

Third, the protocol requiring participants to cease pedaling and to alternate between eyes open and eyes closed phases for 80 s between each step may have influenced both physiological and neural responses. Although this design was intended for a separate research focus, it could have impacted the results. Minimizing such interruptions in future studies could provide more continuous and reliable data.

Moreover, while the study focused on EEG measures, incorporating additional neuroimaging techniques, such as fMRI or near-infrared spectroscopy, could offer a more comprehensive understanding of the neural mechanisms involved in exercise regulation. Future studies should also consider longitudinal designs to examine how neural and physiological responses evolve over time with training adaptations.

5. Conclusions

This study provides important insights into neural and physiological responses to self-paced versus externally controlled exercise protocols. We found that during the SPV test, the beta PSD initially increased but then stabilized, as participants managed their effort effectively. This aligns with the Central Governor Model, which suggests that the brain modulates effort to prevent overexertion. In contrast, the IET test showed a continuous increase in beta activity, indicating greater neural strain and a potentially earlier onset of fatigue. These findings highlight the potential benefits of incorporating self-paced exercise into training programs, especially for improving exercise adherence and optimizing performance.

Additionally, the consistent cardiovascular parameters across both protocols reinforce the reliability of the RPE scale in guiding exercise intensity. The higher cadence observed during self-paced exercise suggests that participants intuitively adopt strategies to optimize energy efficiency and delay fatigue, supporting the utility of self-paced exercise in various settings. The observed stability and slight increases in the EEG/$\dot{V}O_2$ ratio also suggest that this metric could serve as a valuable marker for monitoring neural engagement and fatigue during exercise, potentially informing more personalized training protocols.

Future research should consider larger, more diverse samples and longitudinal designs to further explore the impact of self-paced exercise across different populations and over extended periods. Additionally, minimizing interruptions during exercise testing and integrating additional neuroimaging techniques could provide a more comprehensive understanding of the brain's role in exercise regulation.

Author Contributions: Conceptualization, L.P. and V.B.; methodology, L.P., F.P., I.S.H. and V.B.; software, F.P. and I.S.H.; formal analysis, L.P. and V.B.; data curation, L.P., F.P. and I.S.H.; writing—original draft preparation, L.P. and V.B.; writing—review and editing, L.P., F.P. and I.S.H.; supervision, V.B. All authors have read and agreed to the published version of the manuscript.

Funding: This research received no external funding.

Institutional Review Board Statement: The study was conducted in accordance with the Declaration of Helsinki and approved by the ethics committee of the university Hospital of Brugmann (Brussels, Belgium; reference: B0772022000014; date of approval: 8 February 2022).

Informed Consent Statement: Informed consent was obtained from all subjects involved in the study.

Data Availability Statement: The data presented in this study are available upon request from the corresponding author.

Conflicts of Interest: The authors declare no conflicts of interest.

References

1. Lee, H.H.; Emerson, J.A.; Williams, D.M. Age Moderates the Effect of Self-Paced Exercise on Exercise Adherence among Overweight Adults. *J. Aging Health* **2020**, *32*, 154–161. [CrossRef] [PubMed]
2. Williams, D.M.; Dunsiger, S.; Miranda, R., Jr.; Gwaltney, C.J.; Emerson, J.A.; Monti, P.M.; Parisi, A.F. Recommending self-paced exercise among overweight and obese adults: A randomized pilot study. *Ann. Behav. Med.* **2015**, *49*, 280–285. [CrossRef] [PubMed]
3. Williams, D.M. Exercise, affect, and adherence: An integrated model and a case for self-paced exercise. *J. Sport. Exerc. Psychol.* **2008**, *30*, 471–496. [CrossRef] [PubMed]
4. Hamlyn-Williams, C.C.; Williams, D.M. Comparing exercise adherence strategies: The efficacy of self-paced versus prescribed intensity exercise among inactive adults. *Behav. Med.* **2014**, *40*, 61–69.
5. Chidnok, W.; Dimenna, F.J.; Bailey, S.J.; Wilkerson, D.P.; Vanhatalo, A.; Jones, A.M. VO2max is not altered by self-pacing during incremental exercise. *Eur. J. Appl. Physiol.* **2013**, *113*, 529–539. [CrossRef]
6. Molinari, C.A.; Palacin, F.; Poinsard, L.; Billat, V.L. Determination of Submaximal and Maximal Training Zones From a 3-Stage, Variable-Duration, Perceptually Regulated Track Test. *Int. J. Sports Physiol. Perform.* **2020**, *15*, 853–861. [CrossRef]
7. Mauger, A.R.; Sculthorpe, N. A new VO2max protocol allowing self-pacing in maximal incremental exercise. *Br. J. Sports Med.* **2011**, *46*, 59–63. [CrossRef]
8. Faulkner, J.; Mauger, A.R.; Woolley, B.; Lambrick, D.M. The efficacy of self-paced VO2max testing in athletes. *Med. Sci. Sports Exerc.* **2015**, *47*, 982–987. [CrossRef]
9. Hogg, J.S.; Hopker, J.G.; Mauger, A.R. The self-paced VO2max test to assess maximal oxygen uptake in highly trained runners. *Int. J. Sports Physiol. Perform.* **2015**, *10*, 172–177. [CrossRef]
10. Tucker, R.; Noakes, T.D. The physiological regulation of pacing strategy during exercise: A critical review. *Br. J. Sports Med.* **2009**, *43*, e1. [CrossRef]
11. Borg, G.A. Psychophysical bases of perceived exertion. *Med. Sci. Sports Exerc.* **1982**, *14*, 377–381. [CrossRef] [PubMed]
12. Robertson, C.V.; Marino, F.E. A role for the prefrontal cortex in exercise tolerance and termination. *J. Appl. Physiol.* **2016**, *120*, 464–466. [CrossRef] [PubMed]
13. Palacin, F.; Poinsard, L.; Mattei, J.; Berthomier, C.; Billat, V. Brain, Metabolic, and RPE Responses during a Free-Pace Marathon: A Preliminary Study. *Int. J. Environ. Res. Public. Health* **2024**, *21*, 1024. [CrossRef] [PubMed]
14. Giovanelli, N.; Scaini, S.; Billat, V.; Lazzer, S. A new field test to estimate the aerobic and anaerobic thresholds and maximum parameters. *Eur. J. Sport. Sci.* **2020**, *20*, 437–443. [CrossRef]
15. Billat, V.; Poinsard, L.; Palacin, F.; Pycke, J.R.; Maron, M. Oxygen Uptake Measurements and Rate of Perceived Exertion during a Marathon. *Int. J. Env. Res. Public. Health* **2022**, *19*, 5760. [CrossRef]
16. Noakes, T.D.; St Clair Gibson, A. Logical limitations to the "catastrophe" models of fatigue during exercise in humans. *Br. J. Sports Med.* **2004**, *38*, 648–649. [CrossRef]
17. Amann, M. Central and peripheral fatigue: Interaction during cycling exercise in humans. *Med. Sci. Sports Exerc.* **2011**, *43*, 2039–2045. [CrossRef]
18. Noakes, T.D. Fatigue is a Brain-Derived Emotion that Regulates the Exercise Behavior to Ensure the Protection of Whole Body Homeostasis. *Front. Physiol.* **2012**, *3*, 82. [CrossRef]
19. Bixby, W.R.; Spalding, T.W.; Hatfield, B.D. Temporal Dynamics and Dimensional Specificity of the Affective Response to Exercise of Varying Intensity: Differing Pathways to a Common Outcome. *J. Sport. Exerc. Psychol.* **2001**, *23*, 171–190. [CrossRef]
20. Thompson, T.; Steffert, T.; Ros, T.; Leach, J.; Gruzelier, J. EEG applications for sport and performance. *Methods.* **2008**, *45*, 279–288. [CrossRef]
21. Brümmer, V.; Schneider, S.; Abel, T.; Vogt, T.; Strüder, H.K. Brain cortical activity is influenced by exercise mode and intensity. *Med. Sci. Sports Exerc.* **2011**, *43*, 1863–1872. [CrossRef] [PubMed]
22. Cheron, G.; Petit, G.; Cheron, J.; Leroy, A.; Cebolla, A.; Cevallos, C.; Petieau, M.; Hoellinger, T.; Zarka, D.; Clarinval, A.M.; et al. Brain Oscillations in Sport: Toward EEG Biomarkers of Performance. *Front. Psychol.* **2016**, *7*, 246. [CrossRef] [PubMed]
23. Enders, H.; Nigg, B.M. Measuring human locomotor control using EMG and EEG: Current knowledge, limitations and future considerations. *Eur. J. Sport. Sci.* **2016**, *16*, 416–426. [CrossRef] [PubMed]
24. Bigliassi, M.; Filho, E. Functional significance of the dorsolateral prefrontal cortex during exhaustive exercise. *Biol. Psychol.* **2022**, *175*, 108442. [CrossRef] [PubMed]

25. Bailey, S.P.; Hall, E.E.; Folger, S.E.; Miller, P.C. Changes in EEG during graded exercise on a recumbent cycle ergometer. *J. Sports Sci. Med.* **2008**, *7*, 505–511. [PubMed] [PubMed Central]
26. Edwards, A.M.; Deakin, G.B.; Guy, J.H. Brain and Cardiorespiratory Responses to Exercise in Hot and Thermoneutral Conditions. *Int. J. Sports Med.* **2016**, *37*, 779–784. [CrossRef]
27. Mechau, D.; Mücke, S.; Weiss, M.; Liesen, H. Effect of increasing running velocity on electroencephalogram in a field test. *Eur. J. Appl. Physiol. Occup. Physiol.* **1998**, *78*, 340–345. [CrossRef]
28. Maceri, R.M.; Cherup, N.P.; Hanson, N.J. EEG Responses to Incremental Self-Paced Cycling Exercise in Young and Middle Aged Adults. *Int. J. Exerc. Sci.* **2019**, *12*, 800–810. [CrossRef]
29. Brümmer, V.; Schneider, S.; Strüder, H.K.; Askew, C.D. Primary motor cortex activity is elevated with incremental exercise intensity. *Neuroscience* **2011**, *181*, 150–162. [CrossRef]
30. Robertson, C.V.; Marino, F.E. Prefrontal and motor cortex EEG responses and their relationship to ventilatory thresholds during exhaustive incremental exercise. *Eur. J. Appl. Physiol.* **2015**, *115*, 1939–1948. [CrossRef]
31. Dykstra, R.M.; Hanson, N.J.; Miller, M.G. Brain activity during self-paced vs. fixed protocols in graded exercise testing. *Exp. Brain Res.* **2019**, *237*, 3273–3279. [CrossRef] [PubMed]
32. Bassett, D.R.; Howley, E.T. Limiting factors for maximum oxygen uptake and determinants of endurance performance. *Med. Sci. Sports Exerc.* **2000**, *32*, 70–84. [CrossRef] [PubMed]
33. Utter, A.; Austin, M.; Nieman, D.; Dew, D.; Bowles, E.; Moody, A.; Cartner, B. Validation of Cosmed'S Quark Cpet and Mixing Chamber System: 782. *Med. Sci. Sports Exerc.* **2011**, *43*, 80. [CrossRef]
34. Nieman, D.C.; Austin, M.D.; Dew, D.; Utter, A.C. Validity of COSMED's Quark CPET Mixing Chamber System in Evaluating Energy Metabolism During Aerobic Exercise in Healthy Male Adults. *Res. Sports Med.* **2013**, *21*, 136–145. [CrossRef]
35. Merrigan, J.J.; Stovall, J.H.; Stone, J.D.; Stephenson, M.; Finomore, V.S.; Hagen, J.A. Validation of Garmin and Polar Devices for Continuous Heart Rate Monitoring During Common Training Movements in Tactical Populations. *Meas. Phys. Educ. Exerc. Sci.* **2022**, *27*, 234–247. [CrossRef]
36. Charloux, A.; Lonsdorfer-Wolf, E.; Richard, R.; Lampert, E.; Oswald-Mammosser, M.; Mettauer, B.; Geny, B.; Lonsdorfer, J. A new impedance cardiograph device for the non-invasive evaluation of cardiac output at rest and during exercise: Comparison with the "direct" Fick method. *Eur. J. Appl. Physiol.* **2000**, *82*, 313–320. [CrossRef]
37. Richard, R.; Lonsdorfer-Wolf, E.; Charloux, A.; Doutreleau, S.; Buchheit, M.; Oswald-Mammosser, M.; Lampert, E.; Mettauer, B.; Geny, B.; Lonsdorfer, J. Non-invasive cardiac output evaluation during a maximal progressive exercise test, using a new impedance cardiograph device. *Eur. J. Appl. Physiol.* **2001**, *85*, 202–207. [CrossRef]
38. Enders, H.; Cortese, F.; Maurer, C.; Baltich, J.; Protzner, A.B.; Nigg, B.M. Changes in cortical activity measured with EEG during a high-intensity cycling exercise. *J. Neurophysiol.* **2016**, *115*, 379–388. [CrossRef]
39. Ferree, T.C.; Luu, P.; Russell, G.S.; Tucker, D.M. Scalp electrode impedance, infection risk, and EEG data quality. *Clin. Neurophysiol.* **2001**, *112*, 536–544. [CrossRef]
40. Delorme, A.; Makeig, S. EEGLAB: An open source toolbox for analysis of single-trial EEG dynamics including independent component analysis. *J. Neurosci. Methods.* **2004**, *134*, 9–21. [CrossRef]
41. Makeig, S.; Bell, A.; Jung, T.P.; Sejnowski, T. Independent Component Analysis of Electroencephalographic Data. *Adv. Neural Inf. Process. Syst.* **1996**, *8*, 145–151.
42. Nielsen, B.; Hyldig, T.; Bidstrup, F.; González-Alonso, J.; Christoffersen, G.R. Brain activity and fatigue during prolonged exercise in the heat. *Pflug. Arch.* **2001**, *442*, 41–48. [CrossRef] [PubMed]
43. Miller, E.K.; Cohen, J.D. An integrative theory of prefrontal cortex function. *Annu. Rev. Neurosci.* **2001**, *24*, 167–202. [CrossRef] [PubMed]
44. Aron, A.R.; Robbins, T.W.; Poldrack, R.A. Inhibition and the right inferior frontal cortex. *Trends Cogn. Sci.* **2004**, *8*, 170–177. [CrossRef] [PubMed]
45. Nielsen, J.B. How we walk: Central control of muscle activity during human walking. *Neuroscientist* **2003**, *9*, 195–204. [CrossRef]
46. Grafton, S.T. The cognitive neuroscience of prehension: Recent developments. *Exp. Brain Res.* **2010**, *204*, 475–491. [CrossRef]
47. Hikosaka, O.; Nakamura, K.; Sakai, K.; Nakahara, H. Central mechanisms of motor skill learning. *Curr. Opin. Neurobiol.* **2002**, *12*, 217–222. [CrossRef]
48. Ruddy, K.L.; Carson, R.G. Neural pathways mediating cross education of motor function. *Front. Hum. Neurosci.* **2013**, *7*, 397. [CrossRef]
49. Lambert, E.V.; St Clair Gibson, A.; Noakes, T.D. Complex systems model of fatigue: Integrative homoeostatic control of peripheral physiological systems during exercise in humans. *Br. J. Sports Med.* **2005**, *39*, 52–62. [CrossRef]
50. Garcin, M.; Vandewalle, H.; Monod, H. A new rating scale of perceived exertion based on subjective estimation of exhaustion time: A preliminary study. *Int. J. Sports Med.* **1999**, *20*, 40–43. [CrossRef]
51. Garcin, M.; Billat, V. Perceived exertion scales attest to both intensity and exercise duration. *Percept. Mot. Skills.* **2001**, *93*, 661–671. [CrossRef] [PubMed]
52. St Clair Gibson, A.; Noakes, T.D. Evidence for complex system integration and dynamic neural regulation of skeletal muscle recruitment during exercise in humans. *Br. J. Sports Med.* **2004**, *38*, 797–806. [CrossRef] [PubMed]

53. Billat, V.; Berthomier, C.; Clémençon, M.; Brandewinder, M.; Essid, S.; Damon, C.; Rigaud, F.; Bénichoux, A.; Maby, E.; Fornoni, L.; et al. Electroencephalography Response during an Incremental Test According to the $\dot{V}O_2$max Plateau Incidence. *Appl. Sci.* **2024**, *14*, 5411. [CrossRef]
54. Bentley, D.J.; Newell, J.; Bishop, D. Incremental exercise test design and analysis: Implications for performance diagnostics in endurance athletes. *Sports Med.* **2007**, *37*, 575–586. [CrossRef] [PubMed]
55. Midgley, A.W.; Carroll, S.; Marchant, D.; McNaughton, L.R.; Siegler, J. Evaluation of true maximal oxygen uptake based on a novel set of standardized criteria. *Appl. Physiol. Nutr. Metab.* **2009**, *34*, 115–123. [CrossRef]
56. Hillman, C.H.; Erickson, K.I.; Kramer, A.F. Be smart, exercise your heart: Exercise effects on brain and cognition. *Nat. Rev. Neurosci.* **2008**, *9*, 58–65. [CrossRef]
57. Voelcker-Rehage, C.; Niemann, C. Structural and functional brain changes related to different types of physical activity across the life span. *Neurosci. Biobehav. Rev.* **2013**, *37*, 2268–2295. [CrossRef]
58. Prakash, R.S.; Voss, M.W.; Erickson, K.I.; Kramer, A.F. Physical activity and cognitive vitality. *Annu. Rev. Psychol.* **2015**, *66*, 769–797. [CrossRef]
59. Colcombe, S.; Kramer, A.F. Fitness effects on the cognitive function of older adults: A meta-analytic study. *Psychol. Sci.* **2003**, *14*, 125–130. [CrossRef]
60. Anderson, A.J.; Perone, S. Developmental change in the resting state electroencephalogram: Insights into cognition and the brain. *Brain Cogn.* **2018**, *126*, 40–52. [CrossRef]
61. Swain, D.P.; Franklin, B.A. Comparison of cardioprotective benefits of vigorous versus moderate intensity aerobic exercise. *Am. J. Cardiol.* **2006**, *97*, 141–147. [CrossRef] [PubMed]
62. Neubauer, A.C.; Fink, A. Intelligence and neural efficiency. *Neurosci. Biobehav. Rev.* **2009**, *33*, 1004–1023. [CrossRef] [PubMed]
63. Astorino, T.A.; McMillan, D.W.; Edmunds, R.M.; Sanchez, E. Increased cardiac output elicits higher VO2max in response to self-paced exercise. *Appl. Physiol. Nutr. Metab.* **2015**, *40*, 223–229. [CrossRef] [PubMed]
64. Marsh, A.P.; Martin, P.E. The association between cycling experience and preferred and most economical cadences. *Med. Sci. Sports Exerc.* **1993**, *25*, 1269–1274. [CrossRef] [PubMed]
65. Takaishi, T.; Yasuda, Y.; Ono, T.; Moritani, T. Optimal pedaling rate estimated from neuromuscular fatigue for cyclists. *Med. Sci. Sports Exerc.* **1996**, *28*, 1492–1497. [CrossRef]
66. Takaishi, T.; Yamamoto, T.; Ono, T.; Ito, T.; Moritani, T. Neuromuscular, metabolic, and kinetic adaptations for skilled pedaling performance in cyclists. *Med. Sci. Sports Exerc.* **1998**, *30*, 442–449. [CrossRef]
67. Rose, E.A.; Parfitt, G. A quantitative analysis and qualitative explanation of the individual differences in affective responses to prescribed and self-selected exercise intensities. *J. Sport. Exerc. Psychol.* **2007**, *29*, 281–309. [CrossRef]
68. Ekkekakis, P.; Hall, E.E.; Petruzzello, S.J. The relationship between exercise intensity and affective responses demystified: To crack the 40-year-old nut, replace the 40-year-old nutcracker! *Ann. Behav. Med.* **2008**, *35*, 136–149. [CrossRef]
69. Warburton, D.E.; Nicol, C.W.; Bredin, S.S. Health benefits of physical activity: The evidence. *Can. Med. Assoc. J.* **2006**, *174*, 801–809. [CrossRef]
70. Lee, I.M.; Shiroma, E.J.; Lobelo, F.; Puska, P.; Blair, S.N.; Katzmarzyk, P.T. Effect of physical inactivity on major non-communicable diseases worldwide: An analysis of burden of disease and life expectancy. *Lancet* **2012**, *380*, 219–229. [CrossRef]
71. The Top 10 Causes of Death. Available online: https://www.who.int/news-room/fact-sheets/detail/the-top-10-causes-of-death (accessed on 17 October 2024).

applied sciences

MDPI

Perspective

Is Precision Therapy in Infantile-Onset Epileptic Encephalopathies Still Too Far to Call Upon?

Raffaele Falsaperla [1], Vincenzo Sortino [2] and Piero Pavone [3,*]

1 Department of Medical Science-Pediatrics, University of Ferrara, 44124 Ferrara, Italy; raffaele.falsaperla@unife.it
2 Unit of Clinical Pediatrics, AOU "Policlinico-San Marco", PO "San Marco", University of Catania, 95123 Catania, Italy; sortino.vinci@gmail.com
3 Unit of Clinical Pediatrics, AOU "Policlinico", PO "G. Rodolico", University of Catania, 95123 Catania, Italy
* Correspondence: ppavone@unict.it; Tel.: +39-33-89487072

Featured Application: This research highlights the application of precision medicine in treating epileptic and developmental encephalopathies (EDEs) through early genetic diagnostics and tailored therapies. By enabling gene-specific treatments, this approach offers the potential to improve outcomes and prevent neurodevelopmental impairments in affected patients.

Abstract: Epileptic and developmental encephalopathies (EDEs) are a group of severe, genetically various neurological conditions characterized by early-onset seizures and developmental impairments. Recent advances in molecular genetics and diagnostic tools have led to the development of precision therapies, aiming to address the deep causes of these disorders. Examples, such as pyridoxine for pyridoxine-dependent epilepsy and the ketogenic diet for GLUT1 deficiency syndrome illustrate the potential of presumed tailored treatments. However, challenges persist, as current therapies often fail to fully mitigate neurodevelopmental impairments. Moreover, traditional phenotype-based management strategies, while effective for seizure control, do not address the root causes of these disorders, underscoring the limitations of existing approaches. This article explores the evolving landscape of precision medicine in EDEs, emphasizing the importance of genetic insights in therapy design and the need for a multidisciplinary approach. It also highlights the barriers to widespread implementation, including diagnostic delays, accessibility, and a lack of robust clinical evidence. To fully realize the potential of precision therapies, comprehensive genetic integration, innovation in treatment, and global collaboration are essential. The future of EDE management lies in therapies that not only control symptoms but also correct genetic and molecular defects, offering a more effective, individualized approach to care.

Keywords: epilepsy; encephalopathies; precision; therapy

1. Introduction

Epileptic and developmental encephalopathies (EDEs) are severe neurological disorders marked by early-onset seizures, developmental delays, and cognitive and motor impairments [1]. These disorders, predominantly manifesting in infancy or early childhood, involve multisystemic complications including psychiatric, behavioral, sleep, gastrointestinal, musculoskeletal, and movement disorders [2]. Despite phenotypic similarities, EDEs are genetically heterogeneous, encompassing syndromes such as Early Myoclonic Encephalopathy, Ohtahara Syndrome, West Syndrome, and Dravet Syndrome [3]. Recent advancements in molecular genetics, neurodiagnostic technologies including functional study such as molecular dynamics and patch clamps have refined our understanding of these disorders.

Precision therapies have emerged as targeted approaches to address the root causes of specific EDEs, marking a pivotal shift from conventional treatments. For example, pyridoxine (Vitamin B6) is an effective treatment for pyridoxine-dependent epilepsy, a rare autosomal recessive disorder caused by mutations in the ALDH7A1 gene. Early administration of pyridoxine not only prevents seizures but also mitigates long-term neurodevelopmental damage [4]. Similarly, the ketogenic diet, a high-fat, low-carbohydrate dietary regimen, serves as a precision therapy for GLUT1 Deficiency Syndrome (GLUT1-DS) by providing ketones as an alternative energy source for the brain [5]. This diet has demonstrated remarkable efficacy in reducing seizure frequency and improving cognitive function, offering a tailored solution that directly targets the metabolic dysfunction underlying the condition.

However, it is important to note that not all therapies classified as "precision" share the same degree of specificity or mechanistic insight [6]. While the ketogenic diet directly addresses the metabolic needs of GLUT1-DS patients by bypassing the impaired glucose transport system, it remains a broader intervention compared to gene-specific therapies, such as antisense oligonucleotides or CRISPR-based approaches targeting specific mutations [7]. Similarly, pyridoxine therapy, while life-changing for individuals with pyridoxine-dependent epilepsy, relies on metabolic supplementation rather than directly correcting the genetic mutation itself [8]. These examples illustrate that precision therapies can vary significantly in their level of refinement, ranging from metabolic interventions to advanced gene editing technologies.

The aim of this perspective articles lies in its comprehensive exploration of the current state of precision therapy for EDEs, bridging the gap between emerging research and clinical application. By highlighting the potential and challenges of precision medicine, this work serves as a valuable resource for clinicians, researchers, and policymakers aiming to improve outcomes for individuals with these complex disorders. Moreover, it underscores the need for multidisciplinary collaboration and innovation to expand the scope of precision therapies, ensuring that advances in molecular and genetic science translate into tangible benefits for patients. This article not only emphasizes the transformative potential of precision medicine but also advocates for its integration into routine clinical practice, paving the way for a more tailored and effective approach to managing EDEs.

2. Advances in Diagnostic Approaches

The diagnostic landscape of EDEs has transformed with tools such as next-generation sequencing (NGS), whole-exome sequencing (WES), and whole-genome sequencing (WGS). These technologies have identified over 900 monogenic causes of EDEs, including mutations in SCN1A, KCNQ2, and CDKL5 [9]. Complementary functional assays and neuroimaging techniques, such as MRI spectroscopy and functional connectivity analyses, enable detailed correlations between genetic and brain abnormalities [10]. Integration of biomarkers, such as CSF neurofilament light chain levels and EEG patterns, further enhances diagnostic precision [10].

Advanced functional methods have further revolutionized the understanding of genetic mutations in EDEs. For instance, the patch-clamp technique is a powerful electrophysiological method that allows for the analysis of ion channel mutations by assessing their functional properties. This method has been particularly impactful in characterizing gain-of-function or loss-of-function mutations in genes such as SCN1A and SCN2A, which encode sodium channels critical for neuronal excitability [11]. By elucidating the precise effects of mutations on channel behavior, patch-clamp studies provide essential data for tailoring specific therapeutic strategies.

Additionally, molecular dynamics simulations have emerged as a crucial tool for investigating the interaction between mutated proteins and potential therapeutic compounds. This computational approach models the structural and dynamic behavior of proteins at an atomic level, enabling researchers to predict how specific drugs bind to and modulate the activity of mutated targets. For example, in GNAO encephalopathy, a rare condition caused by pathogenic variants in the GNAO1 gene, molecular dynamics has been employed to analyze how mutations in the Gαo protein affect its interaction with guanosine triphosphate (GTP) and downstream signaling pathways [12]. In this article Falsaperla et al. [12] showed that the molecular dynamics approach was pivotal in identifying tetrabenazine as a candidate drug capable of modulating the dysfunctional Gαo protein and mitigating movement disorders and seizures associated with the condition. This discovery highlights the potential of combining computational insights with clinical pharmacology to develop targeted treatments for rare genetic disorders.

Despite these advancements, conventional antiepileptic drug (AED) treatments remain seizure-focused and disregard specific genetic pathophysiologies. This limitation often leads to delayed interventions, continued developmental regression, and suboptimal patient outcomes. The integration of functional assays, such as patch-clamp and molecular dynamics, into diagnostic workflows holds the potential to bridge this gap, facilitating more accurate diagnoses and personalized treatments.

3. Limitations of Phenotype-Based Therapy

Current management strategies for EDEs rely heavily on phenotype-based antiepileptic drug (AED) selection. This "one-size-fits-all" approach is inadequate for genetically diverse conditions [13]. For instance, sodium channel blockers exacerbate seizures in SCN1A-related Dravet Syndrome but may be beneficial in SCN8A gain-of-function mutations. Similarly, targeted interventions such as mTOR inhibitors for TSC1/TSC2 mutations or antisense oligonucleotides (ASOs) for CDKL5 [14,15] deficiency illustrate the potential of etiology-driven therapies. However, trial-and-error methodologies often delay effective treatments, especially in conditions requiring early neurodevelopmental interventions. While precision therapies aim to overcome these limitations, they are not without their challenges. For example, pyridoxine (Vitamin B6) is considered a precision therapy for pyridoxine-dependent epilepsy (PDE), a condition caused by mutations in the ALDH7A1 gene [16]. Despite the early and timely administration of pyridoxine, nearly 75% of patients with PDE still experience intellectual disability, highlighting that metabolic supplementation alone cannot fully address the complex neurodevelopmental impairments associated with the disorder [17]. Similarly, in GLUT1 Deficiency Syndrome (GLUT1-DS), the ketogenic diet is a cornerstone of precision therapy, providing ketones as an alternative energy source for the brain [18]. However, even with strict adherence to this dietary regimen, many patients continue to experience residual neurological deficits, including learning difficulties and movement disorders [19]. These cases underscore the limitations of precision therapies in fully mitigating the effects of underlying genetic abnormalities. This reliance on phenotype-based therapy reflects a broader issue in epilepsy management: the lack of widespread integration of genetic and molecular insights into routine clinical practice [20]. Phenotypic treatments, while sometimes effective in controlling seizures, fail to address the underlying pathophysiological mechanisms of the disorder. This can lead to suboptimal outcomes, including continued developmental regression, behavioral disturbances, and worsening quality of life.

Furthermore, the absence of robust biomarkers to predict therapeutic response complicates the timely identification of effective treatments. In many cases, patients endure multiple failed therapies before achieving seizure control. Ultimately, therapies based solely on clinical and neurological signs, such as seizure frequency, remain highly limited because they do not target the root cause of the disorder. Instead, they primarily focus on reducing seizure frequency, leaving the underlying genetic or molecular defects unaddressed. Reflecting this limitation, even the terminology in epilepsy treatment is evolving. Drugs previously referred to as antiepileptic drugs (AEDs) are increasingly being described as anti-seizure medications (ASMs) to emphasize their primary role in seizure reduction rather than addressing the broader epileptic pathology [21]. This shift underscores the urgent need for treatments that correct the fundamental pathophysiological mechanisms, enabling not only seizure control but also the prevention of long-term developmental and cognitive impairments.

4. Precision Therapy: A Paradigm Shift

Precision therapy aims to tailor medical interventions to the unique genetic and molecular characteristics of epileptic and developmental encephalopathies (EDEs) [22]. This strategy offers a transformative approach, moving beyond conventional seizure control to directly address the underlying causes of these disorders. The implementation of precision therapy leverages a variety of innovative mechanisms and approaches, including:

1. Gene-Specific Targeting: Approaches such as sodium channel modulators for SCN2A mutations, antisense oligonucleotides (ASOs) for Angelman Syndrome, and CRISPR-based gene editing for mutation correction exemplify how genetic insights can directly inform therapeutic strategies [23].
2. Pathway-Specific Interventions: Examples include mTOR inhibitors for tuberous sclerosis complex (TSC) [24], AMPA receptor modulators for GRIN2A-related encephalopathies [25]. These interventions target disrupted molecular pathways to restore cellular function. Additionally, pyridoxine (Vitamin B6) is a cornerstone therapy for pyridoxine-dependent epilepsy (PDE), effectively reducing seizures by addressing the metabolic defect caused by ALDH7A1 mutations. Similarly, the ketogenic diet, a high-fat, low-carbohydrate regimen, is a precision therapy for GLUT1 Deficiency Syndrome (GLUT1-DS), providing an alternative energy source (ketones) for the brain and mitigating seizures caused by impaired glucose transport.
3. Early Diagnosis and Intervention: Early identification of genetic mutations through newborn screening programs facilitates the timely initiation of treatments during critical neurodevelopmental periods, potentially mitigating irreversible damage. For instance, recognizing ALDH7A1 mutations in PDE [26] or SLC2A1 mutations in GLUT1-DS allows for the immediate application of therapies like pyridoxine or the ketogenic diet, which are most effective when initiated early.
4. Biomarker-Driven Outcomes: Biomarkers, such as CSF neurofilament light chain levels and EEG patterns, enable real-time monitoring of disease progression and treatment efficacy, providing valuable feedback for optimizing interventions [27].
5. Multidisciplinary Care: Precision therapy necessitates collaboration among geneticists, neurologists, metabolic specialists, and dieticians to develop and implement comprehensive care plans that address both the genetic and metabolic needs of patients [28].

Although these advancements represent a paradigm shift, significant challenges remain in implementing precision therapy on a broader scale. The cost of gene-specific treatments, limited availability of advanced diagnostic tools, and the need for robust clinical evidence to validate emerging therapies hinder their widespread adoption. Moreover, even precision therapies like pyridoxine for PDE or the ketogenic diet for GLUT1-DS, while impactful, do not fully mitigate the neurological and developmental challenges posed by these conditions, highlighting the need for ongoing innovation and refinement.

5. Barriers to Implementation

Despite the transformative potential of precision therapy to revolutionize the treatment landscape for epileptic and developmental encephalopathies (EDEs), several critical barriers impede its widespread implementation. Addressing these obstacles is essential to unlock the full promise of precision medicine and ensure equitable access to these life-changing interventions:

1. Diagnostic Delays: In pediatric patients, the lack of access to advanced genetic testing can prolong the diagnostic odyssey, delaying effective interventions during critical neurodevelopmental periods. For example, in a child presenting with early-onset seizures, timely identification of a pathogenic SCN1A mutation could prevent inappropriate treatment with sodium channel blockers, which may exacerbate symptoms. Similarly, delays in identifying ALDH7A1 mutations in Pyridoxine-Dependent Epilepsy (PDE) or SLC2A1 mutations in GLUT1-Deficiency Syndrome (GLUT1-DS) can postpone the initiation of targeted therapies such as pyridoxine supplementation or the ketogenic diet, respectively. These delays not only exacerbate seizure activity but also contribute to long-term neurodevelopmental impairments [29].
2. Therapeutic Accessibility: The financial burden and regulatory complexities surrounding novel therapies pose significant barriers. For instance, treatments like antisense oligonucleotides (ASOs) for CDKL5 deficiency or gene-specific therapies for rare epilepsies often remain prohibitively expensive, limiting availability to families with sufficient financial resources [30]. Even less costly interventions, such as the ketogenic diet for GLUT1-DS, can be challenging to implement due to the need for specialized dietary planning and ongoing monitoring. Similarly, while pyridoxine therapy for PDE is more accessible, its effectiveness is limited by the availability of specialists who can diagnose and monitor these patients effectively. The complexity of personalized treatments often requires advanced technologies, extensive testing, and ongoing monitoring, all of which contribute to elevated costs. Additionally, regulatory hurdles can slow the introduction innovative therapies to clinical practice. To address these challenges, healthcare systems could explore innovative solutions such as public–private partnerships to share costs, adopt more flexible regulatory pathways for faster approval, and implement reimbursement models that prioritize value over volume. International collaboration between researchers, rulers, and the private sector will be essential to overcoming these barriers and making precision therapies more accessible.
3. Clinical Evidence: The scarcity of randomized controlled trials (RCTs) for rare genetic epilepsies limits evidence-based treatment options [31]. Diseases like West Syndrome, Dravet Syndrome, PDE, and GLUT1-DS often rely on small-scale studies or case reports, which hinder the development of standardized therapeutic guidelines. For instance, while the ketogenic diet is widely recognized as a cornerstone therapy for GLUT1-DS, its long-term impact on cognitive outcomes and quality of life remains incompletely understood. Similarly, although pyridoxine supplementation for PDE is effective in reducing seizures, its role in preventing neurodevelopmental delay requires further investigation.
4. Healthcare Education: Knowledge gaps in genomic medicine impede clinical integration. A pediatrician encountering a child with developmental delays and refractory seizures may lack sufficient training to interpret genetic testing results or recognize gene-specific treatment opportunities [32]. For example, clinicians unfamiliar with

the metabolic basis of GLUT1-DS might not consider the ketogenic diet as a therapeutic option, or they might fail to promptly initiate pyridoxine therapy in suspected PDE cases. Comprehensive training programs in genomic medicine, integrated into pediatric neurology curricula, are essential to bridge these gaps. These programs should include case-based learning and real-world applications to equip clinicians with the tools to personalize care effectively. The following table, Table 1, provides a concise summary of key aspects discussed in this paper, highlighting the tools, interventions, and strategies integral to advancing precision therapy in pediatric epileptic and developmental encephalopathies (EDEs).

Table 1. Precision Therapy for EDEs.

Category	Examples	Impact
Diagnostic Tools	NGS, WES, WGS, MRI spectroscopy	Enhanced identification of genetic mutations and structural abnormalities
Gene-Specific Targeting	Sodium channel modulators (SCN2A), ASOs (Angelman), CRISPR	Addressing underlying genetic causes
Pathway-Specific Interventions	mTOR inhibitors (TSC), AMPA modulators (GRIN2A), neuroprotective agents (POLG)	Improved seizure control and neuroprotection
Metabolic Interventions	Pyridoxine (Vitamin B6) for PDE, Ketogenic diet for GLUT1-DS	Reducing seizures and addressing metabolic dysfunctions
Biomarkers	CSF neurofilament light chain, EEG patterns	Real-time monitoring of disease progression and treatment efficacy
Therapeutic Challenges	Limited access, high costs, lack of RCTs	Hinder widespread adoption and equitable implementation
Future Strategies	Genetic testing standardization, global collaboration, education	Accelerating transition to precision medicine

NGS: next-generation sequencing; WES: whole-exome sequencing; WGS: whole-genome sequencing; MRI: Magnetic Resonance Imaging; SCN2A: sodium voltage-gated channel alpha subunit 2; ASOs: antisense oligonucleotides; CRISPR: clustered regularly interspaced short palindromic repeats; mTOR: mechanistic target of rapamycin; TSC: tuberous sclerosis complex; AMPA: α-Ammino-3-idrossi-5-Metil-4-isossazol-Propionic Acid; GRIN2A: glutamate ionotropic receptor NMDA type subunit 2A; POLG: DNA polymerase gamma; PDE: Pyridoxine-Dependent Epilepsy; GLUT1-DS: Glucose transporter type 1 deficiency syndrome; CSF: Cerebrospinal fluid; EEG: electroencephalogram; RCTs: Randomised controlled trials.

6. The Path Forward

To fully harness the transformative potential of precision therapy, a multifaceted approach is necessary. Addressing current barriers requires a comprehensive framework that integrates genetic insights into clinical workflows, drives innovation in treatment modalities, and fosters collaboration across global networks. These strategies must be complemented by robust educational initiatives and patient-centered approaches to ensure meaningful and equitable implementation. The key elements include:

1. Integration of Genetic Testing: Establishing genetic testing as a diagnostic standard is particularly critical in pediatric care. For example, identifying a pathogenic KCNQ2 mutation in an infant presenting with neonatal seizures can facilitate targeted treatment strategies, such as the use of sodium channel modulators, significantly improving developmental outcomes and reducing seizure burden [33]. Early genetic insights also help avoid ineffective or harmful therapies, enabling a precision-guided approach from the outset of care.
2. Studies on the impact of commonly used anti-epileptic drugs on gene expression: There are a few studies in the Literature on the impact of commonly used anti-epileptic drugs on the expression of genes linked to epiepileptogenesis. For example, in human embryonic stem cell based targeted to neural differentiation, valproic acid and

carbamazepine exposure during differentiation determinate concentration-dependent reduced expression of βIII-tubulin, Neurogin1 and Reelin. Valproate caused an increased gene expression of Map2 and Mapt which is possibly related to the neural protective effect [34]. Of interest, Levetiracetam, one of the most widely used and safest anti-epileptic drugs, seems modular, in mice model, epileptogenesis by acting on the adenosine pathway with an increasing gene expression of A1Rs and Kir3.2 in the brain and a reduction in the gene expression of ENT1 central nervous system [35]. Over the past two decades, drug repositioning strategies have become increasingly significant due to their lower failure rates and reduced economic costs. Drugs with comparable side effect profiles may act through a shared mechanism, allowing them to be applied to the treatment of other diseases [36]. This knowledge is particularly important for countries with limited resources where 'old' drugs can be used for 'new' therapeutic targets, representing a variety of precision medicine.

3. Therapeutic Innovation: Expanding pipelines for gene-specific therapies is particularly impactful in pediatric care. For example, in the context of epileptic disorders, therapies targeting SCN1A mutations in Dravet Syndrome have shown transformative potential [37]. Precision treatments such as fenfluramine have demonstrated efficacy in significantly reducing seizure frequency and improving quality of life in affected children [38]. These innovations underscore the critical role of tailoring interventions to the genetic basis of pediatric epileptic conditions, fostering both better seizure management and enhanced developmental outcomes.
4. Global Collaboration: Enhancing data sharing and standardizing guidelines is vital in addressing rare pediatric epileptic disorders. For example, global initiatives like the International League Against Epilepsy's (ILAE) collaborative networks enable clinicians and researchers worldwide to pool genetic data, share clinical insights, and standardize treatment protocols [39]. This collective approach ensures that even children in resource-limited settings can benefit from the latest advancements in precision therapies, improving diagnostic and therapeutic outcomes across diverse healthcare systems.
5. Education and Advocacy: Empowering healthcare providers and patients through education is critical for the effective implementation of precision medicine [40]. For instance, in pediatric epilepsy, equipping clinicians with genomic knowledge allows them to recognize actionable mutations such as SCN1A in Dravet Syndrome, enabling timely and appropriate treatment adjustments. Additionally, educating families about genetic findings fosters informed decision-making and engagement in innovative therapies, ensuring the child receives optimal care.
6. Patient and Family Empowerment: Involving families in therapeutic decision-making and research is particularly impactful in pediatric epilepsy [41]. For instance, educating parents about the potential benefits of genetic testing can lead to earlier identification of mutations such as SCN1A, allowing them to advocate for targeted therapies like fenfluramine for Dravet Syndrome. Additionally, involving families in clinical trials fosters a deeper understanding of available treatments and enhances engagement in the child's care plan, ultimately improving outcomes and quality of life.
7. Strategies to increase affordability and improve equitable access: To increase the affordability and improve equitable access to precision medicine and therapy, global collaborations could play a pivotal role. By fostering partnerships between governments, healthcare providers, pharmaceutical companies, and research institutions, the cost burden of developing and implementing precision therapies can be shared. Such collaborations could enable pooled resources for research and development,

making therapies more affordable and widely accessible, particularly in low- and middle-income countries. Through shared knowledge, economies of scale, and harmonized policies, these global efforts could address disparities in access, ensuring that the benefits of precision medicine reach diverse populations around the world.

7. Conclusions

Precision medicine heralds a transformative era in managing pediatric epileptic and developmental encephalopathies (EDEs) by transcending phenotype-based approaches and addressing the root causes of these disorders. By integrating genetic and molecular insights into diagnostic and therapeutic workflows, precision therapies have the potential to significantly improve clinical outcomes and quality of life for affected individuals. Despite its promise, the path to widespread implementation of precision therapy remains fraught with challenges. Diagnostic delays, therapeutic accessibility, limited clinical evidence, and knowledge gaps in genomic medicine continue to hinder progress. Furthermore, even established precision therapies, such as pyridoxine (Vitamin B6) for Pyridoxine-Dependent Epilepsy (PDE) and the ketogenic diet for GLUT1-Deficiency Syndrome (GLUT1-DS), highlight the limitations of current approaches [42], as they fail to completely mitigate the neurodevelopmental and neurological consequences of these conditions. Addressing these challenges requires a comprehensive, collaborative approach that includes standardizing genetic testing, expanding therapeutic innovation, fostering global collaboration, and equipping healthcare providers with genomic expertise. Equally important is the need to involve patients and families in every step of the diagnostic and therapeutic journey, empowering them to make informed decisions and participate actively in research and care. As the field of precision medicine continues to evolve, the focus must shift toward developing therapies that not only control symptoms, such as seizures, but also correct the underlying genetic or molecular defects. Achieving this vision will necessitate robust investment in research, policy reform, and international cooperation to ensure equitable access to these life-changing interventions. By embracing this paradigm shift, precision medicine can unlock new opportunities for individualized care, paving the way for a future where no patient is left behind.

Author Contributions: Conceptualization, methodology, writing—original draft preparation, R.F. and P.P.; writing—review and editing, V.S.; supervision, R.F. and P.P. All authors have read and agreed to the published version of the manuscript.

Funding: This research received no external funding.

Institutional Review Board Statement: Not applicable.

Informed Consent Statement: Not applicable.

Data Availability Statement: No new data were created or analyzed in this study.

Conflicts of Interest: The authors declare no conflicts of interest.

References

1. Scheffer, I.E.; Zuberi, S.; Mefford, H.C.; Guerrini, R.; McTague, A. Developmental and epileptic encephalopathies. *Nat. Rev. Dis. Primers* **2024**, *10*, 61. https://doi.org/10.1038/s41572-024-00546-6; Erratum in *Nat. Rev. Dis. Primers* **2024**, *10*, 66. [CrossRef] [PubMed]
2. Khan, S.; Al Baradie, R. Epileptic encephalopathies: An overview. *Epilepsy Res. Treat.* **2012**, *2012*, 403592. [CrossRef] [PubMed] [PubMed Central]
3. Pavone, P.; Polizzi, A.; Marino, S.D.; Corsello, G.; Falsaperla, R.; Marino, S.; Ruggieri, M. West syndrome: A comprehensive review. *Neurol. Sci.* **2020**, *41*, 3547–3562. [CrossRef] [PubMed] [PubMed Central]

4. Yeghiazaryan, N.S.; Zara, F.; Capovilla, G.; Brigati, G.; Falsaperla, R.; Striano, P. Pyridoxine-dependent epilepsy: An under-recognised cause of intractable seizures. *J. Paediatr. Child Health* **2012**, *48*, E113–E115. [CrossRef] [PubMed]
5. Ramm-Pettersen, A.; Stabell, K.E.; Nakken, K.O.; Selmer, K.K. Does ketogenic diet improve cognitive function in patients with GLUT1-DS? A 6- to 17-month follow-up study. *Epilepsy Behav.* **2014**, *39*, 111–115. [CrossRef] [PubMed]
6. He, S.; Zheng, L.; Li, J.; Liu, S. Epilepsy Treatment and Diagnosis Enhanced by Current Nanomaterial Innovations: A Comprehensive Review. *Mol. Neurobiol.* **2025**, *62*, 946–961. [CrossRef] [PubMed]
7. Mullard, A. Antisense and CRISPR-based drugs build cases for better hereditary angioedema treatments. *Nat. Rev. Drug Discov.* **2024**, *23*, 488. [CrossRef] [PubMed]
8. Falsaperla, R.; Sciacca, M.; Collotta, A.D.; Tardino, L.G.; Marino, S.; Marino, S.D.; Privitera, G.F.; Vitaliti, G.; Ruggieri, M. PYRIDOXINE-dependent epilepsy (PDE): An observational study of neonatal cases on the role of pyridoxine in patients treated with standard anti-seizure medications. *Seizure* **2024**, *118*, 156–163. [CrossRef] [PubMed]
9. Oliver, K.L.; Scheffer, I.E.; Bennett, M.F.; Grinton, B.E.; Bahlo, M.; Berkovic, S.F. Genes4Epilepsy: An epilepsy gene resource. *Epilepsia* **2023**, *64*, 1368–1375. [CrossRef] [PubMed] [PubMed Central]
10. Cediel, E.G.; Duran, E.A.; Laux, J.; Reuther, W.; Leggio, O.; Robinson, B.; Boerwinkle, V.L. Pre- and post-therapy functional MRI connectivity in severe acute brain injury with suppression of consciousness: A comparative analysis to epilepsy features. *Front. Neuroimaging* **2024**, *3*, 1445952. [CrossRef] [PubMed] [PubMed Central]
11. Thompson, C.H.; Potet, F.; Abramova, T.V.; DeKeyser, J.M.; Ghabra, N.F.; Vanoye, C.G.; Millichap, J.J.; George, A.L. Epilepsy-associated SCN2A (NaV1.2) variants exhibit diverse and complex functional properties. *J. Gen. Physiol.* **2023**, *155*, e202313375. [CrossRef] [PubMed] [PubMed Central]
12. Falsaperla, R.; Sortino, V.; Marino, S.D.; Collotta, A.D.; Gammeri, C.; Sipala, F.M.; Volti, G.L.; Ruggieri, M.; Ronsisvalle, S. Molecular Dynamic Simulations to Determine Individualized Therapy: Tetrabenazine for the GNAO1 Encephalopathy E246K Variant. *Mol. Diagn. Ther.* **2024**, *28*, 329–337. [CrossRef] [PubMed]
13. Pavone, P.; Corsello, G.; Ruggieri, M.; Marino, S.; Marino, S.; Falsaperla, R. Benign and severe early-life seizures: A round in the first year of life. *Ital. J. Pediatr.* **2018**, *44*, 54. [CrossRef] [PubMed] [PubMed Central]
14. Crino, P.B. mTOR: A pathogenic signaling pathway in developmental brain malformations. *Trends Mol. Med.* **2011**, *17*, 734–742. [CrossRef] [PubMed]
15. Gerbatin, R.R.; Augusto, J.; Morris, G.; Campbell, A.; Worm, J.; Langa, E.; Reschke, C.R.; Henshall, D.C. Investigation of MicroRNA-134 as a Target against Seizures and SUDEP in a Mouse Model of Dravet Syndrome. *eNeuro* **2022**, *9*, ENEURO.0112-22.2022. [CrossRef] [PubMed] [PubMed Central]
16. Falsaperla, R.; Vari, M.S.; Toldo, I.; Murgia, A.; Sartori, S.; Vecchi, M.; Suppiej, A.; Burlina, A.; Mastrangelo, M.; Leuzzi, V.; et al. Pyridoxine-dependent epilepsies: An observational study on clinical, diagnostic, therapeutic and prognostic features in a pediatric cohort. *Metab. Brain Dis.* **2018**, *33*, 261–269. [CrossRef] [PubMed]
17. Coughlin, C.R., 2nd; Tseng, L.A.; Abdenur, J.E.; Ashmore, C.; Boemer, F.; Bok, L.A.; Boyer, M.; Buhas, D.; Clayton, P.T.; Das, A.; et al. Consensus guidelines for the diagnosis and management of pyridoxine-dependent epilepsy due to α-aminoadipic semialdehyde dehydrogenase deficiency. *J. Inherit. Metab. Dis.* **2021**, *44*, 178–192. [CrossRef] [PubMed]
18. Tang, M.; Park, S.H.; De Vivo, D.C.; Monani, U.R. Therapeutic strategies for glucose transporter 1 deficiency syndrome. *Ann. Clin. Transl. Neurol.* **2019**, *6*, 1923–1932. [CrossRef] [PubMed] [PubMed Central]
19. Falsaperla, R.; Sortino, V.; Striano, P.; Kluger, G.; Ramantani, G.; Ruggieri, M.; Network for Therapy in Rare Epilepsies (NETRE). Is ketogenic diet a 'precision medicine'? Recent developments and future challenges. *Eur. J. Paediatr. Neurol.* **2024**, *48*, 13–16. [CrossRef] [PubMed]
20. Dwivedi, R.; Kaushik, M.; Tripathi, M.; Dada, R.; Tiwari, P. Unraveling the genetic basis of epilepsy: Recent advances and implications for diagnosis and treatment. *Brain Res.* **2024**, *1843*, 149120. [CrossRef] [PubMed]
21. Tandon, N.; Radosavljevic, M.; Vucevic, D.; Radenkovic, M.; Jancic, J.; Samardzic, J. Anti-seizure Medications: Challenges and Opportunities. *CNS Neurol. Disord.-Drug Targets* **2024**, *23*, 1120–1133. [CrossRef] [PubMed]
22. Esterhuizen, A.I.; Tiffin, N.; Riordan, G.; Wessels, M.; Burman, R.J.; Aziz, M.C.; Calhoun, J.D.; Gunti, J.; Amiri, E.E.; Ramamurthy, A.; et al. Precision medicine for developmental and epileptic encephalopathies in Africa-strategies for a resource-limited setting. *Genet. Med.* **2023**, *25*, 100333. [CrossRef] [PubMed]
23. Quilón, P.G.; Volpedo, G.; Cappato, S.; Ferrera, L.; Zara, F.; Bocciardi, R.; Riva, A.; Striano, P. Antisense oligonucleotides as a precision therapy for developmental and epileptic encephalopathies. *CNS Neurosci. Ther.* **2024**, *30*, e70050. [CrossRef] [PubMed] [PubMed Central]
24. Zhao, W.; Xie, C.; Zhang, X.; Liu, J.; Liu, J.; Xia, Z. Advances in the mTOR signaling pathway and its inhibitor rapamycin in epilepsy. *Brain Behav.* **2023**, *13*, e2995. [CrossRef] [PubMed] [PubMed Central]
25. Pierson, T.M.; Yuan, H.; Marsh, E.D.; Fuentes-Fajardo, K.; Adams, D.R.; Markello, T.; Golas, G.; Simeonov, D.R.; Holloman, C.; Tankovic, A.; et al. GRIN2A mutation and early-onset epileptic encephalopathy: Personalized therapy with memantine. *Ann. Clin. Transl. Neurol.* **2014**, *1*, 190–198. [CrossRef] [PubMed] [PubMed Central]

26. Pauly, K.; Woontner, M.; Abdenur, J.E.; Chaudhari, B.P.; Gosselin, R.; Kripps, K.A.; Thomas, J.A.; Wempe, M.F.; Gospe, S.M., Jr.; Coughlin, C.R., 2nd. Feasibility of newborn screening for pyridoxine-dependent epilepsy. *Mol. Genet. Metab.* **2025**, *144*, 109002. [CrossRef] [PubMed]
27. Custers, M.L.; Vande Vyver, M.; Kaltenböck, L.; Barbé, K.; Bjerke, M.; Van Eeckhaut, A.; Smolders, I. Neurofilament light chain: A possible fluid biomarker in the intrahippocampal kainic acid mouse model for chronic epilepsy? *Epilepsia* **2023**, *64*, 2200–2211. [CrossRef] [PubMed]
28. Tumienė, B.; Del Toro Riera, M.; Grikiniene, J.; Samaitiene-Aleknienė, R.; Praninskienė, R.; Monavari, A.A.; Sykut-Cegielska, J. Multidisciplinary Care of Patients with Inherited Metabolic Diseases and Epilepsy: Current Perspectives. *J. Multidiscip. Healthc.* **2022**, *15*, 553–566. [CrossRef] [PubMed] [PubMed Central]
29. Bariş, S.; Kırık, S.; Balasar, Ö. Importance of targeted next-generation sequencing in pediatric patients with developmental epileptic encephalopathy. *Rev. Assoc. Med. Bras.* **2023**, *69*, e20230547. [CrossRef] [PubMed] [PubMed Central]
30. LoPresti, M.A.; Zhang, L.; Lam, S. Disparities in pediatric drug-resistant epilepsy care. *Childs Nerv. Syst.* **2023**, *39*, 1611–1617. [CrossRef] [PubMed] [PubMed Central]
31. Noble, A.; Nevitt, S.; Holmes, E.; Ridsdale, L.; Morgan, M.; Tudur-Smith, C.; Hughes, D.; Goodacre, S.; Marson, T.; Snape, D. *Seizure First Aid Training for People with Epilepsy Attending Emergency Departments and Their Significant Others: The SAFE Intervention and Feasibility RCT*; NIHR Journals Library: Southampton, UK, 2020. [PubMed]
32. Gotlieb, E.G.; Blank, L.; Willis, A.W.; Agarwal, P.; Jette, N. Health equity integrated epilepsy care and research: A narrative review. *Epilepsia* **2023**, *64*, 2878–2890. [CrossRef] [PubMed]
33. Miceli, F.; Millevert, C.; Soldovieri, M.V.; Mosca, I.; Ambrosino, P.; Carotenuto, L.; Schrader, D.; Lee, H.K.; Riviello, J.; Hong, W.; et al. KCNQ2 R144 variants cause neurodevelopmental disability with language impairment and autistic features without neonatal seizures through a gain-of-function mechanism. *EBioMedicine* **2022**, *81*, 104130. [CrossRef] [PubMed] [PubMed Central]
34. Schulpen, S.H.; de Jong, E.; de la Fonteyne, L.J.; de Klerk, A.; Piersma, A.H. Distinct gene expression responses of two anticonvulsant drugs in a novel human embryonic stem cell based neural differentiation assay protocol. *Toxicol. Vitr.* **2015**, *29*, 449–457. [CrossRef] [PubMed]
35. Jamal, M.; Azam, M.; Khan, S.A.; Ul-Haq, Z.; Simjee, S.U. Levetiracetam ameliorates epileptogenesis by modulating the adenosinergic pathway in a kindling model of epilepsy in mice. *Turk. J. Med. Sci.* **2023**, *53*, 1045–1057. [CrossRef] [PubMed] [PubMed Central]
36. Kumar, P.; Sheokand, D.; Grewal, A.; Saini, V.; Kumar, A. Clinical side-effects based drug repositioning for anti-epileptic activity. *J. Biomol. Struct. Dyn.* **2024**, *42*, 1443–1454. [CrossRef] [PubMed]
37. Schneider, M.F.; Vogt, M.; Scheuermann, J.; Müller, V.; Fischer-Hentrich, A.H.L.; Kremer, T.; Lugert, S.; Metzger, F.; Kudernatsch, M.; Kluger, G.; et al. Brain expression profiles of two SCN1A antisense RNAs in children and adolescents with epilepsy. *Transl. Neurosci.* **2024**, *15*, 20220330. [CrossRef] [PubMed] [PubMed Central]
38. Polster, T. Individualized treatment approaches: Fenfluramine, a novel antiepileptic medication for the treatment of seizures in Dravet syndrome. *Epilepsy Behav.* **2019**, *91*, 99–102. [CrossRef] [PubMed]
39. Le, V.T.; Thuy Le, M.A.; Nguyen, D.H.; Tang, L.N.P.; Pham, T.A.; Nguyen, A.M.; Nguyen, M.K.; Van Ngo, T.; Tran, T.T.; Van Le, T.; et al. Epilepsy surgery program in a resource-limited setting in Vietnam: A multicentered collaborative model. *Epilepsia Open* **2022**, *7*, 710–717. [CrossRef] [PubMed] [PubMed Central]
40. De Wachter, M.; Schoonjans, A.S.; Weckhuysen, S.; Van Schil, K.; Löfgren, A.; Meuwissen, M.; Jansen, A.; Ceulemans, B. From diagnosis to treatment in genetic epilepsies: Implementation of precision medicine in real-world clinical practice. *Eur. J. Paediatr. Neurol.* **2024**, *48*, 46–60. [CrossRef] [PubMed]
41. Hrincu, V.; McDonald, P.J.; Connolly, M.B.; Harrison, M.J.; Ibrahim, G.M.; Naftel, R.P.; Chiong, W.; Alam, A.; Ribary, U.; Illes, J. Choice and Trade-offs: Parent Decision Making for Neurotechnologies for Pediatric Drug-Resistant Epilepsy. *J. Child Neurol.* **2021**, *36*, 943–949. [CrossRef] [PubMed] [PubMed Central]
42. Falsaperla, R.; Sortino, V.; Vitaliti, G.; Privitera, G.F.; Ruggieri, M.; Fusto, G.; Pappalardo, X.G. GLUT-1DS resistant to ketogenic diet: From clinical feature to in silico analysis. An exemplificative case report with a literature review. *Neurogenetics* **2024**, *25*, 69–78. [CrossRef] [PubMed]

MDPI AG
Grosspeteranlage 5
4052 Basel
Switzerland
Tel.: +41 61 683 77 34

Applied Sciences Editorial Office
E-mail: applsci@mdpi.com
www.mdpi.com/journal/applsci

www.ingramcontent.com/pod-product-compliance
Lightning Source LLC
Chambersburg PA
CBHW061826210926
51850CB00006B/822
* 9 7 8 3 7 2 5 8 6 6 1 2 0 *